Computational Science, Engineering and Technology Series: 20

# Trends in Engineering Computational Technology

## Computational Science, Engineering and Technology Series

**Computational Methods for Acoustics Problems**
*Edited by: F. Magoulès*

**Mesh Partitioning Techniques and Domain Decomposition Methods**
*Edited by: F. Magoulès*

**Civil Engineering Computations: Tools and Techniques**
*Edited by: B.H.V. Topping*

**Innovation in Engineering Computational Technology**
*Edited by: B.H.V. Topping, G. Montero, R. Montenegro*

**Innovation in Computational Structures Technology**
*Edited by: B.H.V. Topping, G. Montero, R. Montenegro*

**Innovation in Civil and Structural Engineering Computing**
*Edited by: B.H.V. Topping*

**Progress in Engineering Computational Technology**
*Edited by: B.H.V. Topping, C.A. Mota Soares*

**Progress in Computational Structures Technology**
*Edited by: B.H.V. Topping, C.A. Mota Soares*

**Progress in Civil and Structural Engineering Computing**
*Edited by: B.H.V. Topping*

**Computational Mechanics using High Performance Computing**
*Edited by: B.H.V. Topping*

**Engineering Computational Technology**
*Edited by: B.H.V. Topping, Z. Bittnar*

**Computational Structures Technology**
*Edited by: B.H.V. Topping, Z. Bittnar*

**Computational Modelling of Masonry, Brickwork and Blockwork Structures**
*Edited by: J.W. Bull*

**Civil and Structural Engineering Computing: 2001**
*Edited by: B.H.V. Topping*

**High Performance Computing for Computational Mechanics**
*Edited by: B.H.V. Topping, L. Lämmer*

**Computational Mechanics for the Twenty-First Century**
*Edited by: B.H.V. Topping*

**Parallel and Distributed Processing for Computational Mechanics: Systems and Tools**
*Edited by: B.H.V. Topping*

**Innovative Computational Methods for Structural Mechanics**
*Edited by: M. Papadrakakis, B.H.V. Topping*

# Trends in Engineering Computational Technology

*Edited by*

**M. Papadrakakis and B.H.V. Topping**

published 2008 by
**Saxe-Coburg Publications**
Dun Eaglais
Station Brae, Kippen
Stirlingshire, FK8 3DY, UK

*Saxe-Coburg Publications is an imprint of Civil-Comp Ltd*

Computational Science, Engineering and Technology Series: 20
ISSN 1759-3158
ISBN 978-1-874672-40-1

**British Library Cataloguing in Publication Data**
A catalogue record for this book is available from the British Library

Front cover: A car modelled by Tibor Toth using T-splines and rendered by Dan Haring and Tom Finnigan.
Back cover: Control grids and resulting surfaces of a head model using NURBS and T-splines. These images are reproduced courtesy of Tom Sederberg and T-Splines, Inc. For more information see Chapter 1.

Printed in Great Britain by Bell & Bain Ltd, Glasgow

# Contents

**Preface** **iii**

1 Isogeometric Analysis: Toward Unification of Computer Aided Design and Finite Element Analysis 1
Y. Bazilevs, V.M. Calo, J.A. Cottrell, J. Evans, T.J.R. Hughes, S. Lipton, M.A. Scott and T.W. Sederberg

2 Object-Oriented Finite Elements: From Smalltalk to Java 17
D. Eyheramendy and F. Oudin-Dardun

3 Finite Element Software Design for Today's Computers 41
R.I. Mackie

4 Rheology and Simulation of Fresh Concrete Flow 61
B. Patzák and Z. Bittnar

5 Finite Element Modelling of Two- and Three-Dimensional Viscoelastic Polymer Flows 81
R. Tenchev, O. Harlen, P.K. Jimack and M.A. Walkley

6 Particle Swarm Optimization: Fundamental Study and its Application to Optimization and to Jetty Scheduling Problems 103
J. Sienz and M.S. Innocente

7 Optimization Techniques in Human Movement Analysis 127
A. Eriksson

8 Computational Modelling of Reactive Porous Media in Hydrometallurgy 151
C.R. Bennett, D. McBride, M. Cross, T.N. Croft and J.E. Gebhardt

9 An Enrichment-Based Multiscale Partition of Unity Method 173
M. Macri and S. De

10 A Multi-Scale Formulation of Gradient Elasticity and Its Finite Element Implementation 189
H. Askes, T. Bennett, I.M. Gitman and E.C. Aifantis

11 Multiscale Assessment of Low-Temperature Performance of Flexible Pavements 209
E. Aigner, R. Lackner, M. Wistuba, J. Eberhardsteiner and H.A. Mang

12 Advances in the Meccano Technique for Adaptive Tetrahedral Mesh Generation 229
R. Montenegro, J.M. Cascón, E. Rodríguez, G. Cascón and J.M. Escobar

13 Atoms, Molecules and Flows: Recent Advances and New Challenges in their Multi-Scale Numerical Modeling at the Beginning of the Third Millenium 247
F. Chinesta, A. Ammar, H. Lamari and N. Ranc

14 Continuum and Atomic-Scale Modeling of Self-Positioning Microstructures and Nanostructures 271
G.P. Nikishkov and Y. Nishidate

15 Thermomechanical Deformation Processes of Rate Sensitive Solids: Material Issues and Stability 295
I. Doltsinis

16 Computational Biomechanics of Sensory Organs in Spiders 321
F.G. Rammerstorfer, B. Hößl, H.-E. Dechant, H.J. Böhm and F.G. Barth

17 Three-Dimensional Numerical Analysis of a Dynamic Structure, Saturated Soil and Pore Fluid Interaction Problem 335
A.H.C. Chan and J. Ou

**Author Index** **355**

**Keyword Index** **356**

# Preface

This volume comprises the Invited Lectures presented at The Sixth International Conference on Engineering Computational Technology (ECT 2008) held in Athens, Greece from 2 to 5 September 2008. The ECT conference series began in Edinburgh in 1998. The 2008 conference was held concurrently with The Ninth International Conference on Computational Structures Technology (CST 2008). Chapter 1 is the Plenary Lecture presented by Professor T.J.R. Hughes to open both conferences. We are grateful to the authors and co-authors of these Invited Lectures included in this volume. Their contribution both to the Civil-Comp conferences and this book is greatly appreciated.

Other papers presented at the conferences in 2008 are published as follows:

- *The Invited Lectures from CST 2008 are published in:*
  Trends in Computational Structures Technology, B.H.V. Topping and M. Papadrakakis, (Editors), Saxe-Coburg Publications, Stirlingshire, Scotland, 2008.

- *The Contributed Papers from CST 2008 are published in:*
  Proceedings of the Ninth International Conference on Computational Structures Technology, B.H.V. Topping and M. Papadrakakis, (Editors), (Book of Summaries and CD-ROM), Civil-Comp Press, Stirlingshire, Scotland, 2008.

- *The Contributed Papers from ECT 2008 are published in:*
  Proceedings of the Sixth International Conference on Engineering Computational Technology, M. Papadrakakis and B.H.V. Topping, (Editor), (Book of Summaries and CD-ROM), Civil-Comp Press, Stirlingshire, Scotland, 2008.

We should like to thank the members of the ECT 2008 Conference Editorial Board for their help before and during the conference: Dr P.M. Adler, France; Prof. E.C. Aifantis, Greece; Dr H.M. Al-Humaidi, USA; Dr S.J. Antony, UK; Prof. C.J. Anumba, UK; Prof. H.R. Arabnia, USA; Prof. A.J. Baker, USA; Prof D.C. Barton, UK; Prof. S. Baxter, USA; Dr E. Bellenger, France; Prof. A.C. Benim, Germany; Prof. R.B. Bhat, Canada; Prof. H. Bijl, Netherlands; Dr S. Bordas, UK; Prof. R.I. Borja, USA; Prof. B. Brank, Slovenia; Prof. R. Bru, Spain; Prof. T. Burczynski, Poland; Assoc. Prof. O.S.

Carneiro, Portugal; Prof. T.Y. Chen, Taiwan; Dr Q.V. Chen, USA; Dr H.-P. Cheng, USA; Prof. F. Chinesta, France; Prof. M.-H. Chung, Taiwan; Prof. V. Couaillier, France; Prof. A.L.G.A. Coutinho, Brazil; Prof. M. Cross, UK; Prof. L. Damkilde, Denmark; Dr K. Davey, UK; Prof. J.B. de Paiva, Brazil; Prof. G. Degrande, Belgium; Prof. S. Del Giudice, Italy; Prof. W. Desmet, Belgium; Dr C. Di Napoli, Italy; Prof. E. Dick, Belgium; Dr M. Dolenc, Slovenia; Prof. F.J. Elorza Tenreiro, Spain; Dr J.M. Escobar Sánchez, Spain; Dr J.J. Fang, Taiwan; Prof. L. Ferragut, Spain; Prof. W. Frank, Germany; Prof. J.D. Frost, USA; Dr M.M. Furnari, Italy; Prof. U. Gabbert, Germany; Prof. P. González-Vera, Spain; Dr G.A. Gravvanis, Greece; Dr D.F. Griffiths, UK; Dr Y.-M. Guo, Japan; Prof. W.G. Habashi, USA; Prof. F.C. Hadipriono, USA; Dr L. Hadjileontiadis, Greece; Dr F. Hakl, Czech Republic; Dr S.J. Hardy, UK; Prof. A. Hernandez, Mexico; Dr M. Hirokane, Japan; Dr N. Hu, Japan; Prof. T.J.R. Hughes, USA; Dr J. E. Hurtado, Colombia; Prof. M.Y. Hussaini, USA; Prof. S.-F. Hwang, Taiwan; Prof. S.R. Idelsohn, Spain; Prof. K. Ikeda, Japan; Prof. M.H. Imam, Saudi Arabia; Dr M. Isreb, Australia; Prof. Y. Jaluria, USA; Prof. P.K. Jimack, UK; Dr S. Kenjeres, Netherlands; Prof. T. Kerh, Taiwan; Prof. E. Kita, Japan; Prof. A. Klarbring, Sweden; Dr L.X. Kong, Australia; Prof. V.K. Koumousis, Greece; Prof. D. Kulasiri, New Zealand; Prof. B. Kumar, UK; Prof. L. Laemmer, Germany; Dr O. Laghrouche, UK; Prof. F. Lebon, France; Prof. S.W. Lee, USA; Prof. A.W. Lees, UK; Prof. P. Leger, Canada; Prof. L. Lo Bello, Italy; Dr J. Wei-Zhen Lu, Hong Kong; Dr R.I. Mackie, UK; Prof. F. Magoules, France; Prof. M. Malafaya-Baptista, Portugal; Prof. D.C. Marinescu, USA; Prof. N.C. Markatos, Greece; Prof. K. Matsuno, Japan; Prof. R.V.N. Melnik, Canada; Prof. J.C. Miles, UK; Prof. G. Molnarka, Hungary; Prof. R. Montenegro Armas, Spain; Prof. G. Montero, Spain; Prof. F.C. Morabito, Italy; Prof. J. Morrison, Ireland; Prof. M. Mueller-Hannemann, Germany; Prof. U. Naumann, Germany; Prof. P. Neittaanmaki, Finland; Prof. D.T. Nguyen, USA; Dr F. Nicot, France; Prof. P.A. Pagliosa, Brazil; Prof. N.-F. Pan, Taiwan; Prof. P.C. Pandey, India; Prof. M. Papadrakakis, Greece; Prof. C.D. Perez-Segarra, Spain; Prof. S. Pietruszczak, Canada; Prof. F.T. de Pinho, Portugal; Prof. R. Piltner, USA; Dr S.V. Poroseva, USA; Prof. R. Pusch, Sweden; Dr M.Y. Rafiq, UK; Prof. E. Ramm, Germany; Prof. Z. Ren, Slovenia; Dr L.M. Ribeiro, Portugal; Dr C. Romanel, Brazil; Prof. D. Roose, Belgium; Dr F.-X. Roux, France; Dr S. Samarasinghe, New Zealand; Prof. R.L. Sani, USA; Dr J. Sarrate Ramos, Spain; Prof. M. Schaefer, Germany; Dr O. Schenk, Switzerland; Prof. K. Schilling, Germany; Dr G. Seed, UK; Prof. K.N. Seetharamu, India; Prof. A. Shabana, USA; Dr J. Sienz, UK; Dr J. Sobieski, USA; Prof. V.E. Sonzogni, Argentina; Prof. G.E. Stavroulakis, Greece; Prof. P. Steinmann, Germany; Dr I.A. Stroud, Switzerland; Prof. K.K. Tamma, USA; Prof. J.C.F. Telles, Brazil; Prof. T. Tomiyama, Netherlands; Prof. B.H.V. Topping, UK; Prof. J. Trujillo Jacinto del Castillo, Spain; Prof. C.P. Tsai, Taiwan; Prof. A.S. Usmani, UK; Prof. W.S. Venturini, Brazil; Prof. V.R. Voller, USA; Prof. H. Voss, Germany; Dr C. Walshaw, UK; Prof. W.C. Wang, Taiwan; Prof. L.T. Watson, USA; Prof. N.P. Weatherill, UK; Prof. Dr Ing. T. Weiland, Germany; Prof. M. Wolfshtein, Israel; Prof. L.C. Wrobel, UK; Prof. L. Yan, China; and Dr Z.J. Yang, UK.

Finally, we are grateful to Jelle Muylle for designing these conference proceedings and for all his administrative and organisational skills in organising these conferences. We also wish to thank Dawn Sewell (Civil-Comp Press) and Jenny Mitsoura (National Technical University of Athens) for their administrative support before and during the conference.


Prof. M. Papadrakakis
National Technical University of Athens, Greece

Prof. B.H.V. Topping
Heriot-Watt University, Edinburgh, UK
& University of Pécs, Hungary

# Chapter 1

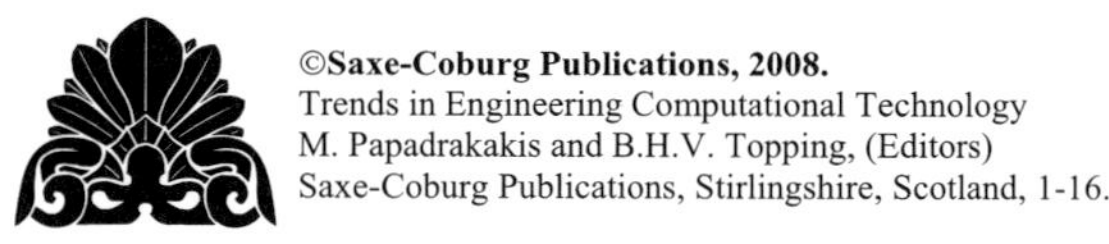

©Saxe-Coburg Publications, 2008.
Trends in Engineering Computational Technology
M. Papadrakakis and B.H.V. Topping, (Editors)
Saxe-Coburg Publications, Stirlingshire, Scotland, 1-16.

# Isogeometric Analysis: Toward Unification of Computer Aided Design and Finite Element Analysis

**Y. Bazilevs[1], V.M. Calo[1], J.A. Cottrell[1], J. Evans[1]**
**T.J.R. Hughes[1], S. Lipton[1], M.A. Scott[1] and T.W. Sederberg[2]**
**[1] Computational and Applied Mathematics**
**University of Texas, Austin TX, United States of America**
**[2] Computer Science**
**Brigham Young University, Provo UT, United States of America**

## Abstract

This chapter discusses the major challenges faced in engineering design and analysis. Going from an initial geometric design to a finite element mesh constitutes the most time-consuming step of the overall design through analysis process. The concept of isogeometric analysis has thus far addressed this issue by utilizing the NURBS functions commonly found in CAD packages as an isoparametric basis for both the geometry and the solution space. We extend the methodology to include T-splines, a technology rapidly gaining favor in the geometry community due to its flexibility and efficiency in representing complex objects. This efficiency stems from its locally refinable basis, a feature ideally suited to finite element analysis where local refinement is frequently desirable. Examples are presented.

**Keywords:** isogeometric analysis, CAD, NURBS, T-splines.

# 1 Introduction

Engineering design and analysis have reached a critical juncture. Each has its own geometric representation, and the design description, embodied in Computer Aided Design (CAD) systems, needs to be translated to an analysis-suitable geometry for mesh generation and use in a Finite Element Analysis (FEA) code. This task is far from trivial. For complex engineering designs it is now estimated to take over 80 percent of overall analysis time, and engineering designs are becoming increasingly more complex. For example, presently, a typical automobile consists of about 3,000 parts, a fighter jet over 30,000, the Boeing 777 over 100,000, and a modern nuclear submarine over 1,000,000 (see Figure 1).

Engineering design and analysis are not separate endeavors. Design of sophisticated engineering systems is based on a wide range of computational analysis and

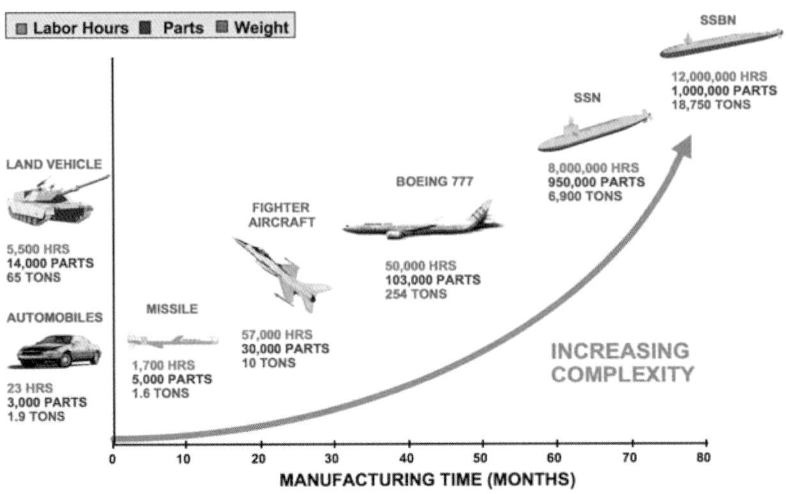


Figure 1: Engineering designs are becoming increasingly complex. As the number of parts comprising an object increases, so too does the amount of time required for it to be manufactured. Such growth in complexity makes analysis a time consuming and expensive endeavor. (Courtesy of S. Gordon, General Dynamics / Electric Boat Division)

simulation methods, such as structural mechanics, fluid dynamics, acoustics, electromagnetics, heat transfer, *etc.* Design speaks to analysis, and analysis speaks to design. However, analysis-suitable models are not automatically created or readily meshed from CAD geometry. Although not always appreciated in the academic analysis community, model generation is much more involved than simply generating a mesh. There are many time consuming, preparatory steps involved. And one mesh is no longer enough. According to Steve Gordon, Principal Engineer, General Dynamics Electric Boat Corporation, "We find that today's bottleneck in CAD-CAE integration is not only automated mesh generation; it lies with efficient creation of appropriate 'simulation-specific' geometry." (In the commercial sector analysis is usually referred to as CAE, which stands for Computer Aided Engineering.)

The anatomy of the process has been studied by Ted Blacker, Manager of Simulation Sciences at Sandia National Laboratories, and is summarized in Figure 2 along with the breakdown in the percentage of time devoted to each task. Note that at Sandia mesh generation accounts for only 20 percent of overall analysis time, whereas creation of the analysis-suitable geometry requires about 57 percent, and only 23 percent of overall time is actually devoted to analysis *per se*. The approximate 80/20 modeling/analysis ratio seems to be a very common industrial experience and there is a strong desire to reverse it, but so far little progress has been made despite enormous effort to do so. The integration of CAD and FEA has proved a formidable problem. It is our opinion that fundamental changes must take place to fully integrate engineering

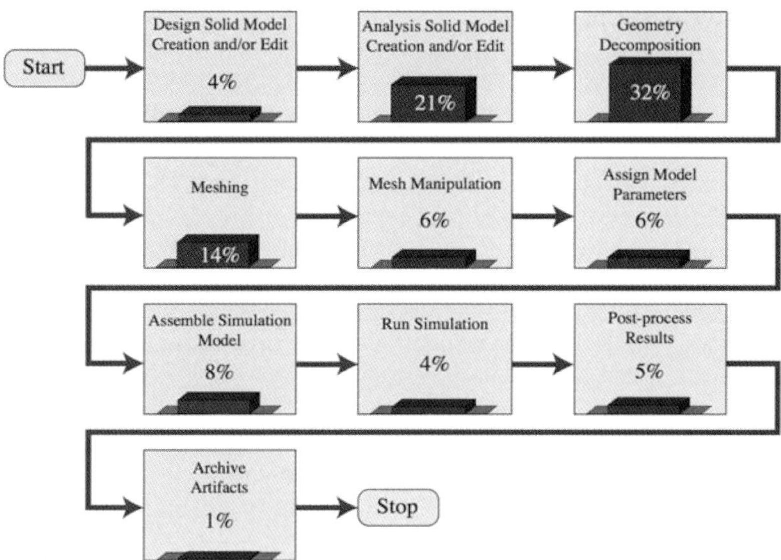


Figure 2: Estimation of the relative time costs of each component of the model generation and analysis process at Sandia National Laboratories. Note that the process of building the model completely dominates the time spent performing analysis. (Courtesy of Ted Blacker, Sandia National Laboratories)

design and analysis.

Recent trends taking place in engineering analysis and high-performance computing are also demanding greater precision and tighter integration of the overall modeling-analysis process. We note that a finite element mesh is only an approximation of the CAD geometry, which we will view as "exact." This approximation can in many situations create errors in analytical results. The following examples may be mentioned: shell buckling analysis is very sensitive to geometric imperfections, boundary layer phenomena and lift and drag are sensitive to precise geometry of aerodynamic and hydrodynamic configurations, and sliding contact between bodies cannot be accurately represented without precise geometric descriptions. Automatic adaptive mesh refinement has not been as widely adopted in industry as one might assume from the extensive academic literature because mesh refinement requires access to the exact geometry, and thus it also requires seamless and automatic communication with CAD, which simply does not exist. Without accurate geometry and mesh adaptivity, convergence and high-precision results are impossible.

Deficiencies in current engineering analysis procedures also preclude successful application of important pace setting technologies, such as design optimization, verification and validation (V&V), uncertainty quantification (UQ), and petascale computing. The benefits of design optimization have been largely unavailable to industry.

The bottleneck is that to do shape optimization the CAD geometry-to-mesh mapping needs to be automatic, differentiable and tightly integrated with the solver and optimizer. This is simply not the case as meshes are disconnected from the CAD geometries from which they were generated. V&V requires error estimation and adaptivity, which in turn requires tight integration of CAD, geometry, meshing, and analysis. UQ requires simulations with numerous samples of models needed to characterize probability distributions. Sampling puts a premium on the ability to rapidly generate geometry models, meshes, and analyses, which again leads to the need for tightly integrated geometry, meshing, and analysis. The era of petaflop computing is on the horizon. Parallelism keeps increasing, but the largest unstructured mesh simulations have stalled because no one truly knows how to generate and adapt massive meshes that keep up with increasing concurrency. To be able to capitalize on the era of $O(10^5)$ core parallel systems, CAD, geometry, meshing, analysis, adaptivity, and visualization all have to run in a tightly integrated way, in parallel and scalably.

Our vision is that the only way to break down the barriers between engineering design and analysis is to reconstitute the entire process. We believe that the fundamental step is to focus on one, and only one, geometric model, which can be utilized directly as an analysis model, or from which geometrically exact analysis models can be automatically built. This will require a change from classical finite element analysis to an analysis procedure based on exact CAD representations. This concept is referred to as Isogeometric Analysis, and was first proposed in Hughes, Cottrell, and Bazilevs [1] and further developed in [2–7]. Isogeometric analysis is based on the same computational geometry representation as the CAD model.

There are a number of candidate computational geometry technologies that may be used in isogeometric analysis. The most widely used in engineering design is NURBS (non-uniform rational B-splines), the industry standard (see, *e.g.*, [8–11]). The major strengths of NURBS are that they are convenient for free-form surface modeling, can exactly represent all conic sections, that is, circles, cylinders, spheres, ellipsoids, *etc.*, and that there exist many efficient and numerically stable algorithms to generate NURBS objects. They also possess useful mathematical properties, such as the ability to be refined through knot insertion, $C^{p-1}$-continuity for order $p$ NURBS, and the variation diminishing and convex hull properties. NURBS are ubiquitous in CAD systems, representing billions of dollars in development investment. The major deficiencies of NURBS are that gaps and overlaps at intersections of surfaces cannot be avoided, complicating mesh generation, and that they utilize a tensor product structure making the representation of detailed local features inefficient. T-splines are a recently developed generalization of NURBS technology (see [12]). T-splines correct the deficiencies of NURBS in that they permit local refinement and coarsening, and a solution to the gap/overlap problem. Commercial T-spline capabilities have been recently introduced in Maya [13] and Rhino [14], two NURBS-based design systems.

A NURBS surface is defined using a set of control points, which lie, topologically, in a rectangular grid, as shown in Figure 3a. This means that a large percentage of NURBS control points are superfluous in that they contain no significant geometric

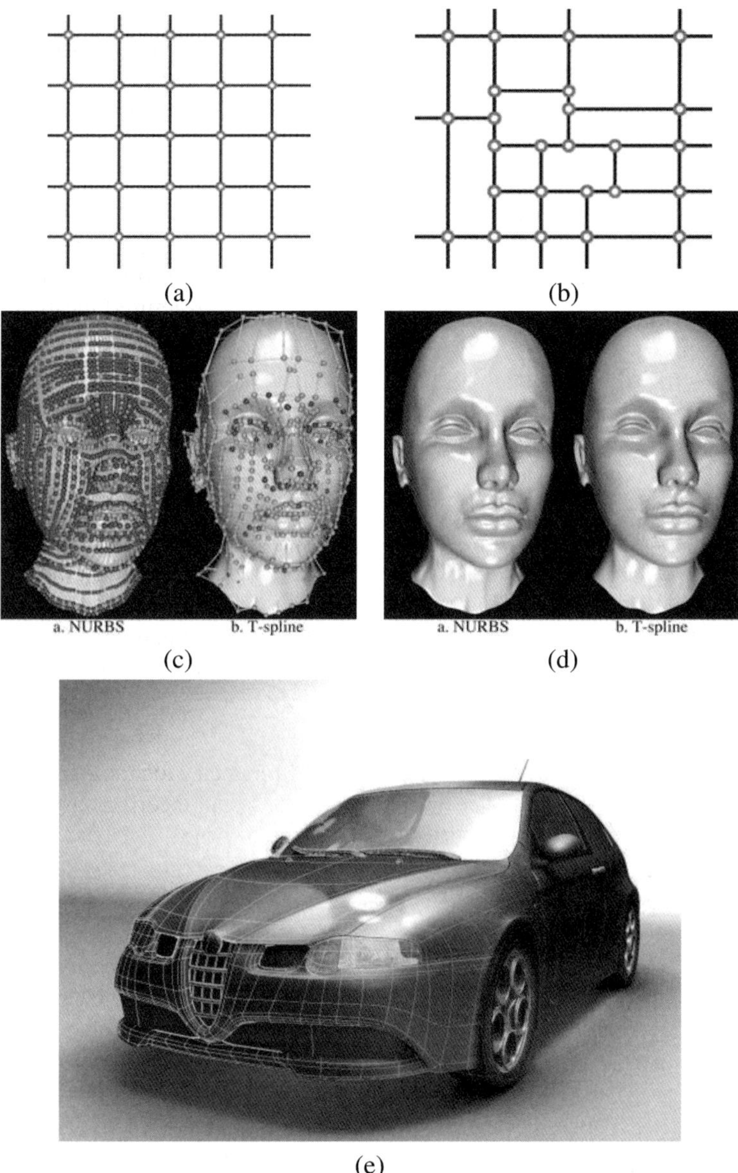


Figure 3: NURBS control points lie in a rectangular grid. Rows of T-spline control points can be incomplete. (a) Topology of NURBS control grid. (b) Topology of sample T-spline control grid. (c) NURBS and T-spline control grids for head model. 4800 NURBS control points; 1100 T-spline control points. (d) Head modeled as NURBS and T-spline. (e) Car modeled using T-splines; one quarter of the modeling time compared with NURBS

information, but merely are needed to satisfy the topological constraints. In Figure 3c, 80 percent of the NURBS control points are superfluous. By contrast, a T-spline control grid is allowed to have partial rows of control points, as shown in Figure 3b. A partial row of control points terminates in a T-junction, hence the name T-splines. In Figure 3c, the highlighted T-splines control points are T-junctions. For the head modeled by NURBS and T-splines in Figures 3c and 3d, the T-splines model requires only 24 percent of the control points compared to the NURBS model. For a designer, fewer control points means faster modeling time. The artist that created the T-splines model of the car in Figure 3e estimated that it took one-quarter the time it would have taken using NURBS. Refinement, the process of adding new control points to a control mesh without changing the surface, is an important basic operation used by designers. A limitation of NURBS is that refinement requires the insertion of an entire row of control points. T-junctions enable T-splines to be locally refined. As shown in Figure 4, a single control point can be added to a T-splines control grid. Another limitation of NURBS is that because a single NURBS surface must have a rectangular topology, most objects must be modeled using several NURBS surfaces. The hand in Figure 5 is modeled using seven NURBS patches, one for the forearm, one for the hand, and one for each finger. It is difficult to join multiple NURBS surfaces in a single watertight model, as illustrated in Figures 5a and 5b, especially if corners of valence other than four are introduced. T-junctions make it possible to merge together several NURBS surfaces into a single gap-free T-spline, as shown in Figure 5c.

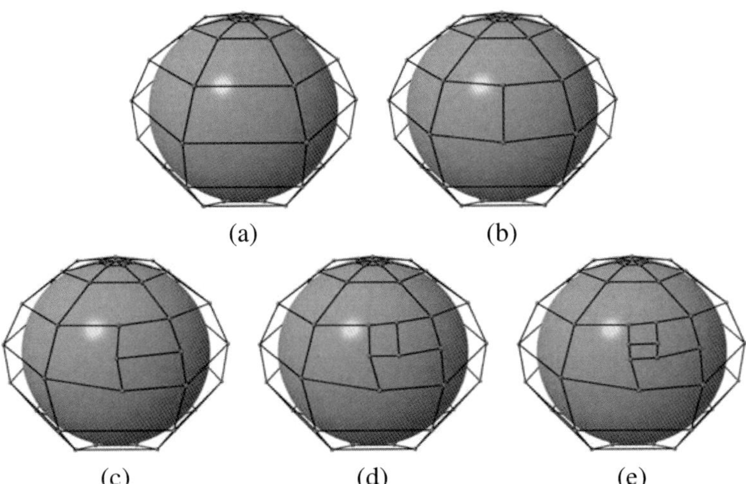


Figure 4: Local refinement using T-splines. The added control points do not change the surface

The most serious problem inherent in NURBS is that it is mathematically impossible for a trimmed NURBS to accurately represent the intersection of two NURBS surfaces without introducing gaps in the model. Figure 6 illustrates how these gaps

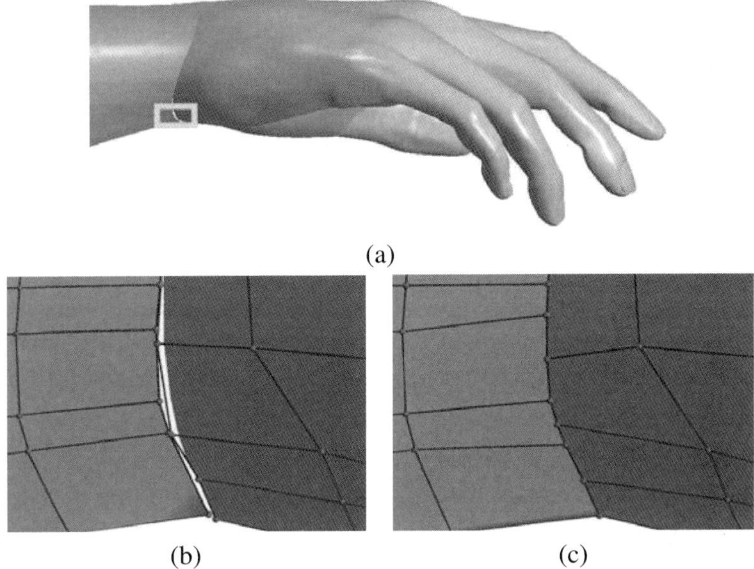

(a)

(b) (c)

Figure 5: (a) Hand modeled using seven NURBS surfaces. (b) Blowup of highlighted region showing NURBS control grids. (c) T-spline control grid; the gap is closed using T-spline merging

occur. Figure 6a shows the Utah teapot model in which the body and spout are modeled as two distinct NURBS surfaces. Using surface intersection operations that are standard in CAD systems, the spout cuts a hole in the body, and the body cuts away the excess portion of the spout. The white rings in Figure 6b show where the cuts will occur, and Figure 6c shows the resulting cuts. The trimmed teapot body and spout in Figure 6c are represented using trimmed NURBS. In Figure 6d, the trimmed spout is moved back into position relative to the trimmed body. However, they do not form a watertight model, as the blowup in Figure 6e shows.

An NSF-sponsored workshop in 1999 identified the unavoidable gaps in trimmed NURBS as the most pressing unresolved problem in the field of CAD [11]. Downstream engineering departments spend as much as fifty percent of their time dealing with such CAD files. It has been estimated to cost over $600 million annually to the U.S. auto industry alone. One maker of aircraft engines estimates that it spends $24 million per year identifying and resolving such problems. The existing approach is to employ "healing" software, which does not fix the problem, but only reduces the size of gaps. The problem is a significant ingredient in the design-to-analysis bottleneck because a CAD model must be closed in order to generate an analysis-suitable mesh. T-splines provide a way to close the gap. Figure 7 shows how T-splines can represent the NURBS model in Figure 6 as a single, gap-free T-spline, the details of which are described in [19]. In addition to the fact that T-splines solve many of the problems with

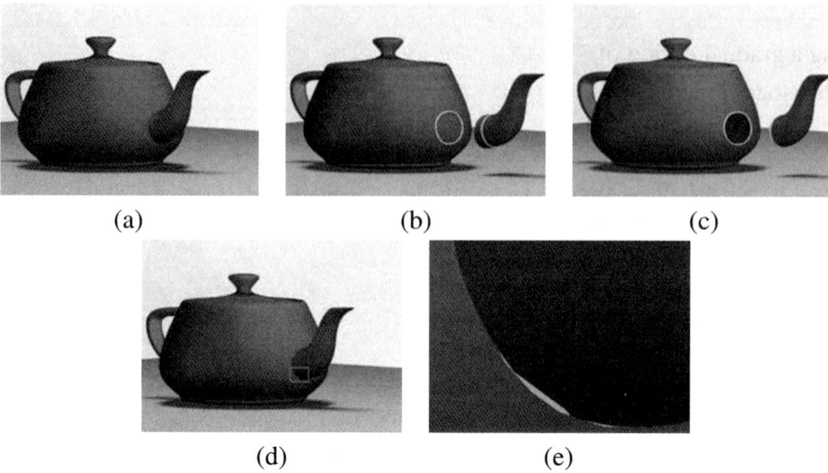

Figure 6: Utah teapot model. (a) The original NURBS surfaces. (b) Intersection curves where the body and spout intersect (the spout has been moved away from the body). (c) Trimmed-NURBS representation in which the hole is cut in the body and the excess part of the spout is trimmed away. (d) Trimmed-NURBS body and spout moved back to their original positions. (e) Blowup showing the gap between the body and spout inside the green rectangle in (d)

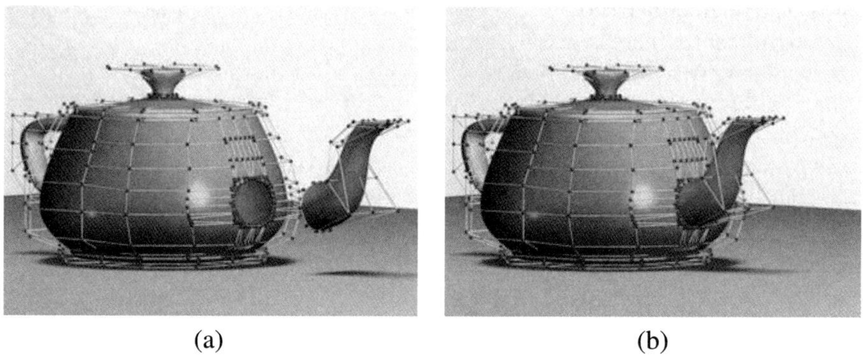

Figure 7: Utah teapot model. (a) The trimmed body and spout from Figure 6c each modeled as a T-spline surface. (b) The two T-splines in (a) merged into a single gap-free T-spline

NURBS that have vexed the CAD community for three decades, T-splines are forward and backward compatible with NURBS. Every NURBS is a special case of a T-spline (*i.e.*, a T-spline with no T-junctions or extraordinary points) and every T-spline can be converted into one or more NURBS surfaces by performing repeated local refinement to eliminate all T-junctions. Compatibility is crucial for commercialization, especially in a mature industry like CAD, and it may allow T-splines and NURBS to coexist during a gradual period of T-splines adoption.

The isogeometric concept would endeavor to use the surface design model emanating from CAD directly in analysis. This would only suffice if analysis only requires the surface geometry, such as in the stress or buckling analysis of a shell. In many cases, if not most, the surface will enclose a solid and an analysis model will need to be created for the solid. The basic problem is to develop a three-dimensional (trivariate) representation of the solid in such a way that the surface representation is exactly preserved. This is far from a trivial problem. Surface differential and computational geometry and topology are now fairly well understood [15], but the three-dimensional problem is still open [16, 17]. The hope is that through the use of new technologies, such as Ricci flows and polycube splines [15], progress will be forthcoming. The polycube spline concept has features in common with the template-based system developed by Zhang *et al.* [5] for modeling patient-specific arterial fluid-structure geometries.

The current state-of-the-art in isogemetric analysis is as follows: A number of single and multiple patch NURBS-based parametric models have been developed and analyzed [1–7]. Various mesh refinement schemes have been investigated, namely, $h$-, $p$- and $k$-refinement (*i.e.*, mesh, $C^0$ order elevation, and $C^{p-1}$ order elevation, resp.). It has been shown that isogeometric analysis preserves geometry at all levels of refinement and that detailed features can be retained without excessive mesh refinement, in contrast with traditional finite element analysis. Superior accuracy to finite element analysis on a degree-of-freedom basis has been demonstrated in all cases, and indications of significantly increased robustness in vibration and wave propagation analysis has been noted (see [6] and [18]). So far, isogeometric analysis has been applied to simple solid parts and thin-walled structures (see Figure 8), and its extension to more complex parts is apparent (see Figure 9). The current range of applicability for isogeometric analysis includes more complex thin-walled structures (see Figure 10). The challenge is to apply it to very complex modules and assemblages. Examples emanating from modern ship design are representative (see Figure 11).

Given the incredible inertia existing in the design and analysis industries, one may reasonably ask why one should believe that the time is right to transform the technology of these industries. Here are our reasons. Recent initiatory investigations of the isogeometric concept have proven very successful. Compatibility with existing design and analysis technologies is attainable. There is interest in both the computational geometry and analysis communities to embark on isogeometric research. Several minisymposia and workshops at international meetings have been held[1]. There is an inexorable march toward higher precision and greater reality. New technologies are being introduced and adopted rapidly in design software to gain competitive advantage. New and better analysis technologies can be built upon these new CAD

[1]For example, the 7th World Congress on Computational Mechanics; the 9th U.S. National Congress on Computational Mechanics and the Seventh International Conference on Mathematical Methods for Curves and Surfaces.

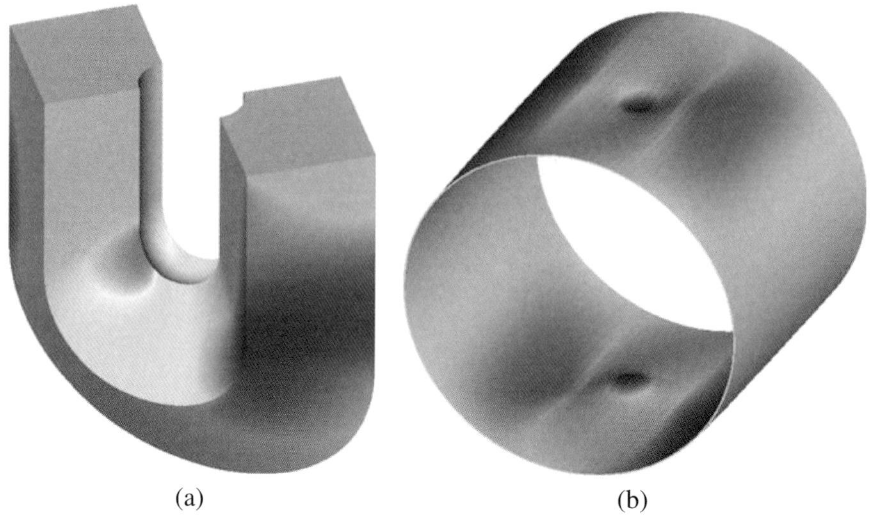

(a) (b)

Figure 8: Examples of some of the simple solid and thin-walled structures to which isogeometric analysis has been applied

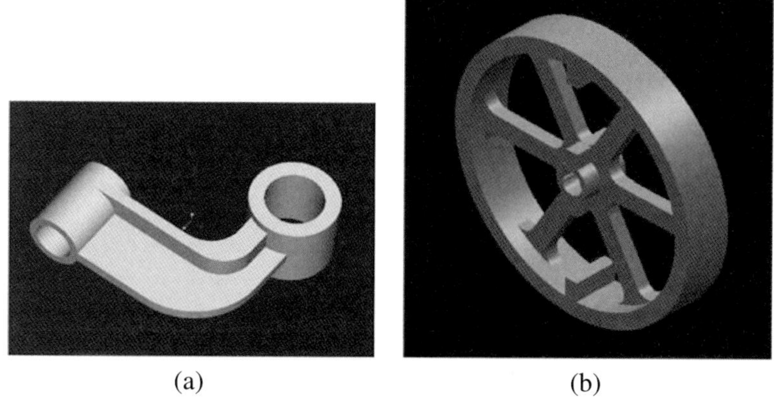

(a) (b)

Figure 9: Extension of the isogeometric analysis methodology to various parts and components is apparent. (Courtesy of S. Gordon, General Dynamics / Electric Boat Division.)

technologies. Engineering analysis can leverage these developments as a basis for the isogeometric concept.

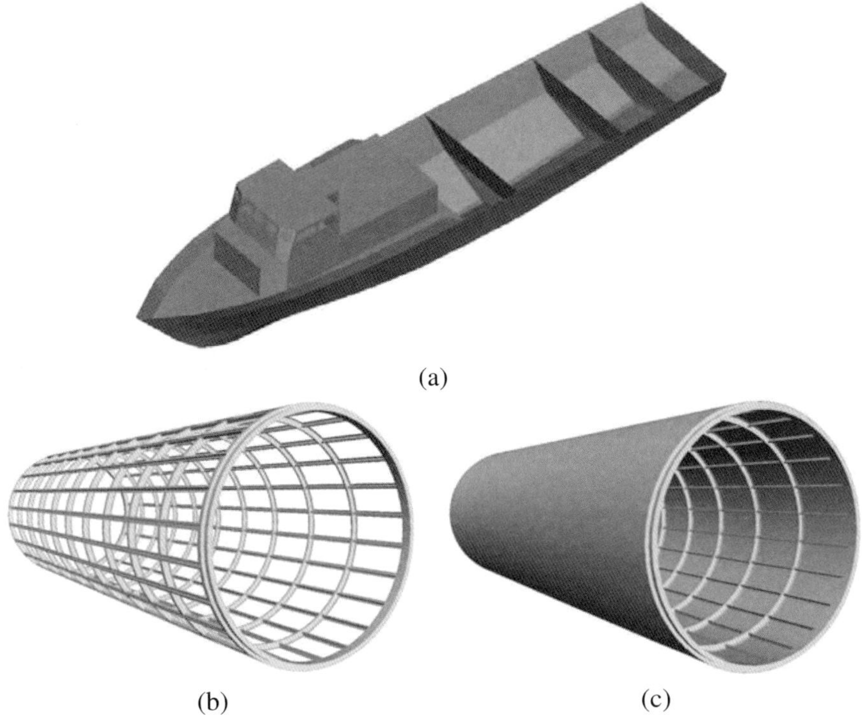

(a)

(b) (c)

Figure 10: Current range of applicability of isogeometric analysis includes more complex thin-walled structures

# 2 Isogeometric Analysis with T-splines

## 2.1 NURBS shortcomings

NURBS have been, and continue to be, widely used by designers. As mentioned, they are a standard in the CAD community and have recently been used with a great deal of success as a basis for isogeometric analysis. However, they have several drawbacks that we would like to avoid. One is that if we assume open knot vectors are always used, then they achieve only $C^0$-continuity across patch boundaries. While there are circumstances under which higher continuity is possible (see [19]), it is not always practical in an analysis code. A related issue is that the joining of two patches that were created separately can be problematic, frequently requiring the insertion of many knots from one patch into the other, and *vice versa*. This is a significant disadvantage of NURBS: knot insertion is a global operation. When we refine by inserting knots into the knot vectors of a surface, the knot lines extend throughout the entire domain (see Figure 12b). In [6], one approach to local refinement was considered that involved multiple patches. This technique, however, required the use of constraint equations

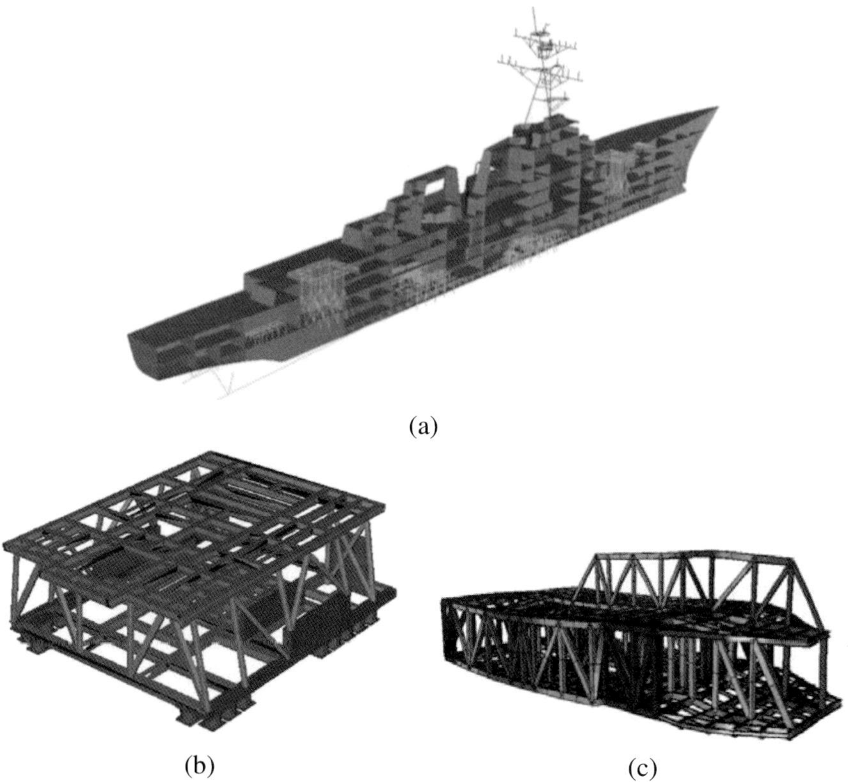

(a)

(b) (c)

Figure 11: Some examples from modern ship design of the types of complex modules and assemblages that present challenges to analysts. (Courtesy of S. Gordon, General Dynamics / Electric Boat Division)

which are inconvenient to implement, and refinement still propagates throughout a given patch. This is a problem in both geometric modeling and in analysis. What we would like is a technology that allows us to use the smooth functions and geometrical flexibility of NURBS, while permitting truly local refinement.

T-splines (see [12]) offer an interesting alternative to NURBS for both geometric modeling and as a basis for isogeometric analysis. They allow us to build bases that are complete up to a desired polynomial order, as smooth as an equivalent NURBS basis, and capable of being locally refined while keeping the original geometry and parameterization unchanged. The properties that make T-splines useful for geometric modeling also make them useful for finite element analysis. As mentioned, T-splines even include NURBS as a special case, making their use in a finite element setting a natural extension of the existing isogeometric analysis technology.

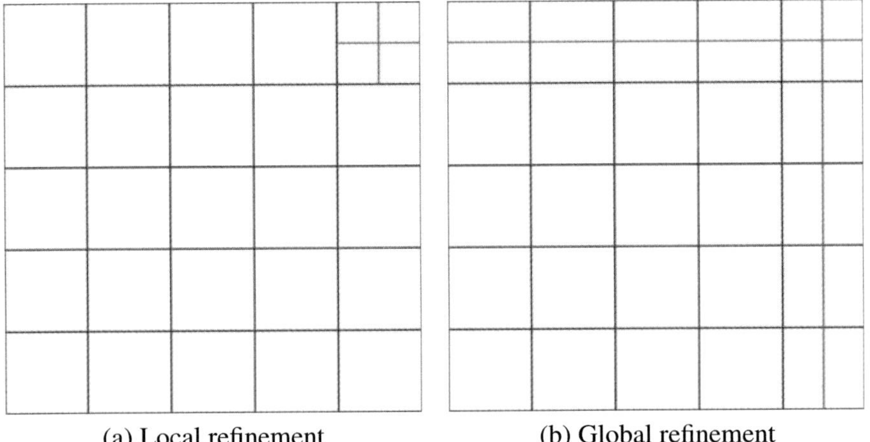


Figure 12: a) There are many instances in which we would like to locally refine an initial NURBS mesh by subdividing an individual element. b) Unfortunately, knot insertion is a global process that necessitates the propagation of the refinement throughout the domain

## 2.2 Isogeometric structural analysis: Pinched cylinder

An isogeometric structural analysis framework based on T-splines preserves many of the desirable properties of its NURBS counterpart. In particular, T-splines also satisfy standard patch tests. In what follows, we present numerical solutions for a linear elastic solid. The Galerkin formulation of linear elasticity is employed. Though the calculations are for a thin-walled structure, it is modeled as a three-dimensional solid and no shell assumptions are employed. A direct algebraic equation solver was employed, and numerical integration was performed using Gaussian quadrature.

In a problem setup adapted from Felippa [20] and Belytschko *et al.* [21], the pinched cylinder is subjected to equal and opposite concentrated forces at its midspan (see Figure 13). The ends are supported by rigid diaphragms. This constraint results in highly localized deformation under the loads. Only one octant of the cylinder is used in the calculations due to symmetry. Two quadratic NURBS elements were utilized in the through-thickness direction and quadratic through quintic orders of T-splines were utilized on the surface, see [1]. Note that these basis functions are at least $C^1$- through $C^4$-continuous everywhere. An example sequence of T-spline meshes is shown in Figure 14. A contour plot of the displacement on the deformed configuration is shown in Figure 15. Convergence of the displacement under the load is plotted in Figure 16.

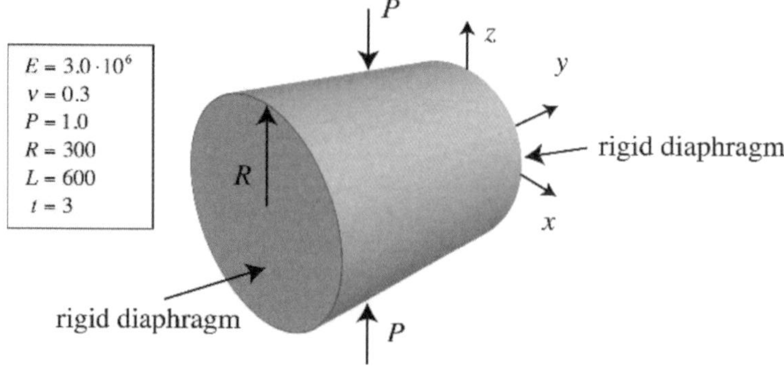


Figure 13: Shell Obstacle Course. Pinched cylinder problem description and data

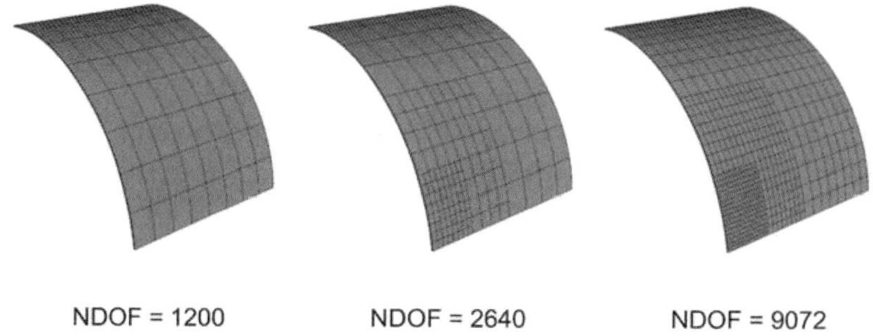


Figure 14: Shell Obstacle Course. Sequence of T-spline meshes for pinched cylinder for $p = 2$

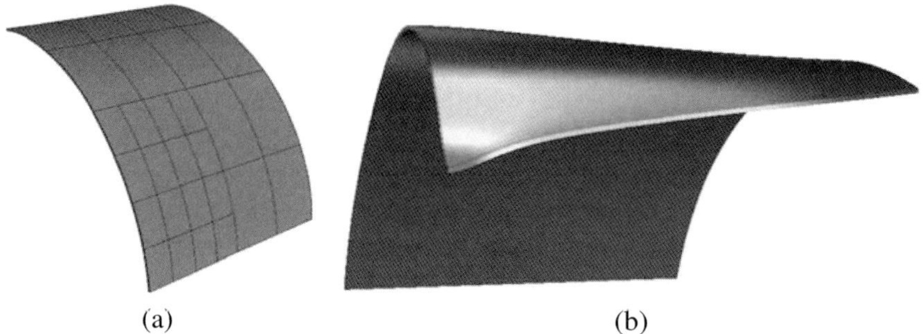


Figure 15: Shell Obstacle Course. Pinched cylinder with $p = 5$, NDOF = 1260: (a) Physical mesh and (b) Displacement contours on deformed configuration (scaling factor of $3 \times 10^6$ used)

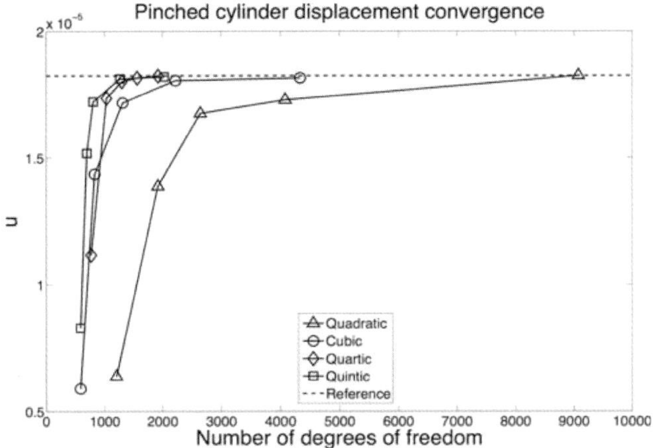


Figure 16: Shell Obstacle Course. Pinched cylinder displacement convergence

# 3 Conclusion

Isogeometric analysis has been extended to include finite element analysis using T-splines. Initial results suggest that the locally refinable T-spline basis may be well-suited to analysis. Highly localized features can be captured without necessitating global refinements of the mesh, as was necessary in the case of NURBS.

# References

[1] T.J.R. Hughes, J.A. Cottrell, Y. Bazilevs, "Isogeometric analysis: CAD, finite elements, NURBS, exact geometry, and mesh refinement", *Computer Methods in Applied Mechanics and Engineering*, 194:4135–4195, 2005.

[2] J.A. Cottrell, A. Reali, Y. Bazilevs, T.J.R. Hughes, "Isogeometric analysis of structural vibrations", *Computer Methods in Applied Mechanics and Engineering*, 195:5257–5296, 2006.

[3] Y. Bazilevs, L. Beirao de Veiga, J.A. Cottrell, T.J.R. Hughes, G. Sangalli, "Isogeometric analysis: approximation, stability and error estimates for $h$-refined meshes", *Mathematical Models and Methods in Applied Sciences*, 16:1031–1090, 2006.

[4] Y. Bazilevs, V.M. Calo, Y. Zhang, T.J.R. Hughes, "Isogeometric fluid-structure interaction analysis with applications to arterial blood flow", *Computer Methods in Applied Mechanics and Engineering*, 38:310–322, 2006.

[5] Y. Zhang, Y. Bazilevs, S. Goswami, C. Bajaj, T.J.R. Hughes, "Patient-specific vascular NURBS modeling for isogeometric analysis of blood flow", *Computer Methods in Applied Mechanics and Engineering*, 196:2943–2959, 2007.

[6] J.A. Cottrell, T.J.R. Hughes, A. Reali, "Studies of refinement and continuity in isogeometric analysis", *Computer Methods in Applied Mechanics and Engineering*, 196:4160–4183, 2007.

[7] H. Gomez, V.M. Calo, Y. Bazilevs, T.J.R. Hughes, "Isogeometric Analysis of the Cahn-Hilliard phase-field model", *Computer Methods in Applied Mechanics and Engineering*, In press.

[8] D.F. Rogers, *An Introduction to NURBS With Historical Perspective*, Academic Press, San Diego, CA, 2001.

[9] L. Piegl, W. Tiller. *The NURBS Book (Monographs in Visual Communication), 2nd ed.*, Springer-Verlag, New York, 1997.

[10] G.E. Farin, *NURBS Curves and Surfaces: from Projective Geometry to Practical Use, Second Edition*, A. K. Peters, Ltd., Natick, MA, 1999.

[11] G. Farin, *Curves and Surfaces for CAGD, A Practical Guide, Fifth Edition*, Morgan Kaufmann Publishers, San Francisco, 1999.

[12] T.W. Sederberg, J. Zheng, A. Bakenov, A. Nasri, "T-splines and T-NURCCSs", *ACM Transactions on Graphics*, 22 (3):477–484, 2003.

[13] T-Splines, Inc., http://www.tsplines.com/maya/

[14] T-Splines, Inc., http://www.tsplines.com/rhino/

[15] H. Wang, Y. He, X. Li, X. Gu, H. Qin, "Polycube splines", *Computer Aided Design*. In press.

[16] W.P. Thurston, "Three-dimensional manifolds, kleinian groups and hyperbolic geometry", *Bulletin of the American Mathematical Society (New Series)*, 6:357–381, 1982.

[17] W.P. Thurston, *Three-dimensional geometry and topology, Vol. 1*, Princeton University Press, 1997.

[18] T.J.R. Hughes, A. Reali, G. Sangalli, "Duality and unified analysis of discrete approximations in structural dynamics and wave propagation: Comparison of $p$-method finite elements with $k$-method NURBS", *Computer Methods in Applied Mechanics and Engineering*, In press.

[19] T.W. Sederberg, G.T. Finnigan, X. Li, H. Lin, H. Ipson, "Watertight Trimmed NURBS", *ACM Transactions on Graphics*, In press.

[20] C.A. Felippa, Course notes for advanced finite element methods, http://caswww.colorado.edu/Felippa.d/FelippaHome.d/Home.html

[21] T. Belytschko, H. Stolarski, W.K. Liu, N. Carpenter, J.S.-J. Ong, "Stress projection for membrane and shear locking in shell finite elements", *Computer Methods in Applied Mechanics and Engineering*, 51:221–258, 1985.

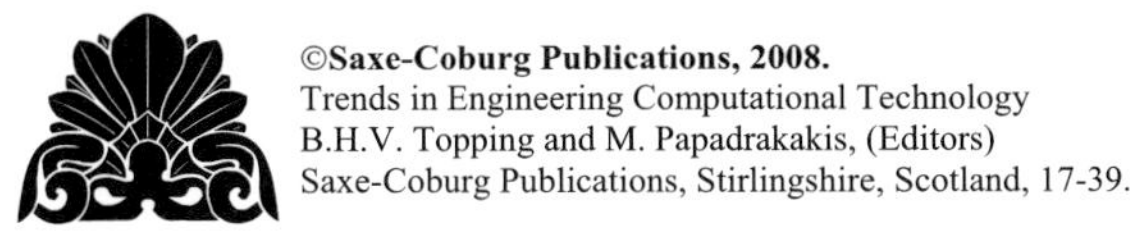
©Saxe-Coburg Publications, 2008.
Trends in Engineering Computational Technology
B.H.V. Topping and M. Papadrakakis, (Editors)
Saxe-Coburg Publications, Stirlingshire, Scotland, 17-39.

# Chapter 2

# Object-Oriented Finite Elements: From Smalltalk to Java

**D. Eyheramendy[1] and F. Oudin-Dardun[2]**
**[1] Ecole Centrale de Marseille - LMA, CNRS UPR7051**
**Marseille, France**
**[2] Institut Camille Jordan / CDCSP / ISTIL, CNRS UMR5208**
**Université de Lyon, Université Lyon 1, Villeurbanne, France**

## Abstract

Since the middle of the 1980s, the object-oriented programming has gained more and more attention in the computational mechanics community. Although the most popular language used today is C++, several languages have been used since the beginning of the object-oriented era. In this chapter, we first try to draw a panel of object-oriented approaches developed in the past years, and second to draw the main lines to design modern computational codes. From an industrial point of view, this choice remains crucial and beyond the technical problems, it may have tremendous economic consequences. The choice of the programming language may have a dramatic impact on the software architecture and cannot be considered as a simple technical choice. Several authors proposed new features based on software engineering techniques such as design patterns. This kind of approach is mandatory to handle the growing complexity of the problems addressed today, usually involving several physics at different space and time scales, and implemented on heterogeneous distributed computer systems. We advocate that a global philosophy involving a single programming paradigm in a complete environment capable of managing hardware (memory, network, GUI,...) is needed to handle this global complexity. Some basic code features are described.

**Keywords:** object-oriented programming, multiphysics, finite elements, nonlinear mechanics, parallel programming.

## 1 Introduction

The problems that are to be solved in modern computational mechanics involve more and more complex phenomena at different space and time scales. This implies in general the growth of the size of the numerical problems to be solved in this context. The use of parallel processing strategies that are to be implemented on heterogeneous computer systems may be an answer to solve this kind of problem.

Despite the developments of both programming languages and associated programming methodologies and tools, and high level commercial computational tools (ABAQUS, Nastran, COMSOL,...), the development of a new computational model in mechanics may still be tremendous. Roughly speaking, commercial tools do not generally offer enough flexibility to introduce new complex problems. Moreover, they are strongly system dependent (computer system, operating system,...). In this chapter, we wish to address the problem of the development of computational tools from a global point of view. The first aspect deals with complexity of the problem to be solved: physical problem, mathematical algorithms, The second one is the complexity of the computer system on which the code is implemented, including the networking aspects. The latter is essential for the development of large scale simulation tools. The programming language chosen for the developments still remains strategic, but is not the only parameter. The object-oriented programming has shown in the past its capability to deal with complexity. However, additional features are needed to achieve reliability and generality in the developments, *i.e.* to ensure portability on complex systems in the context of parallel programming. In Section 2, we give a brief historical point of view of object-oriented programming languages used in the design of computational mechanics applications. We discuss the main features for the design of modern computational tools. Some examples of advanced object-oriented computational codes in Java and C# are then described. High level code structuring features are presented. In Section 3, we present the key ideas for multi-fields applications in the context of finite elements code dedicated to multiphysics applications. The second aspect of the complexity of modern computational tools is described in Section 4 on the example of the development of parallel computations on heterogeneous systems.

# 2 Object-oriented programming of finite elements: improving the design of complex finite elements codes

## 2.1 From the definition of basics objects for finite elements to the solution of global problems

The choice of the programming language remains strategic to the development of computational science applications. The C++ language is today the most popular to design object-oriented finite elements application. Nevertheless, alternative languages have been tested since the 1990's in computational mechanics: Smalltalk (see Zimmermann [1]), LISP (see Miller [2]), CTalk (in the preliminary tests in Zimmermann [1]), Eiffel, Ada (see Lucas [3]), C++ (*e.g.* see Dubois-Pélerin [4] and references therein). These languages have in common to support more or less the object-oriented paradigm. The implementation is in general closely related to the definition of the abstract computational model. This is supposed to enhance the maintainability and extendibility capabilities of finite codes. This approach naturally encompasses concepts for high level architecture design that allows an evolution towards more natural mathematical languages. For the first time in Rehak [5] and Miller [6], object-oriented programming was proposed as a general methodology for finite element implementation. Both implementation examples use a LISP based system. One of the key points of the method to get better structured programs is the very high level of data abstraction capabilities of the approach. In Rehak [5], objects of matrix type appear and in Miller [6] structural objects such as node, degree of freedom, and element are described. The latter is completed in Miller [7] where object-oriented languages are discussed. In Fenves [8], the modularity and the reusability of object-oriented finite element (FE) codes are put forward and the efficiency in the design and the implementation of FE code is emphasized. The same conclusions are drawn in Forde [9]. Here an interesting comparison is performed between a classical FE code (a C program) and an equivalent object-oriented version (a mixed C - Object Pascal program). The size of the OO code is smaller, mainly due to the use of both hierarchical organization and inheritance. Similar remarks can be found in the literature (see references [10] to [23]). In Zimmermann [24] and Dubois-Pélerin [25] and [26], a complete OO environment for linear FE analysis is thoroughly discussed. A new concept is introduced here as a programming rule, "the non-anticipation rule". By never anticipating the state of the object when sending it a message, the code becomes much more robust. The extension of the ideas to nonlinear analysis can be found in Dubois-Pélerin [27] and [28], with additional interesting concepts such as "unassembled matrix" which seems to allow a more flexible implementation of solution schemes by means of alternative storage that does not necessitate the assembling of elemental matrices. This seems to be inspired from Verpeaux [29]. Menetrey [30] proposed an extension of the approach of Dubois-Pélerin [26] for nonlinear constitutive laws, here $J_2$ plasticity. In Besson [31] and Foerch [32], an advanced description of the object "material" is given. The integration of complex constitutive laws in a C++ object-oriented FE code is made easier and more flexible through the use of C++ programming rules permitting dynamic binding and linking of the code. Since then, the object-oriented paradigm has been used in many fields of computational mechanics. The interest of object-oriented programming is highlighted in applications where algorithms are complex and in large scale applications. Among applications are cited e.g.: parallel implementations of the FE code (see *e.g.* in Angus [33], Buffat [34], Hsieh [35]), rapid dynamics (see *e.g.* Potapov [36]), multi-domain analysis for metal cutting, mould filling, composite material forming (see *e.g.* Gélin [37]), in fracture mechanics (see *e.g.* Kawata [38], Nistor [39], Cojocaru [40]), fibre laminates stochastic simulation in Chung [41]... It is interesting to note an application to a symbolic treatment of the finite element method to automate the programming of codes in Zimmermann [42] and Eyheramendy [43] and [44].

## 2.2 Towards global computational approaches

It is obvious that today most of the programming languages available propose high level data structuring schemes, more or less related to object-oriented concepts. But in modern computational mechanics, the application of basic object-oriented programming concepts is probably not sufficient to build large scale easily portable applications. We focus on the development of computational applications based on advanced object-oriented concepts and advanced software technologies. Two main technologies are becoming increasingly relevant in computational mechanics today: Java and C#. The underlying programming concept remains close to one another in both approaches. The technologies supporting both implementations allow similar global developments.

Until now, most of the developments in Java computational mechanics have been considered by the computational science community, usually concerned by parallel computations. In the domain of numerical computations, Java retained attention for its networking capabilities and its Internet easy portability. For example in Nuggehally [45], a trivial application based on a boundary element method has been developed. Similar developments may be found directly on the Internet, including basic finite element applications, such as Nikishkov [46] for an application in fracture mechanics. In the computational mechanics community, Java is often considered as a simple tool to produce applications on the Internet and/or to effectuate computations on the network. For example, Java type technologies can be used to couple traditional codes written in C/C++/Fortran. This permits the developers to use ancient codes or parts of code in coupled applications. A consequence of it is to keep a real computational efficiency. For example in Miller [47], an interactive finite element application based on a coupled C++/Java is described. Comparative tests with Fortran and C are conducted on small problems using direct solvers based on tensor computations; this aims at illustrating the high efficiency computational potential of Java. In Nikishkov [48], the development of GUIs is put forward, such as in Marchand [49] on an unstructured mesh generator. In Padial-Collins [50] and VanderHeyden [51], the Java code CartaBlanca is presented. It is an environment for distributed computations of complex multiphase flows. Based on a finite volume approach, a solution scheme based on a Newton-Krylov algorithm is described. The code exhibits good performances (see numerical tests in Padial-Collins [50]). A similar environment has been developed to simulate electromagnetism problems in Baduel [52]. Both applications show the high potential of the approach to design more complex and general computational tools in mechanics for example. These developments exhibit the networking capabilities provided by Java. A large number of publications show the interest of Java and its computational efficiency: direct solution of linear systems Nikishkov [53], FFT and iterative and direct linear systems solvers on Euler type flows Bull [54], solution of Navier-Stokes flows in Häuser [55] and Riley [56]. Again, the multiprocessor management and networking capabilities of Java are put forward in these papers. In Nikishkov [53], Bull [54], Häuser [55] and Eyheramendy [56], performances of Java are tested on simple matrix/vector products. Compared to C/C++ code, only 20 to 30% of efficiency is lost. The description of a finite element code in Java was recently proposed in Nikishkov [58]. The proposed design remains rather similar to the existing ones based on classical object-oriented approaches. Nevertheless, this paper shows that it is possible to develop a global code based on a pure Java approach. The latter can be applied to design large complex applications in computational mechanics taking into account complexity in modern computational mechanics: multiscale, multiphysics and multiprocessor applications. An example is given in Eyheramendy [60] and in Eyheramendy [59] for material integration.

Firstly, enhanced and homogeneous data organization schemes to deal with complexity are proposed in Java. Secondly, developing a global application in a uniform environment including all the libraries needed is somehow an attractive idea for scientists and industrials. In the following, data organization schemes available in Java are developed.

The development of finite element application based on the language C# (.NET environment) has been investigated more recently (see Mackie [59][63][64]). The C# language possesses similar data structuring capabilities as the Java one. From this point of view, in Mackie [63], the author insists on the use of interfaces to enhance the global code structure. Thus, the specific implementation can be separated from the interface used in general classes. This permits the programmer to minimize the coupling between classes in the implementation of strongly coupled algorithms. The second aspect of the C# language used in Mackie [59] and [62] is the easy and natural programming of distributed applications. An application to modal analysis is given in Heng [66]. The development of distributed applications was made using the MPI and PVM technologies. The main drawback pointed out in Mackie [64] is that they are not mainstream technologies. Similar remarks are developed in Java in Eyheramendy [56] and [65] where the same domain decomposition method is programmed, based first on a simple multi-threaded application, and second on a distributed one. Both codes are based on the same finite element code which is a simple sequential application. The difference between both implementations is the distribution scheme and the communication process.

## 2.3 Advanced data structuring features for the design of finite element codes

Modern programming languages propose enhanced data structuring capabilities. These structures aim at better identifying the data and the associated treatments. We call it "advanced object-oriented concepts", features that help to better structure objects in the finite element codes. In modern programming languages, we can identify several concepts and among them the interface concept, the inner class concept and the package concept. Note that here we use the Java terminology. An interface is a user defined type. This can be viewed as a class specification. Let us define the three concepts: the interface concept, the inner class concept and the package concept. The interface defines a behaviour specification. From a practical point of view, this specification mainly consists of a set of names of methods. A class is said to implement an interface if and only if all the methods specified in the interface are implemented in the class. No implementation of the methods is proposed within the interface. This feature can be really interesting to enhance the robustness of codes by programming independent objects (closely related to the concept of thecomponent). It allows the programmer to impose specifications and rules to a given class. For example, a given algorithm can be used in a given context and not in another context. Generally speaking, the interface clearly separates the use of several objects in a given algorithm from their specific implementation without freezing its implementation into a classical hierarchy of classes. Thus, an object may be involved in a structural class organization and may be involved with different kinds of objects in a given algorithm. Applications of this concept may be found in the context of iterative solvers in [5] and [63] and constitutive law integration in [60]. An inner class is a class that is defined inside another one. The location of the class defines different types of classes and of course different behaviours. These declinations of the concept of inner class are: static member class, member class, local class and anonymous class. All of them share the same basic principle. The advantage of these various kinds of inner classes is that the class can be defined in the right place, where just needed. These schemes allow a very local and partial sharing of data without compromising their global data encapsulation. This kind of organization can be applied to finite elements codes in various situations. In the general description finite elements, many situations in which local and global aspects of data organization are mixed occur. This occurs for example in the context of fields. For a given problem, a discrete field is defined at the global level, *i.e.* on a global domain or subdomain. The definition of this field is based on the definition of elemental fields that may be viewed as local data. The latter defines for example the interpolation based on the nodal description. Finally, a package is a collection of classes and/or interfaces that may partially share data and behaviour. From a practical point of view, a package is a directory which contains files defining classes and interfaces. As matter of fact, the package is named and packages have a hierarchical organization. But the definition of a package is more than this simple directory organisation. The name of a package defines a namespace. This scheme introduces an additional criterion for the scope resolution of data for classes. Partial data sharing may exist in the scope of the package. The simplest application of the package notion is the thematic classification of classes and interfaces. For example, for a finite element code, one could consider global packages to deal with the mesh management, the finite element formulations, the material and their integrators, etc... The three programming features presented here exist or can be simulated more or less in classical programming languages, such as C#, Java. To a lesser extent, these features can be emulated in C++, but note that compiler dependency may still today affect the portability of applications.

# 3 Multi-fields finite elements for multiphysics

## 3.1 Basis for the management of multi-fields applications

Since the origin of object-oriented finite elements, a large number of papers have proposed different kinds of classes' organization. In the approach proposed here, the underlying principle is a clear separation of the geometry description, the mesh description and the different fields needed for a given problem (pressure, velocity, density, stress…). The organization is based on a multi-level data description. Every level is linked to the other one. As shown in Figure 1, the data for a finite element problem is based on a three level description. Geometries such as points and lines (see the Geometry level in Figure 1) permit us to describe the topology of the domain. The description of the mesh (Mesh level in Figure 1) consists of the description of the basic geometries. Different meshes can be based on the same geometry. At this level, the mesh is described in terms of geometric nodes and elemental geometries. Only the basic kinematics of the element are given at this level. Typical features needed for the finite element formulation will be based on the kinematics provided by this elemental geometry. The fields, in the finite element sense (level Field in Figure 1) are defined in a similar way. The field is defined by a set of elemental fields. The discrete description of the field can be made at the level either of a node or of a Gauss point. The description of the elemental description of the field follows the same multilevel description as the field. As shown in Figure 1, the elemental definition of the field is based on the element in the geometric sense. An elemental geometry can be used by several elemental fields. It can have either a set of nodes or a set of gauss points. Information passing between the different levels is made possible by the management of links between the objects of each level (vertical direction in Figure 1). For example, an elemental field knows the underlying elemental geometry on which it is defined. As a matter of fact, it gives access to the kinematics of the element (*i.e.* shape functions and shape function derivatives). It is worth noting that data is never duplicated in this organization. The global organization briefly discussed here permits the developer to focus first on the physics that couple the fields, and second on the solution algorithms. This data organization of fields allows a natural description of multi-fields and multi-physics problems. Moreover, the implementation of solution algorithms is made easier by proposing a global data, the field, and a local data, the description of the field at the element level. Each of them can be involved in the algorithm (*e.g.* global organization of the algorithm leading to a successive solution of linear systems for which the left and right-hand sides of the system are formed by the elemental matrices computed at the local level).

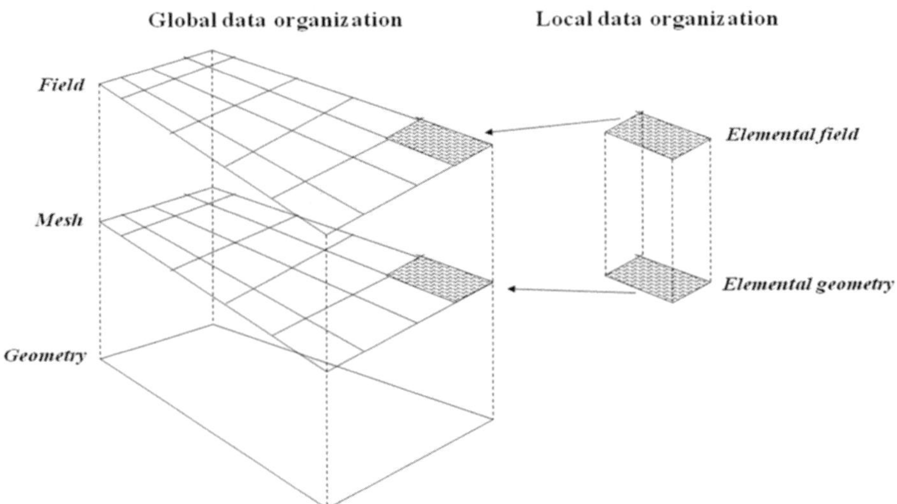


Figure 1: Global data organization versus local data organization.

## 3.2 Application of object-oriented finite elements programming in JAVA to a Seepage problem

In Section 2, typical features of O.O programming have been presented. In this section we give an example of finite element formulation implementation in Java. We give an example of the use of the concept of inner class. As described in the previous section, inner classes allow the programmer to partially hide information to the whole of the code. It may be interesting to share partially global and local aspects of a numerical scheme. The approach is illustrated on an interface tracking scheme for underground water flow that may be found in [66]. The management of the finite element formulation is described in this example.

### 3.2.1 A finite element formulation of a free surface seepage problem

A procedure to locate precisely the free surface of unconfined seepage flow through porous media is described in [66]. Let us briefly recall the problem. We consider the flow of one incompressible and homogeneous fluid into a porous medium. The medium is assumed to be either wet (saturated) or dry. Capillarity, partial saturation and evaporation are neglected. The free surface is defined as the boundary line between the dry and wet soils as shown in Figure 2 (free surface CDE). No flux gets through the free surface and the pressure is zero on the free surface. The domain $\Omega_w$ represents the flow region (the saturated part of the geometric domain $\Omega$ occupied by the earth structure) and $\Omega_d$ the dry part. The equations of the problem are summarized in Table 1.

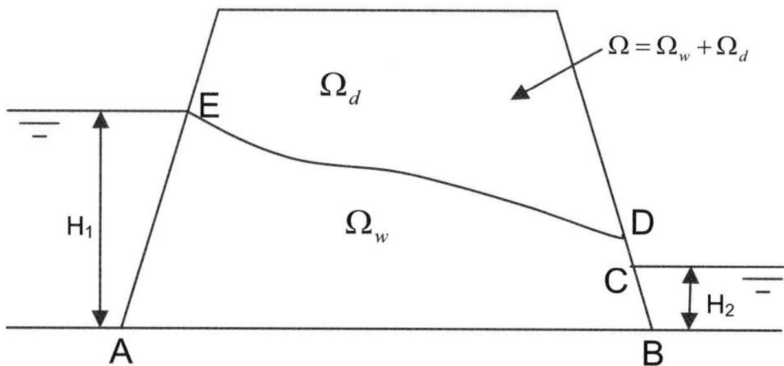


Figure 2: Definition of the free seepage problem

The piezometric head $\Phi$ is defined with respect to the pressure $p$ as:

$$\phi = \frac{p}{\rho g} + y = \frac{p}{\gamma} + y$$

where $\rho$ is the density of the fluid and $\gamma$ the specific weight of the fluid.
The steady state problem may be modeled as follows:

$p \geq 0$ in $\Omega_w$

$p = 0$ in $\Omega_d$

$\nabla . k . \nabla \phi = 0$ in $\Omega$

where $k$ is the Darcy permeability tensor.
The boundary conditions are expressed as follows:

$p = \gamma(H_1 - y)$ on AF

$p = \gamma(H_2 - y)$ on BC

$p = 0$ and $n . k . \nabla \phi = 0$ on ED

$p = 0$ and $n . k . \nabla \phi = 0$ on CD

$n . k . \nabla \Phi = 0$ on AB

where $n$ is the outward unit vector defined on the domain $\Omega_w$.

Table 1: Definition of the seepage problem

The difficulty of the problem is that the location of the free surface is unknown *a priori* and enforcement of the boundary conditions on the free surface may be tremendous. A simple weak formulation was proposed in [67]. The formulation is extended to the entire domain $\Omega = \Omega_w \cup \Omega_d$ by using an extended pressure field *p(x)* for which the $p(x) = 0$ in $\Omega_d$ -the dry soil domain-. Existence and uniqueness of the solution for the extended problem is shown by introducing the following penalized problem:

$$\begin{cases} \text{Find } p_\varepsilon \in P = H^1(\Omega_w) \text{ such that } \forall q \in H_0^1(\Omega_w) \\ \int_\Omega \frac{1}{\gamma} k \nabla p_\varepsilon \nabla q d\Omega = -\int_\Omega k\, H_\varepsilon(p) \nabla y \nabla q d\Omega \end{cases} \quad (1)$$

where $H_\varepsilon(p)$ is an extension of the Heaviside function defined by:

$$H_\varepsilon(p) = \begin{cases} 1 \text{ if } p \geq \varepsilon \\ \frac{p}{\varepsilon} \text{ if } 0 \leq p \leq \varepsilon \\ 0 \text{ if } p \leq 0 \end{cases} \quad (2)$$

Since the location of the free surface is unknown *a priori*, a Newton iterative scheme is required to solve the problem. The same stopping criteria as the one found in [30] are used to solve the penalized problem. A local computation (at the elemental level) is done to determine of the penalty parameter $\varepsilon$ and a cut-off pressure $p_o$ which enforce the pressure to a small negative in the dry soil. The discretized form of the variational problem (5) may be written as follows using a simple linearization technique:

$$\begin{cases} K_t \delta p^n = f - Kp^n \\ p^{n+1} = p^n + \delta p^n \end{cases} \tag{3}$$

The elemental contributions may be written as follows as printed in Table 2 where $A$ denotes the assembling operator and where the $N_a$ are the shape functions.

$$K = K_t = A(k^e)$$
$$p = \sum N_i p_i^e = Np^e$$
$$\nabla p = \begin{bmatrix} N_{1,x} \cdots N_{n,x} \\ N_{1,y} \cdots N_{n,y} \end{bmatrix} \begin{bmatrix} p_1^e \\ \vdots \\ p_n^e \end{bmatrix} = Bp^e$$
$$k^e = \int_{\Omega^e} \frac{K}{\gamma} \nabla p \nabla q d\Omega = \sum_{GP} p^{e^t} B^t DBq^e |J| w_i$$
$$f = A(f^e)$$
$$f^e = \int_{\Omega_e} K H_\varepsilon(p) \nabla y \nabla q d\Omega = \sum_{GP} K H_\varepsilon(p) e_2 Bq^e |J| w_i$$
$$e_2^t = (0,1)$$

Table 2: Elemental matrices for the penalty free surface seepage problem

### 3.2.2 A Java implementation of the seepage problem

As pointed out in Section 3.1, the main object of the application is the object field. The nodal field supports all the nodes and the values of degrees of freedom. The formulation initializes the fields, *i.e.* in this case only one scalar field, the pressure. The definition of the field is automatically built over the discrete domain and its data are involved in the computation of the elemental matrices $k_t^e, k^e, f^e$ at the level of the element supporting the formulation. The implementation of the formulation is made through a class called: **PressureDarcyPenalizedProblem**. The class is posted in Table 3. It belongs to a package called **femjava.fem.formulation.basics**. The class subclasses **Formulation** in which all the basic behavior of the F.E formulation is taken into account, *e.g.* the global way of building fields over all the computational domain. The class embeds an inner class called Darcy; the latter is an inner class, in this case a static member class. The two important methods of the class are:

- *initialize()* which permits us to describe the unknown field, here a scalar field, the pressure, for the problem; this a global method

- o *getElement()* which permits us to instantiate at the local level the finite element called **PressureDarcy** with local elemental fields, numerical quadrature (which may play a crucial role for constitutive law modeling) and material definition

The finite element **PressureDarcy** is defined as an inner static class and exhibits a local behavior which corresponds to the computation of the finite element matrices defined Table 2. The fact that the class is defined as static means that the instantiation of it is intimately linked to the outer class **PressureDarcyPenalizedProblem**. Moreover, this is a subclass of class **Element** which embeds all the behavior mandatory to deal with elemental data such as nodes, attached geometry (the type of element which processes the finite element interpolations), *etc*… The definition of one class and its inner class permits us to completely define a new finite element formulation. The global solution algorithm, *i.e.* the Newton-Raphson procedure, is independent from the definition of the formulation from a programming point of view. The inner class concept available in Java can be seen as a generalization of the O.O concept at the level of a class. This leads to a safer organization of code.

```
package femjava.fem.formulation.basics ;
// Packages inclusion
public class PressureDarcyPenalizedFormulation extends Formulation
{
  public static double Tolerance = 1e-3 ;
  public static double LargeParameter = 1e10 ;

  public String toString() { // … }
  public Material defaultMaterial() { return new PorousMedia () ; }
  public void initialize( Domain domain ) { // … }

  public Element getElement( ElementalGeometry aGeom , Quadrature aQuadrature ,
ElementalField[] flds, int nb , Material m )
  {
    return new PressureDarcy ( aGeom , aQuadrature , flds , nb , m ) ;
  }

  public static class PressureDarcy extends Element
  {
    public PressureDarcy( ElementalGeometry aGeom , Quadrature aQuadrature ,
ElementalField[] flds ,
                              int n , Material m ) { // ... }
    public FullMatrix computeConstitutiveMatrix() { // ... }
    public Hashtable computeElementalMatrices( TimeStep ts ) { // ...}
    protected double computeCutOffPressure( ElementalGeometry aGeometry , double gamma )
{ // ...}
    protected double computeEpsilon( ElementalGeometry aGeometry , double gamma ) { // ...}
    // ... additional non-implemented abstract methods ...
  }
}
```

Table 3: Global description of the implementation of the seepage formulation

### 3.2.3 Numerical application

For all the test cases, the permeability $k$ is set to identity tensor and the specific weight of water $\gamma$ is set to 1.0. Bilinear quadrilateral elements have been used. For all the numerical tests, piezometric heads are indicated in Figure 3. We obtained similar results to the one presented in [66]. The first test shown Figure 3 is a homogeneous dam. The solution exhibits a surface of seepage. In the second test, the dam has a toe drain. In the last test, the dam has an impermeable sheet on the upper part of the right face. On these cases we tested various strategies for the choice of the penalty parameter $\varepsilon$, cut-off pressure definition, penalty parameter to impose the zero-pressure overall the domain. The tests shown here converge in about 10 iterations. The purpose of this work was to rapidly evaluate this formulation and to check its performances. We have noted that the precision of the solution depend drastically on the choice of various parameters involved in the formulation. The way to impose the boundary conditions for the pressure especially on the free surface seepage zone may lead to instable results. These experiments lead us to adopt an alternative strategy for the solution of the free surface seepage problem by introducing a free surface equation and solving the flow using a velocity-pressure formulation.

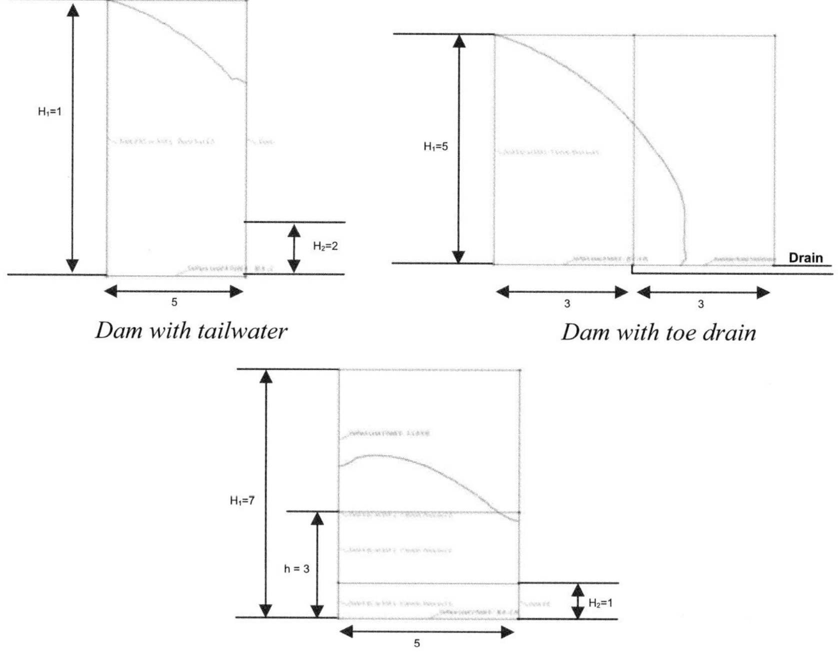


Figure 3: Numerical examples for various kinds of Darcy's flow

# 4 A step forward generalized parallel implementations of finite elements: Application to a domain decomposition method finite element codes

We show that a classical sequential computational code can easily be parallelized by creating adequate classes corresponding first to the global parallel algorithm, and seond to the data exchanges needed in the algorithm. This allows the use of the same finite element formulation either in the context of a sequential application or a parallel one.

## 4.1 A Schwarz domain decomposition method

A domain $\Omega$ is decomposed in two overlapping subdomains $\Omega_1$ and $\Omega_2$ as shown in Figure 4. $\Gamma_1$ is the part of the boundary of $\Omega_1$ which is in $\Omega_2$, and $\Gamma_2$ the part of the boundary of $\Omega_2$ which is in $\Omega_1$. Consider the following boundary value problem to be solved in $\Omega$:

- Find $u$ with appropriate regularity conditions such that $Lu = f$ in $\Omega$, where $u$ satisfies the boundary conditions.
    - $L$ is a differential operator,
    - $f$ a regular function defined on the domain $\Omega$.

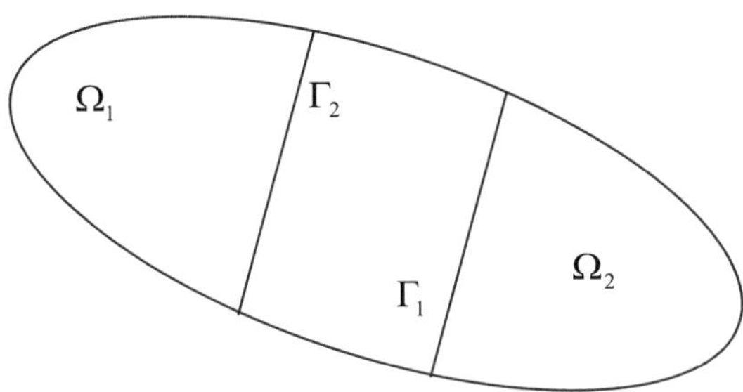


Figure 4: Decomposed domain

The Schwarz alternating method consists of solving in an iterative process the following problems on the subdomains $\Omega_1$ and $\Omega_2$:

At iteration $i$ solve:

- $Lu^{2i} = f$ in $\Omega_1$ with the initial boundary conditions and $u^{2i}\big|_{\Gamma_1} = u^{2i-1}\big|_{\Gamma_1}$
- $Lu^{2i+1} = f$ in $\Omega_2$ with the initial boundary conditions and $u^{2i+1}\big|_{\Gamma_2} = u^{2i}\big|_{\Gamma_2}$

This method can easily be parallelized for *n* subdomains (see Figure 5). Consider the examples of decomposition for a domain. Subdomains are alternatively coloured in black and grey. We consider here a Schwarz multiplicative method:

- At each iteration:
  - o Solve the grey subdomains
  - o Exchange boundary values from the grey subdomains to the black ones
  - o Solve the black subdomains
  - o Exchange boundary values from the black subdomains to the grey ones

This method generally converges.

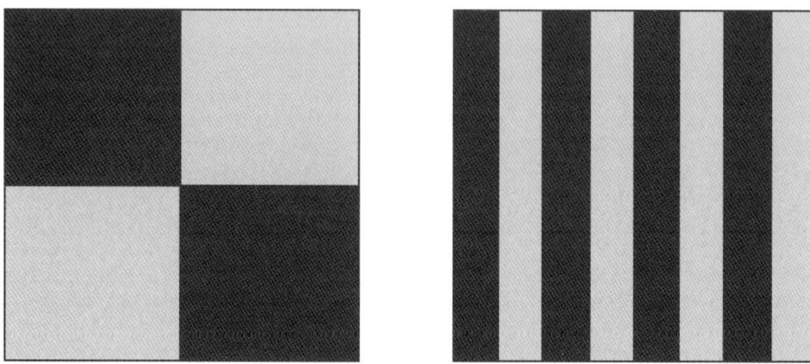

Figure 5: Typical decomposed domains

## 4.2 A Multithreaded implementation of a Schwarz domain decomposition method

The classes for a generic object-oriented domain decomposition method are given in Figure 6. The abstract class **DDAlgorithm** is inherited by the **MultiThreadedSchwarzMultiplicative** class. The latter inherits the attributes *decomposedDomains* (the domains), *epsilon* (the relative residual value) and *maxIteration* (the maximum number of iterations for the algorithm), and the methods needed to manage saving operations during computation. The **MultiThreadedSchwarzMultiplicative** class has no attribute, except the ones inherited from the superclass, and implements the method *solve* which is summarized Figure 7. The structure of the method is given on the left side. Each colour is solved successively. The solution block, which has three steps, an initialization step , a solution step and a synchronization of processes step is given at the top right in Figure 7. The initialisation of the thread for the solution of the domain is given at the bottom right in Figure 7. Note that the thread is instantiated to solve the problem corresponding to a single subdomain to the message called *solveSchwarz()* sent to a given subdomain. This shows the simplicity of the programming of the management of multiple processes in Java which mainly consists of three points: the creation of the thread, the starting of the threads and the global synchronization of threads (similar to MPI barriers).

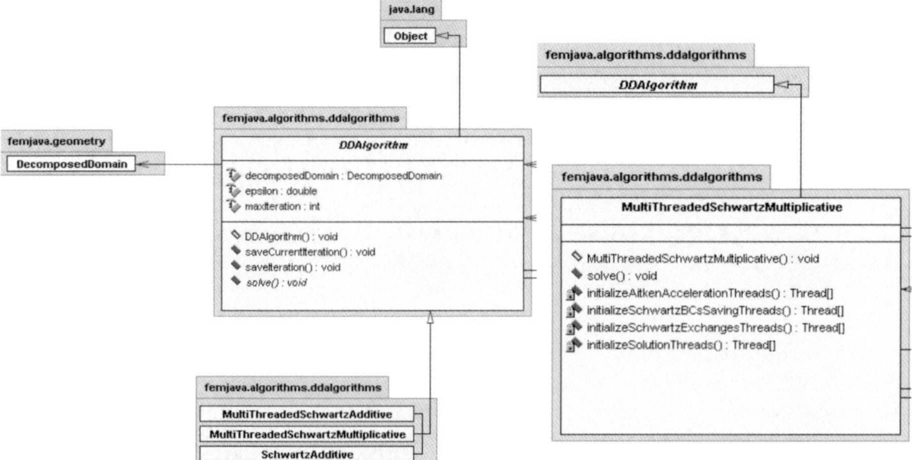


Figure 6: UML diagrams of classes for Domain Decomposition Methods

```
public void solve( )
  {
    // ... INITIALIZING OF THE SOLVER

    for( int it = 0 ; ( it < maxIteration ) && ( ! iteration.converged() ) ; it++ )
    {
       for( int color = 0 ; color < 2 ; color++ )
       {
          int anteColor = ( color + 1 ) % 2 ;


       //*************** solution ****************

       // SOLUTION BLOCK
       // ...

       //*************** exchanges ****************

       // EXCHANGES BLOCK
       // ...

       }

    // FINALIZING THE ITERATIONS
    // ...
    }
  }
```

```
//*************** solution ****************
Thread[] solutionThreads = this.initializeSolutionThreads ( color ) ;

for( int i = 0 ; i < solutionThreads.length ; i++ )
       solutionThreads[i].start () ;

try
{
     for( int i = 0 ; i < solutionThreads.length ; i++ )
            solutionThreads[i].join () ;
}
catch( InterruptedException e )
     System.out.println ( e ) ;
```

```
Thread threadii = new Thread ( new Runnable ()
        {
          int number = Ni ;

          public void run()
          {
            domains[number].solveSchwarz () ;
          }
        });
```

Figure 7: Simplified method *solve*, solution block and threads initialization

## 4.3 Multithreaded versus distributed implementation

The implementation proposed in the previous section for a shared memory system can easily be extended to a distributed memory version by taking into account the communication scheme. The same algorithm can naturally be extended to a distributed objects version using the Remote Method Invocation. Communication between distributed objects can be implemented by using either the Remote Method Invocation scheme or Sockets type communication schemes. The basic management of thread would remain unchanged as shown in Figure 8. Solving both problems is made through the same basic code for the physics and numerical algorithm. The general structure of the code organization is the same. In the distributed version, the global Schwarz algorithm is managed through a thread located on the computer 1. The algorithm consists of two points: the management of the domains located on alternative computers and the communication schemes between these domains located on various computers. The global management is held by way of distributed object. The Java RMI package is used for that purpose. It permits the programmer to keep a natural and homogeneous organization of classes to deal with the domain decomposition algorithm. The consequence of it is to keep exactly the same structure as in both algorithms, managing real objects in the one case, and distributed objects in the other case. The latter remains seamless for the programmer. The drawback in using the RMI package is the lack of efficiency in the communication. This is the reason why communications are programmed using the Socket Java class which is a point to point communication scheme. This scheme is implemented at the level of boundary condition object where the information provided by the neighbor domain is needed. Only a local implementation is needed within the global framework. From a practical point of view, the concerned boundary is associated with a new type of boundary condition object able to manage communications between two geometries (points, lines, surface –for 3D cases-). For illustration purposes, both schemes are applied to a Navier-Stokes flow.

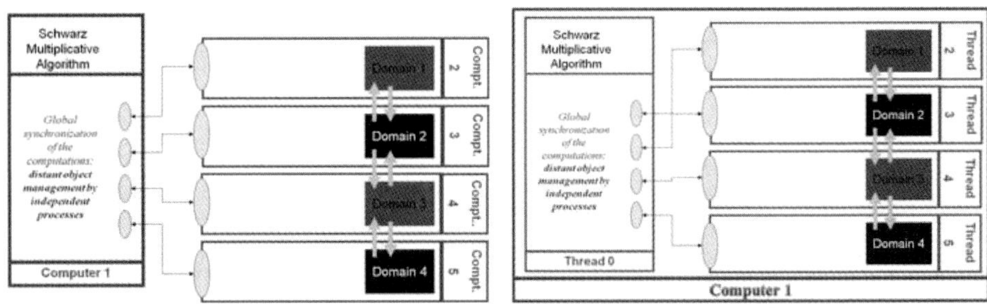


Figure 8: Multi-threaded application versus distributed application

## 4.4 Application to a periodic Navier-Stokes flow

To illustrate both approaches, we study a Galerkin Navier-Stokes formulation stabilized by adding least-squares type terms. The equations of the problem and the formulation are presented in Figure 9. Linearization of the problem is introduced in a Newton like scheme in the Schwarz multiplicative scheme. A direct linear system solver based on a Crout decomposition is used to solve the linear system at each iteration. Note the same code is compiled once and run on all systems. The multi-threaded version of code is run on a SGI Altix 16-processor Itanium2 1.3Ghz, 32 Go of RAM. The distributed version is run on a set of simple PCs linked by a classical network available in a class room. The code used for both applications is exactly the same. The global solution algorithm and the boundary conditions in charge of the communication scheme are of course not the same. The formulation is applied to the computation of a flow through a set of cylinders shown in Figure 10. The domain is a periodic layer of cylinders. Numerical results are given in Figure 10. First, the pressure contour on a typical cell is plotted. Secondly, the mean velocity computed over all the domain is given with respect to the gradient of pressure over the cell in the direction of the main flow. The latter represent the homogenized flow through the cylinders. For low velocities, the relation between the mean velocity and the gradient of pressure is linear. From a global point of view this can be assimilated to a Darcy's flow. But as far as the mean velocity is increasing the linearity disappears. The advection term is no more negligible and the global flow cannot be assimilated to Darcy's flow. It shows the influence of the advection term on the homogenized flow, that cannot any more be considered as a linear Darcy's flow. Filtration laws could be established in such a way. The same computation has been held using both approaches. From a practical point of view, we show that the same code can be run on heterogeneous systems. As both systems are different, we cannot compare the efficiency of both algorithms. This goes beyond the scope of our qualitative test. We show here that we can easily switch from one computer to another one without any problem using exactly the same code without new compilation. This feature is of major interest from an industrial point of view.

$$
\begin{array}{ll}
\sigma_{ij,j} + f_i = \rho(u_i u_{i,j} + u_{i,t}) & \text{on } \Omega \times T \\
u_{i,j} = 0 & \text{on } \Omega \times T \\
\sigma_{ij} n_j = \bar{F}_i & \text{on } \partial_2 \Omega \times T \\
u_i = \bar{u}_i & \text{on } \partial_1 \Omega \times T \\
\sigma_{ij} = -p\delta_{ij} + 2\mu\varepsilon_{ij}(u) & \text{on } \Omega \times T \\
\varepsilon_{ij}(u) = \frac{1}{2}(u_{i,j} + u_{j,i}) & \text{on } \Omega \times T \\
u_i(0) = u_{i0} & \text{on } \Omega \text{ at } t = 0
\end{array}
$$

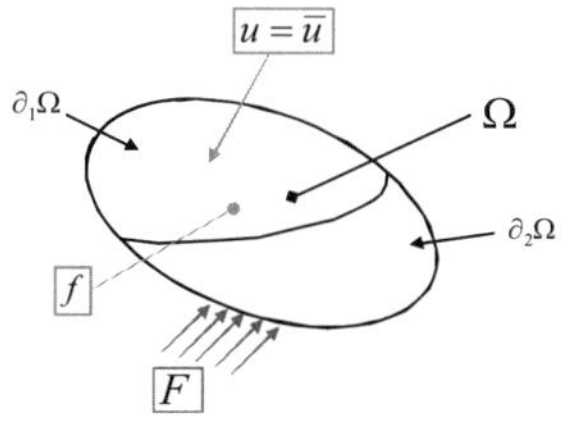

• Given $f$, find $(u^h, p^h) \subset ((\mathcal{S}^h)_n \times (\mathcal{P}^h)_n)$ such that for each $(w^h, q^h) \subset ((\mathcal{W}^h)_n \times (\mathcal{P}^h)_n)$, one has

$$
\int_\Omega \rho u_j^h u_{i,j}^h + \rho u_{i,t}^h dv - \int_\Omega 2\mu \varepsilon_{ij}(u^h) \varepsilon_{ij}(w^h) dv + \int_\Omega p_h w_{i,i}^h dv + \int_\Omega u_{i,i}^h q^h dv - \int_\Omega f_i w_i^h dv
$$

$$
+ \sum_{\Omega^e \in \Omega^h} \left[ \int_{\Omega^e} (\rho u_j^h u_{i,j}^h + \rho u_{i,t}^h - 2\mu \varepsilon_{ij,j}(u^h) + p_{,i}^h - f_i) \tau_{mom} (\rho u_j^h w_{i,j}^h + q_{,i}^h) dv \right] = 0
$$

Figure 9: Navier-Stokes equations and stabilized finite element formulation

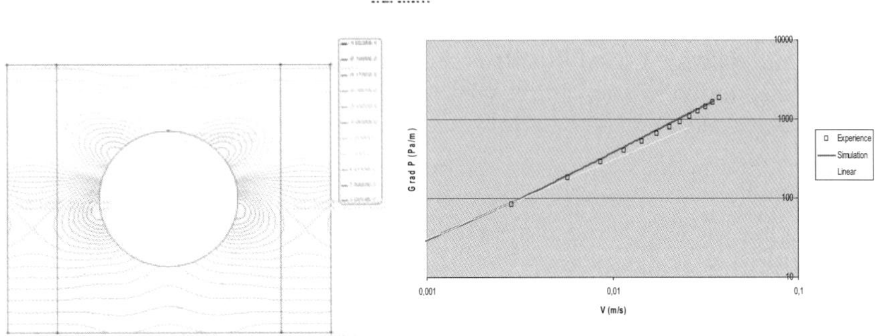


Figure 10: Computational domain – Numerical results

# 5 Conclusion

In this chapter we have briefly recalled the use of different object-oriented programming languages in the context of computational mechanics. We have advocated a global and uniform design of finite element codes, including both a single programming concept based on advanced object-oriented concepts and libraries included in the platform of developments. The first point is illustrated on an example of the use of inner classes in the context of finite element formulations. This allows global and local definitions of the formulation within the same context. The second point is illustrated on two implementations of a Schwarz overlapping domain decomposition method based on Java libraries for multithreading and distributed objects manipulations. We advocated that even the object-oriented paradigm remains indispensable for structuring large scale application, but that additional skills are needed. Global approaches such as the one proposed in Java and C# (.NET) become increasingly relevant in computational mechanics. This kind of approach offers both a high level object-oriented programming language and high level libraries. Programming using a single concept and language, *i.e.* without the requirement of external libraries or tools, somehow simplifies the developments of finite element codes. The homogeneity of codes guarantees a faster extension and easier portability typically in the case of distributed computing. The choice of the programming language remains strategic in the development of computational science applications. It is in general closely related to the definition of the abstract computational model. An important issue in the choice of the language for computational mechanics remains the numerical efficiency but has to be put into perspective with additional feature. Some technical issues in the choice of a programming language or computational tool are: abstraction capabilities of the language, modification and reusability of the codes, efficiency of the language, availability of libraries (GUI, files systems, networking, data persistency, mathematical libraries, data base,…) , user-friendliness of developments tools (team management developments, collaborative developments, large scale application design, design capabilities…), integration of other languages, documentation management,… It is worth noting that design patterns may help the programmer in handling the complexity of application (see Gamma [70] for concepts and Heng [59] for application to finite elements). At last, the Internet may become in the near future an important feature for the deployment of computational tools. The technologies presented in this chapter offer a natural potential for the portability of computational tools.

## References


[1] Th. Zimmermann, Y. Dubois-Pèlerin, P. Bomme, "Object-oriented finite element programming : I. Governing principles", Comput. Methods Appl. Mech. Engrg., 98, 291-303,1992.

[2] G.R Miller., "A LISP-Based Object-Oriented approach to structural analysis", Engr. with Comp.,4, 197-203, 1988.

[3] D. Lucas, B. Dressler, D. Aubry, "Object-oriented finite element programming using the ADA language", Numerical Methods in Engineering '92, C. Hirsh *et al.* (Eds.), 591-598, 1992.

[4] Y. Dubois-Pèlerin, Th. Zimmermann, "Object-oriented finite element programming : III. An efficient implementation in C++", Comput. Methods Appl. Mech. Engrg., 108, 165-183, 1993.

[5] D.R. Rehak, J.W. Baugh Jr., "Alternative Programming Techniques for Finite Element Programming Development", Proceedings IABSE Colloquium on Expert Systems in Civil Engineerings, Bergamo, Italy. IABSE, 1989.

[6] G.R Miller., "A LISP-Based Object-Oriented approach to structural analysis", Engr. with Comp., 4, 197-203, 1988.

[7] G.R Miller., An Object-Oriented Approach to Structural Analysis and Design, Computers & Structures, 40, 75-82, 1991.

[8] G.L. Fenves, Object-Oriented programming for engineering software development, Engr. with Comp., 6, 1-15, 1990.

[9] B.W.R Forde, R.O. Foschi, S.F Stiemer, "Object-Oriented Finite Element Analysis", Computers & Structures, 34, 355-374, 1990.

[10] J.S.R.A. Filho, P.R.B. Devloo, Object-Oriented programming in scientific computations : The beginning of a new area, Engineering Computations, 8, 81-87, 1991.

[11] Y. Dubois-Pèlerin, P. Bomme, Th. Zimmermann, "Object-oriented finite element programming concepts", Proceedings of European conference on new advances in computational structural mechanics, ed. P. Ladevèze, O.C. Zienkiewicz, Elsevier Science Publishers, 95-101, 1991.

[12] J.W. Baugh, D.R. Rehak, "Data abstraction in engineering software development", Comp. Civ. Engr., 6, 282-299, 1992.

[13] P.R.B. Devloo, C.A. Magalhaes, A.T. Noel, "On the implementation of the p-adaptive finite element method using the object oriented programming philosophy", in Numerical methods in engineering and applied sciences, part 1 , CIMNE, Barcelona, 1992.

[14] P.R.B. Devloo, "An object oriented approach to finite element programming (Phase I): a system independent windowing environment for developing interactive scientific programs", Advances in Engineering Software, 14, 41-46, 1992.

[15] J.T. Ross, J.P. Morrow, L.R. Wagner, G.F. Luger, "Two paradigms for OOP models for scientific applications", Proceedings of $8^{th}$ conf. held in conjunction with AEC Systems 92, Dallas, June 7-9 1992, Ed. Barry, Goodno and Wright, ASCE, 535-542, 1992.

[16] J.T. Ross, L.R. Wagner, G.F. Luger, "Object-oriented programming for scientific codes. I: Thoughts and concepts", Comp. Civ. Engr., 6, 480-496, 1992.

[17] J.T. Ross, L.R. Wagner, G.F. Luger, "Object-oriented programming for scientific codes. II: Examples in C++", Comp. Civ. Engr., 6, 480-496, 1992.

[18] R.I. Mackie, Object-oriented programming of the finite element method, Int. J. Num. Meth. Engr., 35, 425-436, 1992.

[19] S.P. Scholz, "Elements of an object-oriented FEM ++ program in C++", Comp. and Struct., 43, 517-529, 1992.

[20] L.O. Nielsen, "A C++ class library for FEM special purpose software", Internal report, Department of Structural Engineering, Technical University of Denmark, Serie R vol. 308, 1994.

[21] G.W. Zeglinski, R.P.S. Han, "Object oriented matrix classes for use in a finite element code using C++", Int. J. Num. Meth. Engr., 37, 3921-3937, 1994.

[22] J. Drolet, "Towards a cross-platform finite element application framework: A toll to simplify finite element simulations", Proceedings of $1^{st}$ Structural Specialty Conference , May $29^{th}$ to June $1^{st}$ 1996, Edmonton,Canada, 1996.

[23] R.I. Mackie, "Using Objects to handle complexity in Finite Element Software", Eng. with computers, 13, 99-111, 1997.

[24] Th. Zimmermann, Y. Dubois-Pèlerin, P. Bomme, "Object-oriented finite element programming : I. Governing principles", Comput. Methods Appl. Mech. Engrg., 98, 291-303, 1992.

[25] Y. Dubois-Pèlerin, Th. Zimmermann, P. Bomme, "Object-oriented finite element programming : II. A prototype program in Smalltalk", Comput. Methods Appl. Mech. Engrg., 98, 361-397, 1992.

[26] Y. Dubois-Pèlerin, Th. Zimmermann, "Object-oriented finite element programming : III. An efficient implementation in C++", Comput. Methods Appl. Mech. Engrg., 108, 165-183, 1993.

[27] Y. Dubois-Pélerin, P. Pegon, "Object-Oriented programming in nonlinear finite element analysis", Computers & Structures, 67(4), 225-241, 1998.

[28] Y.D. Dubois-Pélerin, P. Pegon, "Improving modularity in object-oriented finite element programming", Commun. numer. methods engin., 13, 193-198, 1997.

[29] P. Verpeaux, T. Charras, A. Millard, "CASTEM 2000 : une approche moderne du calcul de structures", Calcul des structures et intelligence artificielle, Vol. 2, J.M. Fouet, P. Ladevèze et R. Oyahon Eds., Pluralis, 1988.

[30] Ph. Menétre, Th. Zimmermann, "Object-Oriented Non-Linear Finite Element Analysis : Application to J2 plasticity", Computers & Structures, vol. 49 n° 5, 767-777, 1993.

[31] J. Besson, R. Foerch, "Large scale object-oriented finite element code design", Comput. Methods Appl. Mech. Engrg., 142, 165-187, 1997.

[32] R. Foerch, "Un environnement orienté objet pour la modélisation numérique des matériaux en calcul des structures", Ph.D. thesis report, Ecole Nationale Supérieure des Mines de Paris, 1996.

[33] I.G. Angus, Parallelism, "Object-Oriented Programming methods, Portable software and C++", Proceedings of 8th conf. held in conjunction with AEC Systems 92, Dallas, June 7-9 1992, Ed. Barry, Goodno and Wright, 506-513, 1992.

[34] M. Buffat, I. Yudiana, C. Leribault, "Parallel simulation of turbulent compressible flows with unstructured domain partitioning. Performance on T3D and SP2 using OOP", in Parallel computational fluid dynamics: Algorithm and results Advanced Computers, Eds. Schiano, A. Ecer, J. Periaux and N. Satofuka, Elsevier, 76-83, 1992.

[35] S.H. Hsieh, E.D. Sotelino, "A message-passing class library C++ for portable parallel programming", Eng. with computers, 13, 20-34, 1997.

[36] S. Potapov, "Un algorithme ALE de dynamique rapide basé sur une approche mixte Eléments finis-Volumes finis", Implémentation en langage orienté objet C++ , PhD thesis report, Ecole Centrale Paris, 1997.

[37] J. C. Gelin, L. Walterthum, Conception d'un logiciel orienté-objets pour la simulation de processus de formage, Actes du 2nd Colloque national en calcul des structures, Giens, Hermès, 552-558, 1995.

[38] H. Kawata, S. Yoshimura, G. Yagawa, H. Kawai, "Object-oriented system for evaluation of fracture mechanics-Parameters of linear and nonlinear 3D cracks", Proceedings of IECS 95, S.N. Atluri, G. Yagawa, T.A. Cruse (Eds.),1, 39-44, 1995.

[39] I. Nistor, O. Pantalé, S. Caperaa, "Numerical implementation of the eXtended Finite Element Method for dynamic crack analysis", Advances in Engineering Software, 39(7) 573-587, 2008.

[40] D. Cojocaru, A.M. Karlsson, "An object-oriented approach for modeling and simulation of crack growth in cyclically loaded structures", Advances in Engineering Software, In Press 2008.

[41] D. B. Chung, M. A. Gutiérrez, R. de Borst, "Object-oriented stochastic finite element analysis of fibre metal laminates", Computer Methods in Applied Mechanics and Engineering, 194(12-16), 1427-1446, 2005.

[42] Th. Zimmermann, D. Eyheramendy, "Object-oriented finite elements : I. Principles of symbolic derivations and automatic programming", Comput. Methods Appl. Mech. Engrg., 132, 277-304, 1996.

[43] D. Eyheramendy, Th. Zimmermann, "Object-oriented finite elements : II. A symbolic environment for automatic programming", Comput. Methods Appl. Mech. Engrg., 132, 259-276, 1996.

[44] D. Eyheramendy, Th. Zimmermann, "Object-oriented finite elements : III. Theory and application of automatic programming", Comput. Methods Appl. Mech. Engrg., 154, 41-68, 1998.

[45] M. Nuggehally, Y.J. Lui, S.B. Chaudhari, P. Thampi, "An internet-based computing plateform for the boundary element method", Adv. In Engrg. Software, 34, 261-269, 2003.

[46] G.P. Nikishkov, H. Kanda, "The development of a Java engineering application for higher-order asymptotic analysis of crack-tip fields", Advances in Engineering Software 30, 469-477, 1999.

[47] G.R. Miller, P. Arduino, J. Jang, C. Choi, "Localized tensor-based solvers for interactive finite element applications using C++ and Java", Comp. & Struct. 81, 423-437, 2003.

[48] G. P. Nikishkov, "Generating contours on FEM/BEM higher-order surfaces using Java 3D textures", Advances in Engineering Softwarc, 34, 469-476, 2003.

[49] R. Marchand, M. Charbonneau-Lefort, M. Dumberry, "ARANEA, A program for generating unstructured triangular meshes with a Java Graphics User interface", Comp. Phys. Communications, 139, 172-185, 2001.

[50] N.T. Padial-Collins, W.B. VanderHeyden, D.Z. Zhang, E.D. Dendy, D. Livescu, "Parallel operation of CartaBlanca on shared and distributed memory computers", Concurrency and Computation: Practice and Experience 16, 61-77, 2004.

[51] W.B. VanderHeyden, E.D. Dendy, N.T. Padial-Collins, "CartaBlanca-a pure-Java, component-based systems simulation tool for coupled nonlinear physics on unstructured grids-an update", Concurency and Computation: Practice and Experience 15, 431-458, 2003.

[52] L. Baduel, F. Baude, D. Caromel, C. Delbé, N. Gama, S. El Kasmi, S. Lanteri, "A parallel object-oriented application for 3-D electromagnetism", ECCOMAS 2004, Jyväskylä, Finland, 2004.

[53] G.P. Nikishkov, Y.G Nikishkov, V.V Savchenko, "Comparison of C and Java performance in finite element computations", Computer & Structures, 81, 2401-2408, 2003.

[54] J.M. Bull, L. A. Schmith, L. Pottage, R. Freeman, "Benchmarking Java against C and Fortran for Scientific Applications", Joint ACM JavaGrande – ISCOPE 2001 Conference, Stanford Universtity, June 2-4, 2001.

[55] J. Häuser, T. Ludewig, R.D. Williams, R. Winkelmann, T. Gollnick, S. Brunett, J. Muylaert, "A test suite for high-performance parallel Java", Advances in Engineering Software, 31, 687-696, 2000.

[56] D. Eyheramendy, "Object-oriented parallel CFD with JAVA", 15th International Conference on Parallel Computational Fluid Dynamics, Eds. Chetverushkin, Ecer, Satofuka, Périaux, Fox, Ed. Elsevier, 409-416, 2003.

[57] C.J. Riley, S. Chatterjee, R. Biswas, "High-performance Java codes for computational fluid dynamics", Concurrency and Computation: Practice and Experience 15, 395-415, 2003.

[58] G.P.Nikishkov, "Object oriented design of a finite element code in Java", Computer Modeling in Engineering and Sciences 11, pp. 81-90, 2006.

[59] D. Eyheramendy, F. Oudin, "Advanced object-oriented techniques for coupled multiphysics", In Civil Engineering Computation: Tools and Techniques, Ed. B.H.V. Topping, ©Saxe-Cobourg Publications, ISBN 978-1-874672-32-6, 37-60, 2007.

[60] D. Eyheramendy, "Advanced object models for mathematical consistency enforcement in scientific computing", WSEAS Transactions on Mathematics, 4(4), 457-463, 2005.

[61] Mackie, R.I., "Distributed finite element analysis and the .NET framework", Proceedings of the ninth international conference on civil and structural engineering computing, Topping B.H.V. (Ed), Civil-Comp Press, Stirling, United Kingdom, 2003.

[62] Mackie, R.I., "Programming Distributed Finite Element Analysis: An Object Oriented Approach", Saxe-Coburg Publications, ISBN 978-1-874672-31-9, 2007.

[63] Mackie, R.I., "Object-oriented programming of distributed iterative equation solvers", Computers & Structures, 86, 511-519, 2008.

[64] Mackie, R.I., "Object-oriented implementation of distributed finite element analysis in .NET", Advances in Software Engineering, 38, 726-737, 2007.

[65] D. Eyheramendy, D. Loureiro, F. Oudin-Dardun, "An integrated object-oriented approach for parallel CFD", 20th International Conference on Parallel Computational Fluid Dynamics, Lyon, May 19th-22nd 2008.

[66] S.J. Lacy and J.H. Prevost, "Flow through porous media: A procedure for locating the free surface", Int. J. Num. An. Meth. Geom., 11, 585-601, 1987.

[67] H. Brezis, D. Kinderlehrer and G. Stampacchia, "Sur une nouvelle formulation du problème de l'écoulement à travers une digue", C.R. Acad. Sci. Paris, 287 (ser. A), 711-714, 1978.

[68] Heng, B. C. P., Mackie, R. I. "Parallel Modal Analysis Using Distributed Objects", In Topping, B. H. V. (Ed.) Proceedings of the Fifteenth UK Conference of the Association of Computational Mechanics in Engineering. Civil-Comp Press, 2007.

[69] Heng, B. C. P., Mackie, R. I. "Design patterns in object-oriented finite element programming", In Topping, B. H. V., Montero, G. & Montenegro, R. (Eds.) Proceedings of the Fifth International Conference on Engineering Computational Technology., Civil-Comp Press, 2006.

[70] E. Gamma, R. Helm, R. Johnson, J. Vlissides, Design Patterns: Elements of Reusable Object-Oriented Software, Addison-Wesley, Reading, 1995.

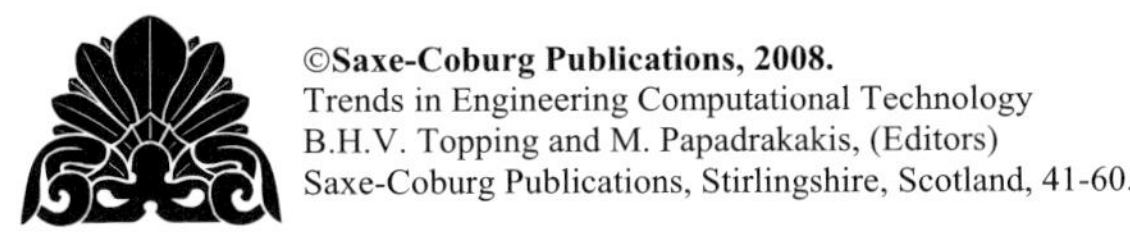
©Saxe-Coburg Publications, 2008.
Trends in Engineering Computational Technology
B.H.V. Topping and M. Papadrakakis, (Editors)
Saxe-Coburg Publications, Stirlingshire, Scotland, 41-60.

# Chapter 3

# Finite Element Software Design for Today's Computers

**R.I. Mackie**
**Civil Engineering**
**School of Engineering, Physics and Mathematics**
**University of Dundee, United Kingdom**

## Abstract

Developments in computers mean that parallel and distributed computing are now mainstream elements. This paper examines the use of component oriented software design to handle the associated complexity. This is applied to finite element software, using domain decomposition methods with the capability to carry out distributed solution. The use of components enables areas of complexity to be separated out from each other. For instance, the equation solver is logically isolated from the finite element program. Object-oriented methods can be combined with component oriented design in designing the implementations of the components. It is demonstrated that component oriented design can be applied at both high and low levels of software design, and both help to handle complexity, and to produce more flexible software.

**Keywords:** distributed computing, finite element analysis, object-oriented, parallel processing, component oriented, iterative solution.

## 1 Introduction

There have been enormous changes in computing over the last fifty years, from early computers that required enormous buildings to contain them and yet had little power, to the notebook computers of today with gigabytes of memory and multiple processors. Similarly, software has transformed from programs that merely implemented numerical algorithms to systems that have sophisticated graphical user interfaces, links to databases, and worldwide interconnectivity. In terms of finite element analysis, finite element packages have changed from being specialist numerical packages requiring specialist users and dedicated hardware, to packages that run quite happily on desktop and laptop computers.

For a long time the improvements in computing power largely consisted of more memory being available, and ever increasing processor speeds. However, in recent years the major change in computing power has been the advent of dual core processors, and now the increasing availability of quad core processors. When increasing power was largely a matter of faster processors software automatically took advantage of the increased speed as the computer architecture was still essentially serial. However, in order to take full advantage of multi-core computers programs need to be specifically designed to use parallel algorithms. Parallel computing has been in use for many years, but has been very much a specialised area of computing. Developments mean it is now a mainstream requirement.

In addition to the changes in processors, all computers are now connected to the internet and most are on a local network. Even homes will commonly have several computers linked by a wireless network. Computing is increasingly a distributed activity, rather than all the activity being focused on a single machine.

The situation is that modern computing is now inherently parallel and distributed. As well as offering many potential advantages, the software design challenges are increased, and additional sources of complexity are introduced. Therefore software design paradigms are needed to meet the challenge. In the last twenty years object-oriented programming has become the dominant paradigm in software design. The prime advantage of object oriented programming is its greater data modelling capabilities. A related development is component oriented programming, which offers even greater program flexibility, and most importantly, greater separation between areas of complexity.

This paper will examine the use of component and object oriented methods in the design of finite element software for distributed analysis. It will be demonstrated how these methods allow both logical and physical separation to be achieved between different components of a finite element system, and provide valuable tools for meeting the challenges of modern computing.

## 2 Literature Review

Object-oriented programming has been applied to finite element analysis for some twenty years now, with early papers focusing on straightforward implementations of finite elements in an object-oriented language [1,2]. The new features were that separate objects were created for degrees of freedom (DOF), nodes, elements, materials etc. A goal common to much of the work on object-oriented approaches to finite elements was the need to produce a new generation of tools to develop easily maintainable and extensible finite element software systems [3]. It was also found that the object oriented approach could be used without sacrificing computational efficiency [4].

Object–oriented languages have richer data modelling capabilities than traditional languages. This suggests that it is sensible to apply them to the complex algorithms, such as node renumbering schemes, arising in finite element work, including nonlinear analysis [5,6].

While the object-oriented approach offers many advantages, particularly in dealing with complexity [7], it has been recognised that there are limitations, especially the ease with which large complex class hierarchies can arise. Accordingly object-oriented design has been complemented with component oriented design [8]. Object-oriented design relies heavily on inheritance and polymorphism; component oriented design uses interfaces and object composition [9]. This generally leads to a more flexible designs and smaller class hierarchies.

The component oriented approach was applied to finite element analysis software by Dolenc [10]. According to Dolenc component-oriented software consists of three parts: interface, implementation and deployment. Component oriented design plays a key role in the work presented in this paper. Java and COBRA allow for component oriented software, and distributed systems. However, in the present work the .NET framework [11] will be used. .NET is component and object oriented, and was designed specifically with distributed computing in mind.

Parallel computing has a long history in scientific computing in general, and finite element analysis in particular, and object-oriented methods have been applied to this [12-14]. Until recent times parallel computing was largely a specialised area, requiring clusters of workstations, and ultimately leading to supercomputers. Now, cheap desktop computers and notebooks have dual core processors and are networked together. This means that parallel and distributed computing is now readily available.

The key difference between parallel and distributed processing is that parallel processing is fine-grained, whereas distributed processing is coarse-grained [15]. With distributed computing the processing is typically carried out on distinct computers. So while communication overheads are always an issue in parallel computing, they are even more so in distributed computing. Moreover, while individual desktop machines now commonly have more than one processor, they are currently limited to two or four processors.

The finite element method lends itself naturally to distributed processing. Activities like stiffness matrix formulation and stress calculation can be carried out at the element level, with the calculations for each element being carried out independently, though this would normally be at too fine a level for efficient distributed processing. More importantly, sub-structuring [16] and domain decomposition methods [17, 18] are commonly used and these are very natural candidates for distributed processing. They break down the calculation process into fairly large batches of work, most of which can be performed independently of one another, and so are suitable for distributed processing.

A number of implementations of distributed finite element solvers have been developed over the years. Chanda and Baugh [15] used domain decomposition methods. Domain decomposition methods can be used for both direct and iterative solvers, and are said to promote better performance of pre-conditioners. They investigated the use of both direct and iterative solvers, and hybrid schemes that used a combination of both methods.

Much of the work has made use of the MPI [19] or PVM [20] frameworks. However, it has been recognised by several workers that the development of a flexible software library would be of great benefit, promoting quicker development [21-23]. As Chow and Heroux [24] stated, "General software for preconditioning the iterative solution of linear systems is greatly lagging behind the literature. This is partly because specific problems need specific matrix and pre-conditioner data structures". A similar motivation lies behind the current work.

Computational efficiency is dependent upon reducing the inter-process communications and effective load balancing. It should be recognised that distributed computing also increases the amount of memory available, which in itself can allow much larger problems to be solved effectively [25]. In PC networks there is likely to be a significant degree of heterogeneity, with machine shaving different processor speeds, numbers of processors and memory. This aspect is not part of the current work, but research in this area has been reported [26, 27].

This paper is focused on developing flexible frameworks and software design methods for iterative equation solvers, with applications to finite element analysis. This involves careful design of matrix and vector components, and several workers have recognised that this is no simple task [28, 29].

As well as networks of computers being an important area of distributed computing in engineering analysis, internet based computing is the object of increasing attention [30-33].

# 3 Component Oriented Design

Following Dolenc [10] there are three elements of component-oriented design:

- Interface
- Implementation
- Deployment

An interface defines what a component can do, but says absolutely nothing about the implementation. Interface design follows Meyer's [34] principle of designing to a contract. The interface defines what a component can do and what it expects. Perhaps the most important aspect of component-oriented design is that one designs to an interface. This makes the design as independent as possible of implementation details. Figure 1 shows the principle.

The overall design is dependent only upon interfaces, so if the implementation of an interface changes, then this has no effect on the system. Other advantages are that existing objects can be readily adapted to implement an interface. For instance, if a class had been designed without knowledge of a particular interface, then it can be adapted to implement the interface, and so fit into a framework. Moreover, classes that implement a given interface do not have to derive from a common base class. The work presented in this paper was implemented using C#, but other languages implement interfaces as well, most notably Java.

Object-oriented design can be used to implement the components themselves, and advantage taken of object-oriented programming principles such as inheritance and polymorphism. Component-oriented design can also be used in the implementation of the interfaces themselves, as will be seen in the examples given below.

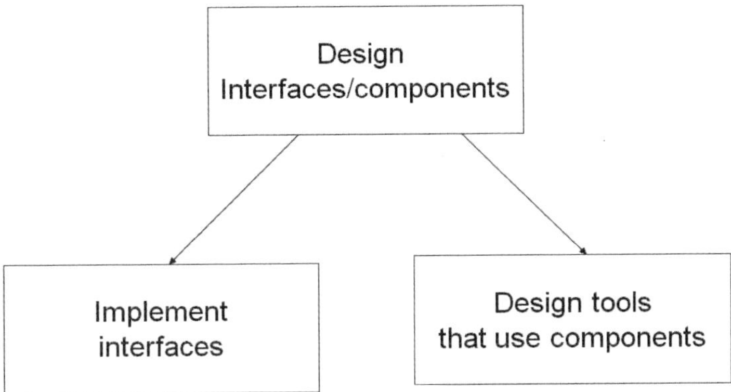


Figure 1: Designing to an interface

Once designed and implemented, the components may be deployed as local objects as part of the main program (i.e. the traditional approach); as objects on the same computer but in a different process; or as remote objects on a networked computer, or on the internet. If component-oriented design principles have been followed, then the client program will be blind logically to whether the object is local or remote.

# 4 Using Components to Handle Complexity

There are many areas of complexity in finite element software; this paper will focus on the following:

- Equation solution
- Finite element system
- Parallel solution methods
- Distributed and remote solution

## 4.1 Separating Equation Solution and the Finite Element Model

One of the keys to handling complexity is to separate the various areas of complexity so that they do not interact with each other. One of the most important areas is the separation of the finite element model and the solution process. More details on this can be found elsewhere [35-37], but the essential details will be outlined here. The focus will be on domain decomposition solvers as these are the most suitable for distributed and parallel processing on multi-core computers and clusters of computers. Using this sort of structure the set of equations has the structure:

$$\begin{pmatrix} \mathbf{K}_{00} & 0 & \cdots & \mathbf{K}_{0n} \\ 0 & \mathbf{K}_{11} & \cdots & \mathbf{K}_{1n} \\ \vdots & \vdots & . & \vdots \\ \mathbf{K}_{0n}^{t} & \mathbf{K}_{1n}^{t} & \cdots & \mathbf{K}_{nn} \end{pmatrix} \begin{pmatrix} \mathbf{d}_0 \\ \mathbf{d}_1 \\ \vdots \\ \mathbf{d}_n \end{pmatrix} = \begin{pmatrix} \mathbf{f}_0 \\ \mathbf{f}_1 \\ \vdots \\ \mathbf{f}_n \end{pmatrix} \tag{1}$$

Each matrix $\mathbf{K}_{ii}$ is a symmetric block matrix associated with the internal degrees of freedom for sub-domain i. $\mathbf{K}_{nn}$ is associated with the interface degrees of freedom. Likewise, $\mathbf{d}_i$ and $\mathbf{f}_i$ represent the internal degrees of freedom and loads for sub-domain i, and $\mathbf{d}_n$ and $\mathbf{f}_n$ represent the interface degrees of freedom and loads for the interface degrees of freedom. The matrices $\mathbf{K}_{in}$ are rectangular blocks and represent the coupling between the internal and interface degrees of freedom. In general, $\mathbf{K}_{ii}$ will be sparse. $\mathbf{K}_{in}$ will also be sparse, and many of its columns will be empty.

Rather than defining a specific solver, an interface is declared. The interface is defined as:

```
public interface ISubDomSolverSym
{
        ISubDomSym Add(AssemblerSubDomSym subDom);
        void Solve();
        IVector RhsExt {get; set;}
        IVector XExt {get;}
        int NoSubDoms {get;}
}
```

It can be seen that the interface itself uses interfaces. `IVector` represents vectors of double precision numbers. Using interfaces increases the flexibility of the software as it is not tied into any one particular implementation. `ISubDomSym` is an interface that represents a sub-domain for the symmetric case and has the definition:

```
public interface ISubDomSym
{
        int IntSize {get;}
        int ExtSize {get;}
        IVector XInt {get; set;}
        IVector RhsInt {get; set;}
        IVector RhsExt {get; set;}
        ...
}
```

In other words, it provides access to the vectors such as $\mathbf{x}_i$, $\mathbf{f}_i$ and $\mathbf{f}_i^{(n)}$, and `IntSize` and `ExtSize` give the number of internal and interface degrees of freedom, respectively. These mean that the client of the solver needs only to know about the interfaces, and needs have no knowledge of any particular interface. For instance, it does not know whether a direct or iterative solver is being used, or whether the solver is using remote or local objects. `AssemblerSubDomSym` is an object that contains the basic data for a sub-domain, i.e. the coefficient matrices $\mathbf{K}_{ii}$, $\mathbf{K}_{in}$, $\mathbf{K}_{nn}^{(i)}$ etc. The data is stored in an efficient manner. The method `Add` uses this data to create an object that implements the interface `ISubDomSym`, and returns a reference to this, possibly remote, object. `Solve` carries out the solution process. `RhsExt` provides access to the vector $\mathbf{f}_n$, and `XExt` provides access to $\mathbf{x}_n$.

An `ISubDomSolverSym` object can be used by a finite element program as follows:

```
foreach (FeSubMesh subMesh in m_FeModel.SubMeshes())
{
  m_Solver.Add(subMesh.SubDom);
}
m_Solver.RhsExt = m_FeModel.RhsExt;
m_Solver.Solve();
```

`m_FeModel` is an object representing the finite element model, and `FeSubMesh` is a sub-domain or sub-structure of the finite element model. The loop iterates through each sub-domain of the model, and `subMesh.SubDom` returns the associated `AssemblerSubDomSym` object. These are added to the `m_Solver` object, which is an `ISubDomSolverSym` object. The important point is that the code is dealing only with solver interface objects, so it is isolated from all implementation details. Hence, implementations and design of solvers can be dealt with completely independently. Changes can be made to the solvers without affecting the design of the finite element program.

## 4.2 Parallel Implementation

There are two aspects to parallel programming in the current context, and it should be noted that the current work is concerned only with work on everyday desktop and notebook computers, not with super-computers. In both cases domain decomposition methods seem to be the most appropriate. A multi-core computer will have a small number of processors, currently two or four. Distributed computing will typically have a number of computers linked by Ethernet. In both cases it is best to use coarse grained parallelism, and this is facilitated by domain decomposition methods. Domain decomposition methods break up the work into relatively large segments, and much of the work on each segment can be carried out independently, with communication required at key points. The mathematical details can be found elsewhere [35].

Parallelism has been commonly implemented on clusters of workstations or PCs using MPI or PVM. Now that parallel processing ability is commonly available, mainstream and relatively easy-to-use technologies are needed. The .NET framework is freely available and provides a number of facilities for multi-threaded and distributed computing, and the use of these will be described here.

Parallelism is achieved using multi-threading and remote objects. Two methods are available in .NET for multi-threading. One is the use of threads, and objects that directly represent the threads. These are similar to multi-threading capabilities provided in C++. They are very flexible and give the programmer the highest degree of control. However, .NET provides an alternative that is very efficient and easy to use. This is achieved using delegates. In .NET a delegate is equivalent to function pointers in C. For instance the following code will carry out a series of operations on the sub-domains using separate threads for each sub-domain:

```
public void DoLotsofSums()
{
        m_NoSubsComplete =  0;
        m_AllSubsComplete = new AutoResetEvent(false);
        foreach(ISubDomain subDom in m_SubDoms)
        {
                DoBigSumDel del = new DoBigSumDel(DoBigSum);
                del.BeginInvoke(subDom,null,null);
        }
        m_AllSubsComplete.WaitOne();
}

delegate void DoBigSumDel(ISubDomain subDom);
private void DoBigSum(ISubDomain subDom)
{
   … Do something with subDom
   BigSumComplete(subDom);
}
```

`m_SubDoms` is a collection of sub-domain objects and `DoBigSum` is a method that does some calculations on a sub-domain. `m_AllSubsComplete` is an event that waits for all the sub-domains to complete their operations. `DoBigSumDel` is the declaration of a delegate type as a function that takes an `ISubDomain` as an argument. `DoBigSum` is the method that will be implemented in parallel, and is used in the creation of a specific instance of the delegate type. .NET automatically supplies delegates with a `BeginInvoke` method, this method causes the method `DoBigSum` to be implemented in a separate thread. Therefore the `foreach` loop causes `DoBigSum` to be implemented simultaneously for all the sub-domains, thus enabling parallel computation. Further details on delegates can be found in many books on C# [38].

.NET allows objects to be created on remote computers very easily, and are called remote objects. Details can be found in general books on C# and books dedicated to remote computing [39]. Reference [35] gives these in the context of finite element programming. In the current context the key feature is that once the object is created it can be treated in the same way as a local object, i.e. the use of remote objects does not affect the overall design of the program. Moreover, interfaces can be used. For instance an interface, `ISolverFactory`, may be created with the following definition:

```
public interface ISolverFactory
{
   ISubDomSolverSym GetSubDomSolver();
}
```

The client program, e.g. a finite element program, would know only that the object implemented the `ISolverFactory` interface. When it calls the `GetSubDomSolver` method to obtain the solver, it knows only that the object implements the `ISubDomSolverSym` interface. This means that if the object exists on a remote computer, then the client computer needs to have no knowledge whatsoever of the implementation. In fact the object code for the implementation does not need to exist at all on the client computer. So there is both logical and physical separation between the client program and the solver system. This idea is explored further elsewhere [40]. However, the logical separation is by far the most important aspect.

## 4.3 Conjugate Gradient Solvers

The previous sections have considered matters at a rather high level, now attention will be given to some of the details in terms of conjugate gradient solvers. Conjugate gradient solvers provide an excellent case study as there are many variations. The mathematical details for the algorithm are well known and so will not be elaborated on here, instead attention will be focused on the software design aspects.

The basic conjugate gradient method works on a single matrix. However, there are a multitude of variations, including:

- In most cases a pre-conditioner is used. This improves the condition number of the matrix and reduces the number of iterations required to achieve convergence. However, there is no one ideal pre-conditioner and there are many variations [41].
- Domain decomposition can be used, and this can be applied in two main ways:
    - Full-matrix methods. With these methods the standard conjugate gradient algorithm is applied, but advantage of the domain decomposition structure is taken to distribute the calculations.
    - Schur complement methods. These condense out the internal degrees of freedom.

In both domain-decomposition approaches, pre-conditioning can be used, usually based upon pre-conditioners for the standard matrix case. The use of object and component oriented methods to conjugate gradient solvers have been covered in detail elsewhere [35]-[37], so an outline of the component-oriented design aspects will be given here.

### 4.3.1 Conjugate Gradient Algorithm

First of all a class `CGSolver` was written that implements the conjugate gradient algorithm, and the class diagram is shown in Figure 2. The key design feature is the interface `ICGMatrix`. `ICGMatrix` represents the set of equations and the `CGSolver` class is supplied with an object that implements the `ICGMatrix` interface. The methods in `ICGMatrix` represent various steps in the conjugate gradient matrix. For instance `UpdateXR` implements the steps:

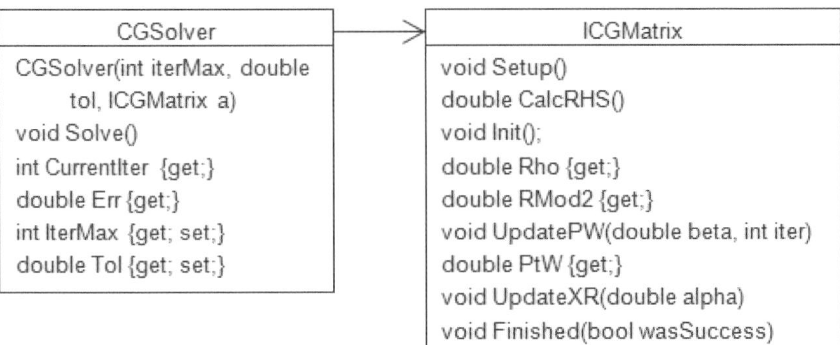


Figure 2: Class diagram for `CGSolver`

$$
\begin{aligned}
&\mathbf{x} = \mathbf{x} + \alpha_k \mathbf{p} \\
&\mathbf{r} = \mathbf{r} - \alpha_k \mathbf{w} \\
&\mathbf{z} = \mathbf{M}^{-1}\mathbf{r} \text{ (if pre-conditioning is used)}
\end{aligned}
\tag{2}
$$

The main steps in the algorithm are delegated to the `ICGMatrix` object. So the algorithm has been coded in such a way that it is independent of the way the data for the right-hand side, the solution vector and the coefficient matrix are stored, whether pre-conditioning is used, whether or not domain decomposition methods are used, and whether or not solution is implemented in a distributed fashion. The advantage is that no further attention need be paid to the algorithm. The use of a component has isolated it from all other details.

### 4.3.2 Implementation of ICGMatrix

The simplest case is for the case where the coefficient data is stored in a single matrix. Figure 3 shows the class diagrams, and illustrates how object-oriented and component-oriented design work hand-in-hand. `ICGMatrixStd` is an extension of

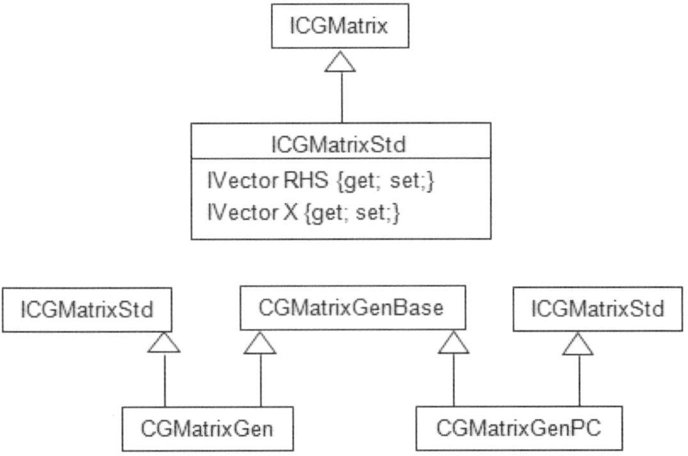


Figure 3: Class diagram for `ICGMatrix` and associated classes

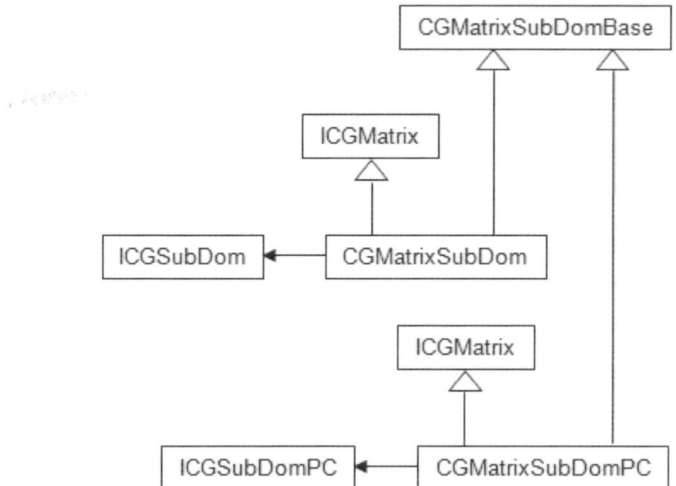


Figure 4: Domain decomposition implementations of `ICGMatrix`

`ICGMatrix` for the standard case. `CGMatrixGen` and `CGMatrixGenPC` implement `ICGMatrix` for without and with pre-conditioning, respectively. Many features are common to the two cases, and these are encapsulated in `CGMatrixGenBase`. Further details can be found elsewhere [36].

Figure 4 shows the class diagram for the domain decomposition case using the full-matrix approach. `CGMatrixSubDomBase` implements common features, and the two descendents implement the features particular to whether or not pre-conditioning is being used. The key feature of domain decomposition is that the data is distributed amongst the sub-domains. The sub-domain objects used by `CGMatrixSubDom` and `CGMatrixSubDomPC` must implement the interfaces `ICGSubDom` and `ICGSubDomPC`, respectively. These two interfaces define the necessary methods that sub-domains must be able to execute. The interfaces are shown in Figure 5.

Using the full-matrix method the conjugate gradient is applied in the same way as for the standard case, except that various operations will be distributed among the sub-domains. Importantly, many of the operations will be implemented in a multi-threaded manner using delegates. These mean that the class can take advantage of multi-core capabilities or distributed processing. The sub-domain objects may or may not be remote objects, the `CGMatrixSubDom` objects are blind to whether or not this is the case.

As previously mentioned, there are numerous ways in which pre-conditioning can be applied. The class `CGMatrixGenPC` has an object that implements the interface `IPreConStd`. This interface has the following definition:

```
public interface IPreConStd
{
   IVector PreCon(IVector r);
   void Setup();
}
```

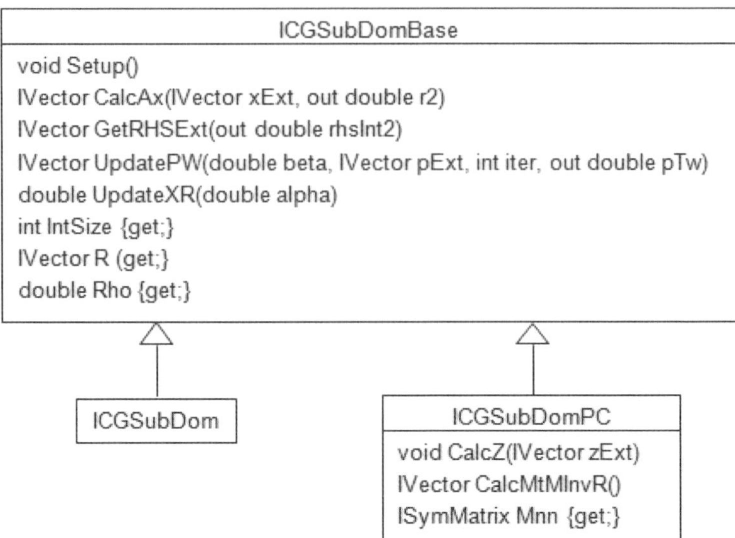


Figure 5: `ICGSubDom` and associated interfaces

Setup is called to initialise the object, and the `PreCon` method is used to apply pre-conditioning steps, and simply specifies that if a vector **r** is supplied, it will return a vector **z**. Different methods of pre-conditioning can be used with `CGMatrixGenPC` by simply supplying it with a different object that implements the `IPreConStd` interface. Similarly, `CGMatrixGenPC` makes use of a `IPreConStd` object. So once an object has been designed that implements `IPreConStd` for a particular pre-conditioning algorithm, then it can be used for either standard conjugate gradient method, or full-matrix domain decomposition method. Furthermore, as will be shown in the next sub-section, the same component can be used in the Schur Complement approach. So the component is pluggable. Plugging in a different pre-conditioner object changes the behaviour of the conjugate gradient solver. This is an example of achieving different behaviour by object composition, rather than polymorphism and inheritance.

### 4.3.3 Schur Complement

A widely used method in domain decomposition is the Schur complement approach. This approach condenses out the internal degrees of freedom and produces a reduced set of equations, and can be used with either direct or iterative solution methods. The Schur complement approach is implemented as follows:

The i'th block matrix equation leads to:

$$\mathbf{d}_i = \mathbf{K}_{ii}^{-1}\left(\mathbf{f}_i - \mathbf{K}_{in}\mathbf{d}_n\right) \tag{3}$$

The n'th equation gives:

$$\mathbf{K}_{nn}\mathbf{d}_{n} + \sum_{i=0}^{n-1} \mathbf{K}_{in}^{t}\mathbf{d}_{i} = \mathbf{f}_{n} \tag{4}$$

Using the result in Equation (3) then gives:

$$\mathbf{K}_{nn}\mathbf{d}_{n} + \sum_{i=0}^{n-1} \mathbf{K}_{in}^{t}\mathbf{K}_{ii}^{-1}\left(\mathbf{f}_{i} - \mathbf{K}_{in}\mathbf{d}_{n}\right) = \mathbf{f}_{n} \tag{5}$$

So the set of equations that needs to be solved in order to obtain $\mathbf{d}_n$ is:

$$\mathbf{S}\mathbf{d}_{n} = \mathbf{c} \tag{6}$$

Where

$$\begin{aligned} \mathbf{S} &= \mathbf{K}_{nn} - \sum_{i} \mathbf{K}_{in}^{T}\mathbf{K}_{ii}^{-1}\mathbf{K}_{in} \\ \mathbf{c} &= \mathbf{f}_{n} - \sum_{i} \mathbf{K}_{in}^{T}\mathbf{K}_{ii}^{-1}\mathbf{f}_{i} \end{aligned} \tag{7}$$

The solutions for the $\mathbf{d}_i$ are then obtained from Equation (3). This approach is known as the Schur complement approach. There are various ways in which the object-oriented design can be carried out for the Schur complement approach. Figure 6 shows one class structure.

`CGMatrixSchur` inherits from `CGMatrixGenBase` and implements the `ICGMatrix` interface. It possesses an object that implements `ISchurSubDoms`. `ISchurSubDoms` specifies the operations relevant to the Schur complement approach, and `CGMatrixSchur` uses this object to carry out most of its operations.

`SchurSubDoms` is an implementation of `ISchurSubDoms`, and is a collection of sub-domain objects that implement `ICGSubDomSchur`. This latter interface defines the Schur complement operations for a sub-domain, and `CGSubDomSchur` is an implementation of the interface. `CGSubDomSchur` uses an `ISolver` object to implement the operations.

`CGMatricSchurPC` inherits from `CGMatrixSchur` and uses pre-conditioning. Pre-conditioning is applied by means of an `IPreConStd` object, using the induced pre-conditioner method.

There are a number of important points to note:

- `CGMatrixSchur` and `CGMatrixSchurPC` implement `ICGMatrix` and so can be used with the `CGSolver` class. The `CGSolver` class would not know whether or not it is using the Schur Complement approach.
- Multi-threaded implementation will be used if the implementation of `ISchurSubDoms` uses multi-threading, as indeed `SchurSubDoms` does.
- The implementation will be used if the `ICGSubDomSchur` objects are remote ones.

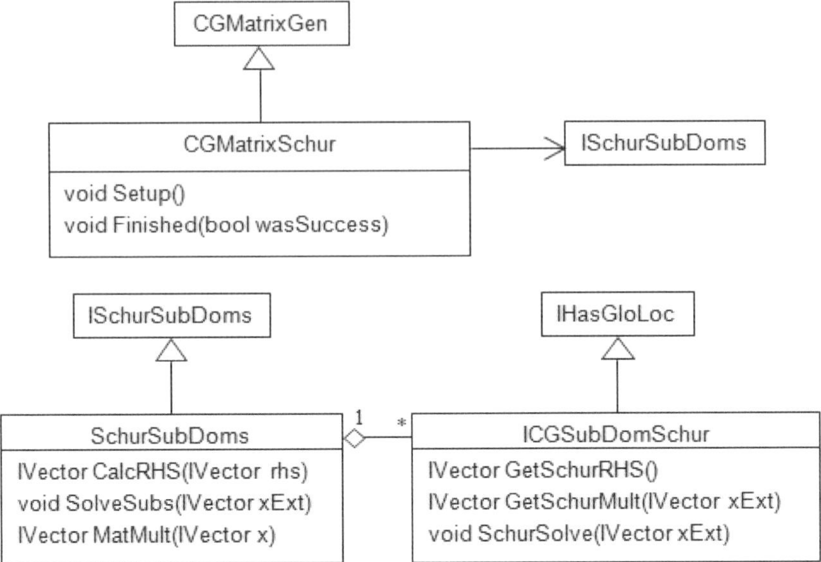


Figure 6: Schur complement class diagrams

- The `CGSubDomSchur` class is not tied to one method of equation solver, it could use a direct or an iterative method. The method is changed by plugging in a different ISolver object.
- The design is not tied to one particular pre-conditioner, different methods can be used by plugging in a different `IPreConStd` object.

This illustrates a key feature of component oriented design and how it isolates sources of complexity. The classes and interfaces described in this section have focused purely on the Schur complement aspects of the implementation of the conjugate gradient method. The design is not dependent upon a particular pre-conditioner, or sub-domain solver method being used, nor on whether the objects are local or remote.

### 4.3.4 Putting it Altogether

The foregoing sub-sections have described how various aspects of a finite element program and equation solver can be designed. This final sub-section will describe how it can all be fitted together.

Suppose that a finite element program is carrying out a linear elastic static solution and is going to use the conjugate gradient method with pre-conditioning. Figure 7 illustrates the process.

The starting point is the structural model. This will typically contain the geometry of the structure and the associated attributes such as loads, restraints etc. A finite element model is created from this. The finite element model consists of a collection of substructures. These could either be created locally or on remote computers.

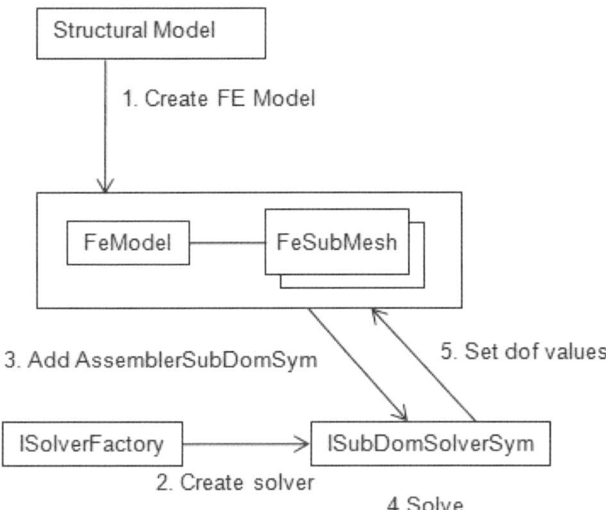


Figure 7: Solution process

The program has an `ISolverFactory` object. This object can return either a standard solver or a domain decomposition solver. The factory object need not exist in the same process as the finite element program. It could either exist in a separate process on the same computer or on a different computer altogether. This separation is illustrated in Figure 8. It is not essential for it not to be part of the program itself, but this demonstrates the complete logical separation between the finite element and solver systems. Typically several `ISolverFactory` objects would be available, one for each different solution method. The server program that provides these factory objects could run either as a normal program, or as a Windows service. In the current example the program uses a factory that supplies conjugate gradient solvers, though this has no logical impact on the finite element program.

Each of the sub-structures in the finite element model supplies an `AssemblerSubDomSym` object containing the information on the stiffness matrix etc. for the sub-structure. These are added to the solver. The finite element program then

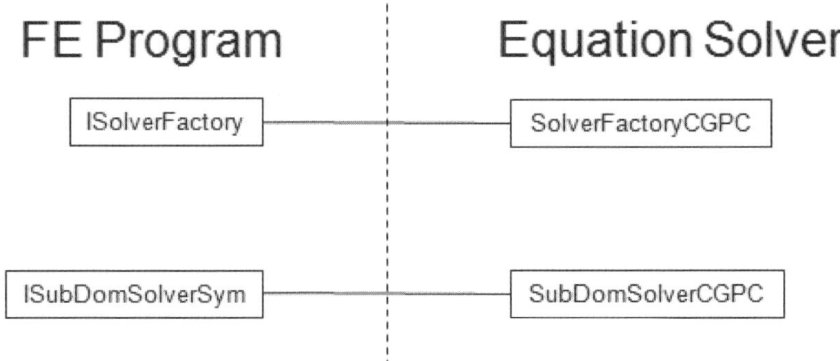


Figure 8: Separation of finite element program and solver system

instructs the solver to carry out the solution. Once complete the finite element program obtains the degree of freedom values and carries out post-processing operations, such as stress calculations.

The above process would be the same regardless of the type of domain decomposition solver that was used; all the details of the process are contained within the solver object itself, the finite element program is completely isolated from them. As the example is using a conjugate gradient solver, the object created by the factory is actually one that implements this method. However, this object exists only in the solver process. The finite element program receives a proxy object that implements the `ISubDomSolverSym` interface. Calls to this object from the finite element program are then routed to the actual object.

So far the description has been limited to what happens on the finite element solver process. Now consider the solver side. Figure 9 shows the key features of what happens in the solver process for the case where it is actually using a conjugate gradient solver with pre-conditioning. When the solver factory is asked to create a domain decomposition solver it creates a pre-conditioner factory. This is an object that creates pre-conditioners, in this case an incomplete $U^tDU$ decomposition preconditioner. Then it creates the solver object itself.

Figure 9b shows what happens when the finite element program adds an `AssemblerSubDomSym` object to the solver. The solver creates a `CGSubDomPC` object from the data contained within the `AssemblerSubDomSym` object, then it adds it to the solver's `CGMatrixSubDomPC` object (which implements `ICGMatrix`).

Figure 9c demonstrates the solve process. The solver creates a `CGSolver` object using its `CGMatrixSubDomPC` object, then tells the `CGSolver` object to carry out the solution process.

# 5 Conclusions

Developments in modern computing, namely the advent of multi-core computers as standard and the interconnected world, mean that parallel and distributed computing are now mainstream parts of software engineering.

This introduces additional sources of complexity. Component and object oriented programming methods can help in handling this complexity. This paper has looked at the application of these principles to finite element and equation solution software.

Component-oriented design can be used to isolate areas of complexity. This has been demonstrated in this paper at various levels of software design, from high level separation between aspects of a software system, to low level parts of solution algorithms. More specifically, examples demonstrated in this paper include:

- The finite element system can be completely separated from the equation solver, so the design and maintenance of each part can be considered independently. The separation can be both logical and physical.
- The .NET framework allows distributed objects and multi-threaded design to be implemented relatively easily.

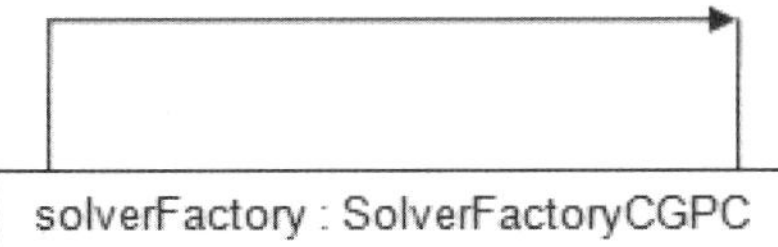


Figure 9(a): Domain-decomposition solver creation

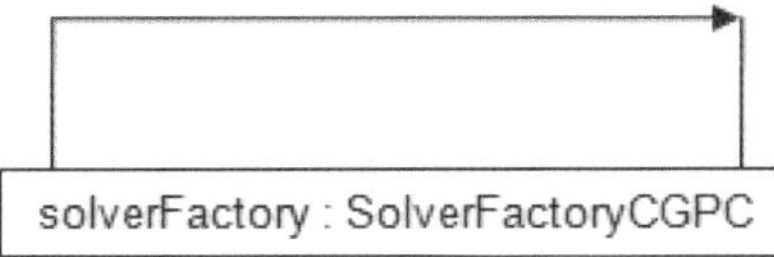


Figure 9(b): Process when an `AssemblerSubDomSym` object is added to the solver

1. Create cgSolver: CGSolver(cgMat)
2. Solve()

solver : subDomSolverCGPC

Figure 9(c): Solution process

- Component and object oriented design can work together, and this has been applied to conjugate gradient solvers.
- Component oriented both simplifies design and produces more flexible software. For instance, in the conjugate gradient solver design of the pre-conditioners is separated from other aspects, and a solver can use a different pre-conditioner by simply plugging in a different component. These features were further demonstrated in the context of the Schur complement approach.
- It was demonstrated how these features can be put together in a complete finite element system.

It is not claimed that the designs presented here are perfect, nor the only or best way of doing things, but they do demonstrate some of the ways in which component –oriented software design can be used in the context of scientific computing.

## References


[1] B.W.R. Forde, R.O. Foschi, S.F. Stiemer, "Object-Oriented Finite Element Analysis". Computers & Structures , 34, 355-374, 1990

[2] R.I. Mackie, "Object Oriented Programming of the Finite Element Method", Int J Num Meth Eng, 35, 425-436,1992.

[3] G. Yu, H. Adeli, "Object-oriented finite element analysis using EER model", Journal of Structural Engineering, 119(9), 2763-2781, 1993.

[4] Y. Dubois-Pelerin, T. Zimmerman, "Object-oriented finite element programming: III.An efficient iimplementation in C++", Comp Meth in App Mech and Eng, 108, 165-183, 1993.

[5] P. Menetrey, T. Zimmerman, "Object-oriented non-linear finite element analysis: application to J2 plasticity", Computers and Structures, 49(5), 767-777, 1993.

[6] Y. Dubois-Pelerin, P. Pegon, "Object-oriented programming in nonlinear finite element analysis", Computers and Structures, 67, 225-241, 1998.

[7] R.I. Mackie, "Using Objects to Handle Complexity in Finite Element Software", Engineering with Computers , 13 99-111, 1997.

[8] R.I. Mackie, "Object Oriented programming for structural mechanics: A Review", in Civil and Structural Engineering Computing 2001, ed B.H.V. Topping, Saxe-Coburg Publications, Stirling, UK, ISBN 1-874672-15-6, 137-160, 2001.

[9] C. Szyperski, "Component software: beyond object oriented programming", Addison-Wesley, Harlow, UK, 1998.

[10] M. Dolenc, "Developing extendible component-oriented finite element software", Advances in Engineering Software, 35, 703-714, 2004.

[11] "Overview of the .NET Framework", .NET Framework Developer's Guide. http://msdn2.microsoft.com/en-us/library/zw4w595w.aspx .

[12] G. Carey, "A prototype scalable, object-oriented finite element solver on multicomputers", Journal of parallel and distributed computing, 20, 357-379, 1994.

[13] A. Bose, G.F. Carey, "A class of data structures and object-oriented implementation for finite element methods on distributed memory systems" Computer Methods in Applied Mechanics and Engineering, 171, 109-121, 1999.

[14] P.R,B. Devloo, "Object oriented tools for scientific computing", Engineering with Computers, 16, 63-72, 2000.

[15] H.S. Chadha, J.W. Baugh, "Network-distributed finite element analysis" Advances in Engineering Software, 25, 267-280, 1996.

[16] R.I. Mackie, "Implementation of Sub-structuring within an object-oriented framework". Advances in Engineering Software 32(10-11), 1-10, 2001.

[17] P.J. Jimack, "Domain decomposition preconditioning for parallel PDE software" in Engineering Computational Technology, B H V Topping and Z Bittnar (Editors), Saxe-Coburg Publications, Stirling, UK, 193-220, 2002.

[18] E.D. Sotelino, Y. Dere, "Parallel and distributed finite element analysis of structures" in Engineering Computational Technology, B H V Topping and Z Bittnar (Editors), Saxe-Coburg Publications, Stirling, UK, 221-250, 2002.

[19] Message Passing Interface, http://www-unix.mcs.anl.gov/mpi/

[20] Parallel Virtual Machine, http://www.csm.ornl.gov/pvm/

[21] G. Mukunda, E. Sotelino, S. Hsieh, "Distributed finite element computations using object-oriented techniques", Engineering with Computers, 14, 59-72, 1998.

[22] R. Wyrzykowski, T. Olas, N. Sczygiol, "Object-oriented approach to finite element modelling on clusters", Lecture Notes in Computer Science, 1947, 250-257, 2000.

[23] J. Kruis, "Domain Decomposition Methods on Parallel Computers", in Progress in Engineering Computational Technology, B.H.V. Topping and C.A. Mota-Soares (editors), Saxe-Coburg Publications, Stirling, UK., 299-322, 2004.

[24] E. Chow, M.A. Heroux, "An object-oriented framework for block pre-conditioning", ACM Transactions on Mathematical Software, 24, 159-183, 1998.

[25] R.M. Rao, T.V.S.R. Appa Rao, B. Dattaguru, "A new parallel overlapped domain decomposition method for nonlinear dynamic finite element analysis", Computers & Structures, 81, 2441-2454, 2003.

[26] P. Rus, B. Stok, N. Mole, "Parallel computing with load balancing on heterogeneous distributed systems", Advances in Engineering Software, 34, 185-201, 2003.

[27] P. Iványi, B.H.V. Topping, "Static Partitioning for Heterogeneous Computational Environments", in Proceedings of the Ninth International Conference on Civil and Structural Engineering Computing, B.H.V. Topping, (Editor), Civil-Comp Press, Stirling, United Kingdom, paper 138, 2003.

[28] E. Noulard, N. Emand, "Object oriented design for reusable parallel linear algebra software", Lecture Notes in Computer Science, 1685, 1395-1392, 1999.

[29] M. Lujan, T.L. Freeman, J.R. Gurd, "OoLaLa: an object-oriented analysis and design of numerical linear algebra", ACM SIGPLAN Notices, 35, 229-252, 2000.

[30] H.M. Chen , Y.C. Lin, "Internet based Analytical Services using a Java based GUI", in Proceedings of the Tenth International Conference on Civil, Structural and Environmental Engineering Computing, Topping BHV, editor, Civil-Comp Press, Stirling, United Kingdom, paper 14, 2005.

[31] M. Nuggehally, Y.S. Liu, S.B. Chaudhari, P. Thampi, "An internet-based computing platform for the boundary element method", Advances in Engineering Software, 34, 261-269, 2003.

[32] J. Peng, K.H. Law, "Building finite element analysis programs in distributed services environment", Computers & Structures 82, 1813-1833, 2004.

[33] Z. Yang, J. Lu, A. Elgamal, "A Web-based platform for computer simulation of seismic ground response". Advances in Engineering Software 35, 249-259, 2004.

[34] B. Meyer, "Object-oriented software construction", Prentice-Hall, 1988.

[35] R.I. Mackie, "Programming Distributed Finite Element Analysis: An Object Oriented Approach", Saxe-Coburg, Stirling, ISBN 978-1-874672-31-9, 2007.

[36] R.I. Mackie, "An Object-Oriented Framework for Programming Iterative Solution Algorithms", in Proceedings of the Tenth International Conference on Civil, Structural and Environmental Engineering Computing, Topping BHV, editor, Civil-Comp Press, Stirling, United Kingdom, paper 60, 2005.

[37] R.I. Mackie, "Object-Oriented Design of Pre-conditioned Iterative Equation Solvers using .NET", in Proceedings of the Fifth International Conference on Engineering Computational Technology, B.H.V. Topping, G. Montero and R. Montenegro, (Editors), Civil-Comp Press, Stirlingshire, United Kingdom, paper 111, 2006.

[38] A. Troelsen, "Pro C# 2500 and the .NET 2.0 Platform", APress, New York, 2005.

[39] I. Rammer, "Advanced .NET Remoting", APress, Berkeley, CA, USA, 2002.

[40] R.I. Mackie, "Design and Deployment of Distributed Numerical Applications Using .NET and Component Oriented Programming", submitted to Advances in Engineering Software, 2007.

[41] M. Benzi, "Preconditioning techniques for large linear systems: A survey", J of Computational Physics, 182, 418-477, 2002.

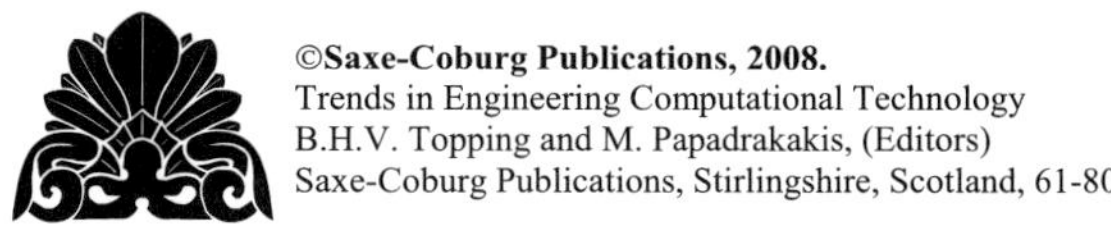

©Saxe-Coburg Publications, 2008.
Trends in Engineering Computational Technology
B.H.V. Topping and M. Papadrakakis, (Editors)
Saxe-Coburg Publications, Stirlingshire, Scotland, 61-80.

# Chapter 4

# Rheology and Simulation of Fresh Concrete Flow

**B. Patzák and Z. Bittnar**
**Faculty of Civil Engineering**
**Czech Technical University, Prague, Czech Republic**

## Abstract

The modeling of fresh concrete flow can significantly contribute to the durability and strength of a structure and it is necessary for design optimization of casting procedure. The purpose of this chapter is to give an overview of fresh concrete rheology and to present a numerical model based on homogeneous approach for the simulation of fresh concrete casting. The application of the model is illustrated on two examples.

**Keywords:** fresh concrete rheology, fresh concrete casting, non-Newtonian flow, interface-capturing.

# 1 Introduction

The rheology of fresh concrete is very important for the construction industry because concrete is usually put into a place in its plastic form. In the construction field, subjective terms like workability, flow-ability, and cohesion are used, sometimes interchangeably, to describe the behavior and flow properties of fresh concrete. These factors depend on flow (rheological) properties of concrete, that have direct influence on the strength and durability of concrete. Concrete that is not properly cast or consolidated may have defects, such as air voids, honeycombs, and aggregate segregation.

The modeling of fresh concrete flow can significantly contribute to the durability and strength of a structure and it is necessary for design optimization of the casting procedure. The aim of this contribution is to show that the reliable numerical modeling of fresh concrete flow is feasible, provided that suitable algorithms and constitutive models are combined together.

Most commonly used tests to describe the flow properties of concrete are limited to the measurement of only one parameter, often not directly related to any physical parameter. They have been typically designed to in-situ comparison between concretes.

Recently, some new tests or modifications of existing tests have been proposed to determine physical parameters that characterize concrete flow. An overview of these experimental techniques will be given, discussing their capabilities and potentials. An extensive review of fresh concrete rheology is given also in [1, 2, 3].

The challenge is also the task of the prediction of fresh concrete properties from its composition. There are several promising models that could predict the rheological parameters (yield stress and viscosity). These include the compressible packing model (CPM) developed by Larrard [4, 5] and simulation of flow suspension developed at NIST [6]. There are also many semi-empirical models that can be used to estimate rheological properties, see for example [7, 8].

Aside from measuring the flow properties of concrete, rheology is concerned with the prediction of the flow. No attempt to predict the flow from the properties of the components has yet been successful, due to the many factors influencing flow properties of concrete, that are more complex than factors influencing flow properties of constituents, and due to the complexity of phenomena taking place. An overview of the computational techniques for fresh concrete modeling is presented in Section 4.

In the rest of the chapter, recent results from numerical modeling of fresh concrete flow, that is treated as a homogeneous non-Newtonian fluid will be presented. As the characteristic flow velocity is very small compared to the speed of sound in fresh concrete, the fluid is treated as incompressible. In the case of incompressible flow, the mass and momentum conservation equations, together with the incompressibility condition and constitutive equation form a complete system. The Bingham model is considered, with the yield stress and plastic viscosity as parameters. In the present chapter, the numerical solution is based on the finite element (FE) method. The solution algorithm is based on a stabilized Eulerian FE formulation to prevent potential numerical instabilities. The stabilization techniques include streamline-upwind / Petrov-Galerkin (SUPG) and pressure-stabilizing / Petrov-Galerkin (PSPG) formulations. These stabilization techniques were introduced by Tezduyar and Hughes, see [9, 10] for further reference.

In analysis of problems with moving boundaries, depending on the complexity of the interface and other aspects of the problem, an interface-tracking or interface capturing techniques are usually used. These techniques will be briefly discussed. The given focus will be on interface-tracking techniques based on Volume-of-Fluid (VOF) and level set method, as they are based on the volume tracking concept to allow natural handling of topological changes of the free surface and straightforward generalization into multiple dimensions (level set method).

The numerical simulations of small scale casting problems will be presented in section 6 as well as their comparison with available experimental data. The numerical results show good agreement with measured data, validating the initial assumptions and mathematical model and illustrate high potential towards simulation of complex casting problems.

# 2 Rheology of fresh concrete

Concrete in a fresh state can be considered as a fluid. This assumption is valid when a certain degree of flow can be achieved and when concrete is homogeneous. This is usually satisfied because concrete is put in place in its plastic form in the majority of industrial applications. It is widely recognized that concentrated suspensions, such as concrete, typically behave as non-Newtonian fluids. The flow of concrete must be defined by at least two parameters, one being the yield stress, because it shows an initial resistance to the flow which should not be neglected, and the other one being plastic viscosity that governs the flow after it was initiated. The yield stress is a consequence of the inter-particle forces that are irreversibly broken by shearing the material. In concentrated dispersions of solid in liquid, the proximity of particles leads to strong mutual interactions, the strength of which depends on many factors, including particle size and shape, concentration, liquid properties, *etc.* Commonly there is a net attraction, causing flocculation. The shape and distribution of flocculated particles play the major role in rheology of the dispersion. In cement systems, the applied shearing reduces the flocculation, indicated by the reduced resistance to the flow, and often followed by link restoration when dispersion is in rest.

Concrete, as a fluid, is most often described using the Bingham model

$$\boldsymbol{\tau} = \boldsymbol{\tau}_0 + \mu\dot{\boldsymbol{\gamma}} \quad \text{if } \dot{\tau} \geq \tau_0 \tag{1}$$

$$\dot{\boldsymbol{\gamma}} = \mathbf{0} \quad \text{if } \dot{\tau} \leq \tau_0 \tag{2}$$

where $\boldsymbol{\tau}$ is the shear stress applied to material, $\dot{\tau} = \sqrt{\boldsymbol{\tau} : \boldsymbol{\tau}}$ is the shear stress measure, $\dot{\boldsymbol{\gamma}}$ is the shear rate, $\boldsymbol{\tau}_0$ is the yield stress, and $\mu$ is the plastic viscosity.

For a certain class of concretes, especially for self-compacting concrete, the linear dependency of the Bingham model is not adequate and better models have been proposed by Larrard *et al.* [11], who used the three-parameter Herschel-Bulkley's model

$$\boldsymbol{\tau} = \boldsymbol{\tau}_0 + a\dot{\boldsymbol{\gamma}}^b \tag{3}$$

The same authors also proposed an alternative, that allows the number of parameters to be reduced to two and to use the simpler Bingham model. The idea is to fit Herschel-Bulkley model on experimental results and then the Bingham model parameters are determined from Herschel-Bulkley curve using least-squares fit. The yield stress is unchanged and viscosity for the Bingham model is calculated as

$$\mu = \frac{3a}{b+2}\dot{\gamma}_{max}^{b-1} \tag{4}$$

where $\dot{\gamma}_{max}$ is the maximum shear strain rate achieved during the test.

# 3 Testing methods

The parameters for the model generally can be determined in two ways: (i) using stress controlled rheometer, when the stress is applied to material and the shear rate is measured, and (ii) using shear rate controlled rheometer, when concrete is sheared and the stress is measured. Most commonly used rheometers are shear rate controlled. However, commercially available rheometers, suitable for homogeneous fluids, are not suitable for concrete, containing solid particles, since the design of the rheometer must take into account the dimensions of the aggregate. It is widely recognized that the smallest gap in the device should be at least three times larger than the largest particle size to prevent interlocking of particles.

Unfortunately, most of the existing tests (slump test, penetrating rod test, *etc*) are unsatisfactory in the sense that they relate only to one parameter, typically to yield stress. A refined version of the standard slump test has been developed for estimating the yield stress and plastic viscosity. The test is based on measuring the time necessary for the upper surface of the concrete cone in the slump to fall a distance 100 mm. Semi-empirical models are then proposed for estimating the yield stress and viscosity based on measured results. The advantage is that this test does not require any special equipment, provided that the one for the standard version is available.

Recently, several rotational rheometers for concrete providing two parameters on output have been designed. The different designs are employed, BML and CEMAGREF-IMG use two coaxial cylinders and BTRHEOM is using parallel rotating plane. For the reliable prediction of parameters, the space between cylinders needs to be relatively small compared to their diameters. The generally agreed ratio of the two cylinder radii is between 1 and 1.1. For concrete the gap needs to be at least three times the aggregate size, the minimum inner cylinder radius turns out to be at least 0.5 m. Because of impractical dimensions, some rheometers are based on a rotating impeller inside cylindrical container (IBB and Two-point rheometers).

A round-robin test has been organized under sponsorship of ACI's Concrete Research Council in Nantes, France. The tests included five available units and the goal was to compare measured data on the same concrete mixtures. As could be expected, they did not produce identical values - absolute rheological parameters depended on the rheometers used. Nevertheless, the results from all rheometers exhibited the same general pattern for both viscosity and yield stress among 12 tested mixtures and the pairwise correlations of measurements were reasonably good, see [12].

# 4 Modeling of concrete flow

Depending on the problem, the computational techniques can be divided into three main classes: homogeneous fluid simulations, discrete particle simulations, and simulations based on particles suspended in the fluid. The choice of particular method depends on the type of problem under consideration.

Homogeneous fluid simulations work on a macro scale, where the fresh concrete can be regarded as homogeneous suspension during the flow. This problem is described by Navier-Stokes equations, typically under incompressible flow assumptions, as the attained velocity is very small compared to the speed of sound. The numerical solution can be based on Eulerian, Lagrangian, or mixed approaches. Eulerian formulation, which is typically used in CFD applications, is characterized by a coordinate system that is either stationary or moving in some prescribed manner in order to accommodate the continually changing solution domain. The mass travels between computational cells even if the grid moves, because grid movements are not related to the motion of the mass. The main advantage is the ability to represent large distortions without the lost of accuracy (in contrast to other methods) and handling of multiple interfaces is relatively straightforward. The main issue is the selection of the optimal strategy for the interface (free surface) tracking. Several authors followed this approach, for example Korokava *et al.* [13], Thrane *et al.* [14], and Roussel [15].

Lagrangian description can naturally handle material interfaces, but this method is not suitable for very large deformation processes, requiring frequent remeshing. Mixed methods are trying to avoid disadvantages of both approaches by trying to combine the best features of both. They typically use fixed computational grid and Lagrangian (integration) points. An example of such an approach, employed in the context of concrete flow, is the work by Dufour and Pijaudier-Cabot [16]. In general, homogeneous approaches can describe complex constitutive laws and can be used to simulate real casting problems. On the other hand, particle blocking and segregation cannot be modeled.

Particle based methods operate typically on a much lower scale, simulating the interactions between particles. Such approaches allow one to take directly into account the influence of mix composition, predict particle interlocking and segregation, and to study the effects of reinforcement. The constitutive parameters could not be directly measured and are typically determined by numerical fitting. The high number of particles required also limits the application to only small problems. Available numerical models are based on discrete element method (DEM) [18, 17] or on Dissipative Particle Dynamics (DPD), recently employed by Martys [19]. A state-of-the-art overview on computational methods in the field of computational modeling of fresh concrete was given by Roussel, *et al.* in [20].

# 5 Modeling of concrete casting based on homogeneous approach

The present solution technique is based on the Eulerian formulation. The numerical solution is based on the finite element method (FEM) and on the interface capturing method to track the position of a free surface. As the characteristic flow velocity will be very small compared to the speed of sound in fresh concrete, the fluid will be treated as incompressible. In the case of incompressible flow, the mass and momentum conservation equations, together with the incompressibility condition and constitutive equation form a complete system. The energy conservation equation could be additionally used to obtain the temperature field.

The solution algorithm is based on a stabilized FEM formulation to prevent potential numerical instabilities. The stabilization techniques include streamline-upwind / Petrov-Galerkin (SUPG) and pressure-stabilizing / Petrov-Galerkin (PSPG) formulations. These stabilization techniques were introduced by Tezduyar and Hughes, see [9, 10] for further reference.

In the analysis of problems with moving boundaries, depending on the complexity of the interface and other aspects of the problem, interface-tracking or interface-capturing techniques are usually used [21, 22, 23, 24]. The interface-tracking technique requires meshes that track or follow the interface. The mesh is updated as the interface evolves. In most cases, an automatic mesh generator needs to be used, sometimes at an overwhelming expense, to generate a new 3D mesh. Furthermore, every time the new mesh is generated, the solution needs to be projected from the old mesh to the new one. This process involves projection errors, and, in 3D, it requires significant computational time.

In the interface-capturing technique the computation is done on fixed spatial domains. An interface function, marking the location of the interface, needs to be computed to track the interface. The interface is captured within the resolution of the finite element mesh, in the sense that the actual physical discontinuity is located near the middle value of characteristic function, which is defined to be equal to one for one reference phase and zero for the other.

Surface tracking methods represent interface as a series of interpolation curves using a discrete set of points on the interface. The points move according to interface evolution. In the simplest case, only a sequence of heights above the reference line is maintained. More general algorithms rely on parametric representation, allowing representation of more complex interface shapes. The advantage is the ability to represent details of the interface that are smaller than cell spacing. On the other hand, it is extremely difficult to handle topological changes of the boundary (merging, folding, etc). Volume tracking methods [22, 25, 26] solve a transport equation for the fraction of the cell occupied by the liquid phase. These methods rely on the ability to advect volume fraction through the grid without smearing. The reconstruction of the interface is done on a cell by cell basis using volume fraction value within the cell and its neighborhood. The advantages include the capability to treat any number of mass constituents, tracing interfaces with large distortions, and implicit handling of interface topological changes.

In this chapter, two methods for tracking the free surface are presented. The first one, based on Volume of Fluid approach, is a typical example of the volume tracking method. It uses a geometrical-based method to advect volume fraction through the grid without smearing. The reconstruction of the interface is done on a cell by cell basis using volume fraction value within the cell and its neighborhood. In general, the resolution of the volume of fluid method is limited by the grid size and it may be difficult to impose boundary conditions on the interface, also the implementation in 3D is usually quite complicated.

The second interface-tracking technique is based on the level set method, originally introduced by Osher and Sethian [27]. The idea is not to track the interface position directly; the interface is defined as a zero level set of a suitable higher-dimensional function (called the level set function). For example, in two dimensions, the level set method represents a given curve in the plane as the zero level set of a two-dimensional auxiliary function. Then the interface is manipulated implicitly through the level set function. The advantages of this approach consist of natural handling of topological changes and straightforward generalization into multiple dimensions.

## 5.1 Stabilized Finite Element Formulation

In this section, the variational formulation with SUPG and PSPG stabilization terms is briefly described. The stabilization provides stability and accuracy in the solution of advection-dominated problems and permits usage of equal-order interpolation functions for velocity and pressure. Furthermore, stabilized formulation significantly improves convergence rate in iterative solution of large nonlinear systems of equations. The governing equations are based on the velocity-pressure formulation of the Navier-Stokes equations

$$\rho\left(\frac{\partial \boldsymbol{u}}{\partial t} + \boldsymbol{u}\cdot\boldsymbol{\nabla}\boldsymbol{u}\right) - \boldsymbol{\nabla}\cdot\boldsymbol{\sigma} = \boldsymbol{0} \text{ on } \Omega_{\mathrm{t}} \tag{5}$$

$$\boldsymbol{\nabla}\cdot\boldsymbol{u} = 0 \text{ on } \Omega_{\mathrm{t}} \tag{6}$$

where $\rho$ is the density and $\boldsymbol{u}$ is the velocity vector. The stress tensor $\boldsymbol{\sigma}$ is decomposed into its deviatoric part $\boldsymbol{\tau}$ and pressure $p$

$$\boldsymbol{\sigma}(p, \boldsymbol{u}) = \boldsymbol{\tau} - p\boldsymbol{I} \tag{7}$$

The essential and natural boundary conditions are

$$\boldsymbol{u} = \boldsymbol{g} \text{ on } \Gamma_{\mathrm{g}} \tag{8}$$

$$\boldsymbol{n}\cdot\boldsymbol{\sigma} = \boldsymbol{h} \text{ on } \Gamma_{\mathrm{h}} \tag{9}$$

where $\Gamma_g$ and $\Gamma_h$ are complementary subsets of the boundary $\Gamma$, $\boldsymbol{n}$ is the unit outward normal vector, and $\boldsymbol{g}$ and $\boldsymbol{h}$ are given functions. A divergence free velocity field is specified as the initial condition.

If the problem does not involve any moving boundary, the spatial domain does not need to change in time. This may be even the case for flows with moving boundaries, provided that the spatial domain is not defined to be the part of space occupied by the fluid. For example, one can have a fixed domain and by assuming that the domain is occupied by two immiscible fluids, the problem can be modeled by simultaneous tracking of mutual interface between the two fluids.

When modeling a free-surface problem, where one fluid is irrelevant, one can assign a sufficiently low density to that fluid. This is also the approach adopted in this work.

Provided that trial solution and test function spaces for velocity and pressure ($S_u^h$, $V_u^h$, $S_p^h$, and $V_p^h = S_p^h$) are defined, then the stabilized finite element formulation of equations (5) and (6) can be written as follows: find $\boldsymbol{u}^h \in S_u^h$ and $p^h \in S_p^h$ such that $\forall \boldsymbol{w}^h \in V_u^h$ and $\forall q^h \in V_p^h$

$$\int_\Omega \boldsymbol{w}^h \cdot \rho \left( \frac{\partial \boldsymbol{u}^h}{\partial t} + \boldsymbol{u}^h \cdot \boldsymbol{\nabla} \boldsymbol{u}^h \right) d\Omega + \int_\Omega \boldsymbol{\varepsilon}(\boldsymbol{w}^h) : \boldsymbol{\sigma}(p^h, \boldsymbol{u}^h)\, d\Omega$$
$$+ \int_{\Gamma_h} \boldsymbol{w}^h \cdot \boldsymbol{h}\, d\Gamma + \int_\Omega q^h \boldsymbol{\nabla} \cdot \boldsymbol{u}^h\, d\Omega + \sum_{e=1}^{n_{el}} \int_{\Omega_e} (\delta^h + \varepsilon^h)$$
$$\cdot \left[ \rho \left( \frac{\partial \boldsymbol{u}^h}{\partial t} + \boldsymbol{u}^h \cdot \boldsymbol{\nabla} \boldsymbol{u}^h \right) - \boldsymbol{\nabla} \cdot \boldsymbol{\sigma}(p^h, \boldsymbol{u}^h) \right] d\Omega = 0 \tag{10}$$

The first three terms correspond to the standard Galerkin formulation of (5)-(9). The last term contains two additional terms. The one with $\delta^h$ is the SUPG term, and the one with $\varepsilon^h$ is the PSPG term. The functions $\delta^h$ and $\varepsilon^h$ are defined as

$$\delta^h = \tau_{SUPG}\, \boldsymbol{u}^h \cdot \boldsymbol{\nabla} \boldsymbol{w}^h, \quad \varepsilon^h = \tau_{PSPG}\, \frac{1}{\rho} \nabla q^h \tag{11}$$

The parameters $\tau_{SUPG}$ and $\tau_{PSPG}$ depend on element Reynolds numbers, which are based on the local and global scaling velocities and element length measures. Their definitions can be found, for example, in [28, 29, 10]. The spatial discrctization of equation (10) leads to the following set of nonlinear differential equations

$$(\boldsymbol{M} + \boldsymbol{M}_\delta)\boldsymbol{a} + \boldsymbol{N}(\boldsymbol{v}) + \boldsymbol{N}_\delta(\boldsymbol{v}) + (\boldsymbol{K} + \boldsymbol{K}_\delta)\boldsymbol{v} - (\boldsymbol{G} + \boldsymbol{G}_\delta)\boldsymbol{p} = \boldsymbol{F} + \boldsymbol{F}_\delta \tag{12}$$
$$\boldsymbol{G}^T\boldsymbol{v} + \boldsymbol{M}_\varepsilon\boldsymbol{a} + \boldsymbol{N}_\varepsilon(\boldsymbol{v}) + \boldsymbol{K}_\varepsilon\boldsymbol{v} + G_\varepsilon\boldsymbol{p} = \boldsymbol{E} + \boldsymbol{E}_\varepsilon \tag{13}$$

where $\boldsymbol{v}$ is the vector of unknown nodal velocities, $\boldsymbol{a}$ is the vector of nodal accelerations, and $\boldsymbol{p}$ is the vector of nodal pressure values. The matrices $\boldsymbol{M}$, $\boldsymbol{N}$, $\boldsymbol{K}$, and $\boldsymbol{G}$ are derived from the time-dependent, advective, viscous, and pressure terms. The right-hand side vectors $\boldsymbol{F}$ and $\boldsymbol{E}$ are due to the boundary conditions. The subscripts $\delta$ and $\varepsilon$ identify the SUPG and PSPG contributions.

By considering the following time discretization

$$\boldsymbol{a}^{t+\Delta t} = \boldsymbol{a}^t + \Delta\boldsymbol{a} \tag{14}$$
$$\frac{\boldsymbol{v}^{t+\Delta t} - \boldsymbol{v}^t}{\Delta t} = \alpha\boldsymbol{a}^{t+\Delta t} + (1-\alpha)\boldsymbol{a}^t \tag{15}$$
$$\boldsymbol{p}^{t+\Delta t} = \boldsymbol{p}^t + \Delta\boldsymbol{p} \tag{16}$$

we can formulate the following solution algorithm:

1. Evaluate velocity vector predictor at time $t + \Delta t$ as

$$\boldsymbol{v} = \boldsymbol{v}^t + \Delta t\boldsymbol{a}^t$$
$$\boldsymbol{a} = \boldsymbol{a}^t$$
$$\boldsymbol{p} = \boldsymbol{p}^t$$

2. From (12) and (13) written in an incremental form solve the pressure and acceleration increments

$$\begin{aligned}\boldsymbol{M}^*\Delta\boldsymbol{a} - \boldsymbol{G}^*\Delta\boldsymbol{p} &= \boldsymbol{R}\\ (\boldsymbol{G}^T)^*\Delta\boldsymbol{a} + \boldsymbol{G}_\varepsilon\Delta\boldsymbol{p} &= \boldsymbol{Q}\end{aligned}$$

where

$$\begin{aligned}\boldsymbol{M}^* &= \boldsymbol{M} + \boldsymbol{M}_\delta + \alpha\Delta t\left(\left.\frac{\partial\boldsymbol{N}}{\partial\boldsymbol{v}}\right|_{\boldsymbol{v}} + \left.\frac{\partial\boldsymbol{N}_\delta}{\partial\boldsymbol{v}}\right|_{\boldsymbol{v}} + \boldsymbol{K} + \boldsymbol{K}_\delta\right)\\ \boldsymbol{G}^* &= \boldsymbol{G} + \boldsymbol{G}_\delta\\ (\boldsymbol{G}^T)^* &= \boldsymbol{M}_\varepsilon + \alpha\Delta t\left(\left.\frac{\partial\boldsymbol{N}_\varepsilon}{\partial\boldsymbol{v}}\right|_{\boldsymbol{v}} + \boldsymbol{K}_\varepsilon + \boldsymbol{G}^T\right)\\ \boldsymbol{R} &= \boldsymbol{F} + \boldsymbol{F}_\delta - [(\boldsymbol{M} + \boldsymbol{M}_\delta)\boldsymbol{a} + \boldsymbol{N}(\boldsymbol{v}) + \boldsymbol{N}_\delta(\boldsymbol{v})\\ &\quad +(\boldsymbol{K} + \boldsymbol{K}_\delta)\boldsymbol{v} - (\boldsymbol{G} + \boldsymbol{G}_\delta)\boldsymbol{p}]\\ \boldsymbol{Q} &= \boldsymbol{E} + \boldsymbol{E}_\varepsilon - \left[\boldsymbol{G}^T\boldsymbol{v} + \boldsymbol{M}_\varepsilon\boldsymbol{a} + \boldsymbol{N}_\varepsilon(\boldsymbol{v}) + \boldsymbol{K}_\varepsilon\boldsymbol{v} + \boldsymbol{G}_\varepsilon\boldsymbol{p}\right]\end{aligned}$$

3. Update velocity, pressure and acceleration as

$$\begin{aligned}\boldsymbol{a} &= \boldsymbol{a} + \Delta\boldsymbol{a}\\ \boldsymbol{v} &= \boldsymbol{v} + \Delta t\alpha\Delta\boldsymbol{a}\\ \boldsymbol{p} &= \boldsymbol{p} + \Delta\boldsymbol{p}\end{aligned}$$

4. Repeat steps 2-3 until the convergence is reached.

By changing the value $\alpha$, different methods from "Generalized mid-point family" can be chosen, i.e., Forward Euler ($\alpha = 0$), Midpoint rule ($\alpha = 0.5$), Galerkin ($\alpha = 2/3$), and Backward Euler ($\alpha = 1$). Except for the first one, all the methods are implicit and require matrix inversion for the solution. Some results from the energy method analysis suggest unconditional stability for $\alpha \geq 0.5$ for the generalized mid-point family. As far as accuracy is concerned, the midpoint rule generally is to be preferred.

In order to avoid numerical difficulties caused by the existence of the sharp angle in material model response of the Bingham model at $\tau = \tau_0$, the numerical implementation uses smoothed relation (see [30]). The following smooth relation for the viscosity is adopted

$$\mu = \mu_0 + \frac{\tau_0}{\dot\gamma}(1 - e^{-m\dot\gamma}) \tag{17}$$

where $m$ is so called stress growth parameter. The higher value of parameter $m$, the closer approximation of the original constitutive equation (1) is obtained.

As an alternative approach, the fractional-step family of methods (introduced by Chorin [31]) has to be mentioned, where the velocity and pressure variables are uncoupled, as opposed to the mixed methods already described. Advantages of this method are mainly due to treating the advective terms explicitly and that only the linear system, coming from standard discretization of second-order elliptic and parabolic partial differential equations, has to be solved. This original idea has been the basis for the extensive development of many new efficient algorithms (see, for example [32, 33]).

## 5.2 Volume of fluid based Capturing Algorithm

The first interface tracking method is based on Volume of Fluid approach, that introduces another unknown - fluid volume fraction in each grid cell. In principle, if we know the amount of fluid in each cell it is possible to locate surfaces, as well as to determine surface slopes and surface curvatures. Surfaces are easy to locate because they lie in cells partially filled with fluid or between cells full of fluid and cells that have no fluid. To compute the evolution of surfaces in time, a technique is needed to move volume fractions through the grid in such a way that the step-function nature of the distribution is retained. A straightforward numerical approximation cannot be used, because numerical diffusion and dispersion errors destroy the sharp, step-function nature of the VOF distribution. An important advantage of the VOF method is the fact that it is based on a volume tracking concept, and this means that it is robust enough to handle the breakup and coalescence of fluid masses.

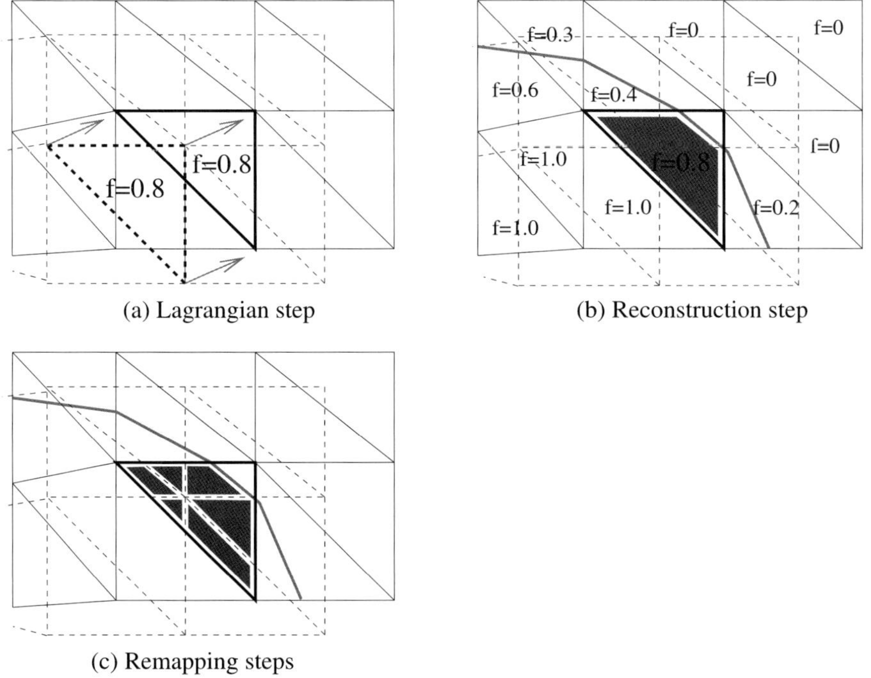

(a) Lagrangian step

(b) Reconstruction step

(c) Remapping steps

Figure 1: Three phases of Interface Capturing Algorithm

The volume tracking algorithm used in this work is based on the paper by Shahbazi *et al.* (see [34]) and consists of three parts (see also Figure 1):

1. Lagrangian part, in which original mesh is projected along trajectories.

2. Reconstruction part. In this step the volume materials are reconstructed on the updated grid, assuming that volume fractions remain constant during the Lagrangian phase. This includes the calculation of interface segment normals, line constants, and material volume truncation at each Lagrangian cell.

3. Remapping part, consisting of assembling the polygonal representation of a reference fluid material for each cell on the updated grid and their deposition to the target (original) grid to obtain updated volume fractions. This phase is performed by a series of polygon intersection procedures between polygon representing reference fluid on updated grid and polygons representing original mesh cells, yielding the contributions to updated reference fluid volumes in individual cells.

The details of numerical implementation can be found in [35, 36].

A similar interface tracking technique, called Edge-Tracked Interface Locator Technigue (ELILT), was introduced in [24], with numerical tests reported in [37, 38]. In this technique, interfaces are represented by a collection of positions along element edges crossed by the interface. To update the interface position, the nodal representation of the interface function is computed using the least-squares projection, which is then updated by solving stabilized advection equation. Edge based representation is then recovered using the least-squares projection combined with the volume conservation enforcement, leading to iterative correction procedure. Compared to the adopted approach, the volume conservation is strictly enforced at the cost of a numerically demanding update.

## 5.3 Level Set Representation of Free Surface

The central idea behind the level set method is the representation of an interface as the zero level set of a suitably defined higher-dimensional function $\phi$. The interface is manipulated implicitly, through the level set function, which is typically defined as a signed distance function from the interface. The interface position can always be recovered as the zero level set, as well as other geometrical characteristics, such as normal direction or curvature. Figure 2 illustrates the embeddment of free surface as the zero level set of the signed distance function for the classical breaking dam problem.

The level set equation for the moving interface advected by a velocity field $\boldsymbol{u}$ can be written in the form

$$\phi_t + \boldsymbol{u}\cdot\nabla\phi = 0 \tag{18}$$

where the level set function $\phi = \phi(t, x)$ describes the evolving interface implicitly by its zero level set. This can be reformulated into a form of Hamilton-Jacobi equation as

$$\begin{gathered} \phi_t + F(\nabla\phi, x)|\nabla\phi| = f(x) \\ F(\nabla\phi, x) = \frac{\boldsymbol{u}\cdot\nabla\phi}{|\nabla\phi|},\ f(x) = 0 \end{gathered} \tag{19}$$

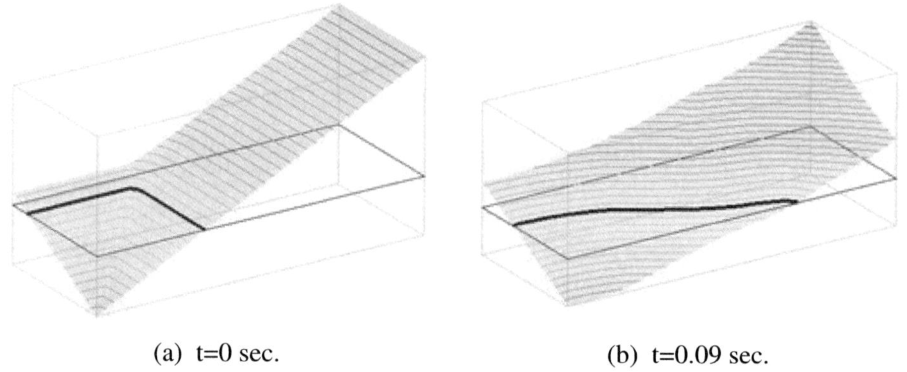

(a) t=0 sec. (b) t=0.09 sec.

Figure 2: Level set representation of a free surface for breaking dam problem: profiles of the level set function at different times, the zero values of level set function representing interface are in bold

The above problem is solved using the explicit positive coefficient scheme, as proposed by Barth and Sethian [39] for unstructured triangular meshes. The algorithm is summarized in Table 1.

As the information propagates in an upwind direction, then for each node there should be a source triangle surrounding this node. This element can be located by tracking back along the gradient of $\nabla F$ at a particular node. The above mentioned algorithm can be used for any node for which at least one source element exists, which produces a nonzero $\alpha_i$ coefficient. This is satisfied for all domain nodes except inflow nodes on boundary, where special update procedure is needed. In this work, we use simple differential-based update formula for inflow nodes

$$\phi_i^{n+1} = \phi_i^n - \Delta t\, \boldsymbol{u}\cdot\nabla\phi^n$$

One of the drawbacks of level set based methods is that they are not conservative, if applied to passive transport having divergence free velocity. For such applications, some combinations of level set methods with other techniques were proposed to improve mass preservation property. This includes re-initialization techniques to recover signed distance property of the level set, hybrid particle level set [40], or high resolution flux-based level set [41]. In this work, techniques from the first category are used to recover signed distance property. The first technique is based on the solution of the following problem to steady state (see [42])

$$\begin{aligned} d_\tau &= S(\phi)(1 - |\nabla d|),\ \tau > 0 \\ d(x,0) &= \phi(x,t) \end{aligned} \tag{20}$$

where $S$ is the sign function and $\tau$ is an artificial time. It can be shown, that the steady state solution of Equation (21) corresponds to the signed distance function.

1 Initialize $\phi_i^* = w_i = 0, \ \ i = 1, 2, ..., M$

2 For each triangle T, i=1,2,3:

$$N_i(x) \in P_1, N_i(x) = \delta_{ij}, \ j = 1, 2, 3, \ x \in T,$$

$$F = \frac{1}{A_i} \int_T \frac{\boldsymbol{u} \cdot \nabla\phi}{|\nabla\phi|} dx, \ \boldsymbol{n}_i = 2A_i \nabla N_i, \ \nabla\phi = \sum_{j=1}^{3} \nabla N_i \phi_j,$$

$$K_i = \frac{F \nabla\phi \cdot \boldsymbol{n}_i}{2|\nabla\phi|}, \ \delta\phi = \sum_{l=1}^{3} K_l \phi_l, \ \delta\phi_i = K_i^+ \sum_{l=1}^{3} K_l^- (\phi_i - \phi_l)(\sum_{l=1}^{3} K_l)^{-1},$$

$$\alpha_i = \frac{\max(0, \delta\phi_i/\delta\phi)}{\sum_{l=1}^{3} \max(0, \delta\phi_l/\delta\phi)}, \ \phi_i^* = \phi_i^* + \alpha_i \delta\phi, \ w_i = w_i + \alpha_i A_i.$$

3 For each $i \in M$:

$$\phi_i^{n+1} = \phi_i^n - \Delta t \phi_i^* / w_i.$$

Table 1: The algorithm of the explicit positive coefficient scheme; M is number of nodes, $A_i$ is area of triangle $T_i$, and $N_i$ is the shape function corresponding to i-th vertex. The superscript "+" means positive part, "-" negative part of given value

The fact, that $S(0) = 0$ implies that $d$ has the same zero level set as $\phi$. This problem is solved using iterative solution technique, also based on the explicit positive coefficient scheme [39]. Another and efficient way to solve this equation is to employ the fast marching method.

# 6 Examples

In this section, two examples of numerical simulations of laboratory experiments (slump and L-box tests) are presented. In both cases, the Bingham material model is adopted. In the first example, VOF technique has been used, while in the second, the Level-set method has been utilized to track the position of free surface. All the computations were performed in a free, object-oriented FEM solver OOFEM [43, 44], which is distributed under GNU Public License.

## 6.1 Slump test

This is a classical test used in many countries to determine the "workability" of concrete. The concrete is put (by a prescribed procedure) into a conical form-work, which is then suddenly released and concrete starts spreading under the gravity loading, see Figure 3. In its standard form, only the final value of the slump is measured. For our purposes, the spreading and slump have been recorded as a function of time. The mea-

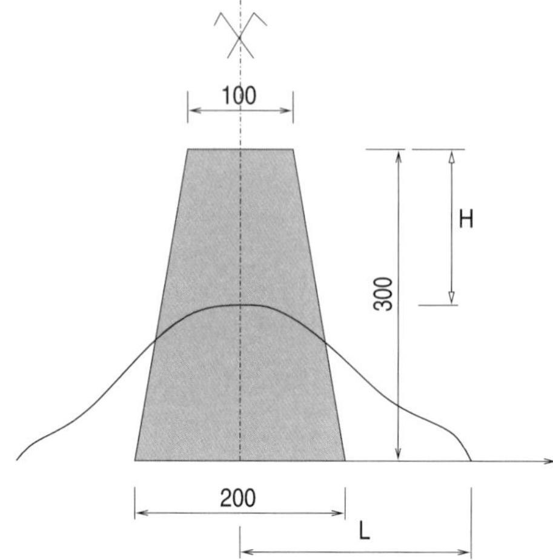


Figure 3: Slump test geometry

sured slump history (not the spreading history) has been used to calibrate the model parameters in this example. Then the numerically obtained results are compared to experimental values for slump as well as spreading history (see Figure 4) to illustrate the ability of the developed model to describe fresh concrete flow. Numerical analysis of this test requires at least axisymmetric analysis, which has been also used in this case.

The trial and error approach has been used to determine the constitutive parameters from slump history. The fitted parameters lead to a very good agreement with experimental measurements, illustrating the good capability of the constitutive model to describe concrete flow. The comparison of results on the spreading history reveals some differences, caused by the existence of an experimentally observed friction on the horizontal base, which was included in the numerical model only as the constant tangential traction. For practical purposes, more work on suitable boundary conditions should be done. Nevertheless, this example shows a very good agreement between experimental and numerical results.

## 6.2 LBOX test

This test is an example of a real casting procedure, for which experimental data are available. The concrete is confined in an L-shaped reservoir with a vertical gate (see Figure 5a). After the removal of the gate, concrete starts spreading into the form-work. Frictionless boundary conditions are assumed on walls forming the reservoir as well as form-work. The problem is solved as two-dimensional. Experimental results were taken from [16], where the constitutive properties of concrete mixture (summarized in

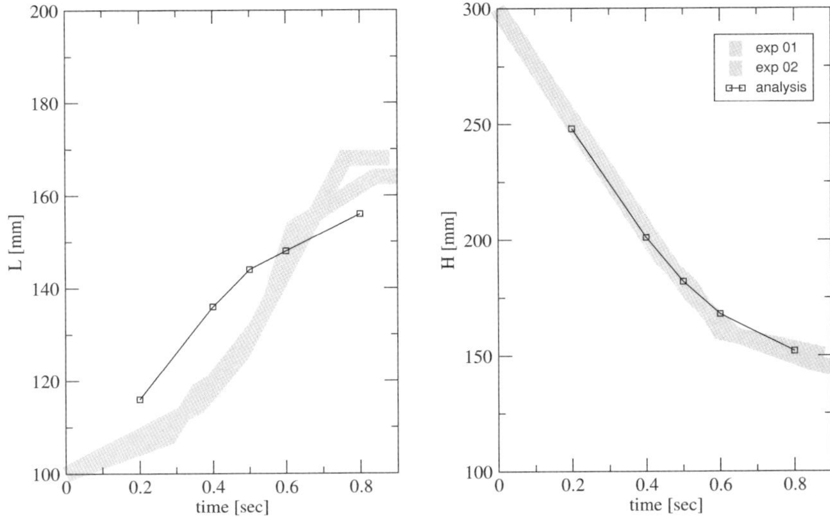


Figure 4: Slump test - comparison of the measured and computed slump (H) and spreading values (L) as the function of time

| | |
|---|---|
| Aggregate 6/10 (kg/m$^3$) | 838.7 |
| Sand (kg/m$^3$) | 806.3 |
| Cement (kg/m$^3$) | 350 |
| Limestone filler (kg/m$^3$) | 250 |
| Superplasticizer (kg/m$^3$) | 6.13 |
| Water (kg/m$^3$) | 197.5 |

Table 2: Concrete composition for LBOX test.

Table 2) have been determined by using *Bétonlab* code [45], based on the compressible packing model and calibrated using parallel plate rheometer (BT RHEOM). The constitutive parameters determined using this technique are: $\tau_0 = 244$ Pa, $\mu = 200$ Pa.s. Available experimental data consist of values of measured heights of concrete at specific locations at the final stage. The unstructured, triangular grid consists of 3652 nodes and 6927 triangles. The numerically resolved profiles of cast concrete in different times are shown in Figure 6. Figure 6 (d) shows the final profile, compared with experimental results indicated as bold cross marks. A good agreement between experimental and numerical results has been obtained. The level set representation of free surface is illustrated in Figure 5b, corresponding to the solution time t=0.3 seconds.

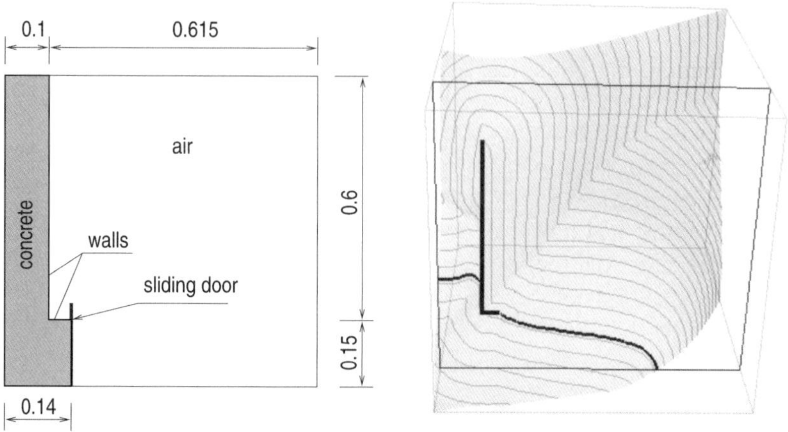


Figure 5: LBOX test: (a) geometry and (b) level set representation at t=0.3 sec.

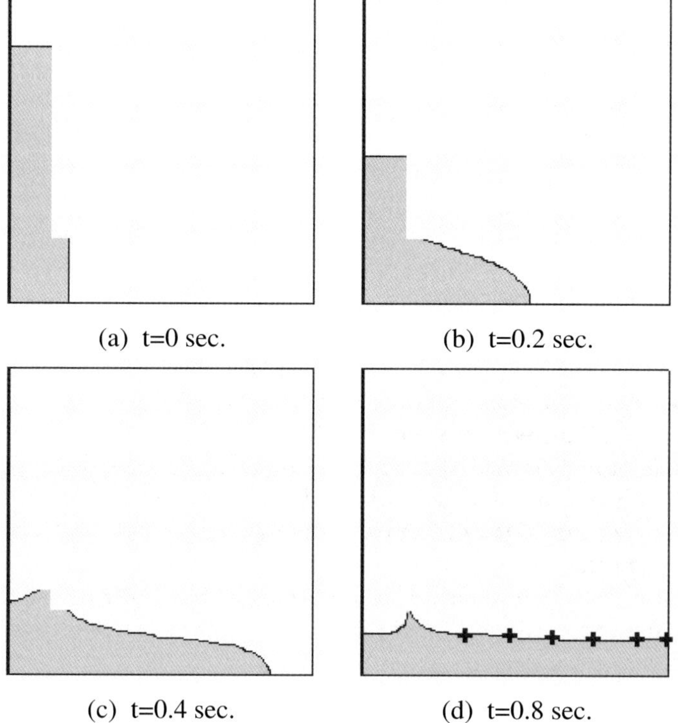


Figure 6: LBOX test: profiles of spreading concrete at different times

# 7 Conclusions

In this contribution, the numerical aspects of fresh concrete flow modeling are addressed. Particular attention is given to algorithms that can be used to extend existing engineering codes. The fresh concrete is considered as a non-Newtonian fluid. Eulerian formulation is adopted, which is typically used in CFD applications. To track the position of the free surface, an interface-capturing approaches are presented. They are based on the Volume-of-Fluid and Level-set approaches. The advantage of the above approaches is the fact that they are based on the volume tracking concept, which can implicitly handle the breakup and coalescence of fluid masses. The presented simulations show, that obtained results are in good agreement with experimental data.

# Acknowledgments

This work was supported by the Grant Agency of the Czech Republic - Project No.: 103/06/1845.

# References

[1] C. Ferraris, "Measurement of the Rheological Properties of High Performance Concrete: State of the Art", Report. J. Res. Natl. Inst. Stand. Technol, 104, 1999.

[2] C. Ferraris, F. Larrard, "Testing and Modelling of Fresh Concrete Rheology", National Institute of Standards and Technology, report NISTIR 6094, 1998.

[3] P.F.G. Banfill, "The Rheology of Fresh Cement and Concrete - A Review", Proc. 11th International Cement Chemistry Congress, Durban, 2003.

[4] F. de Larrard, T. Sedran, "Mixture-proportioning of high-performance concrete", Cement and Concrete Research, Vol. 32, 11, 1699-1704, 2002.

[5] F. de Larrard, "Concrete Mixture Proportioning", Taylor & Francis, 1999.

[6] S.T. Erdogan et al., "Influence of the shape and roughness of inclusions on the rheological properties of a cementitious suspension", Cement Concrete Comp (2008), in press.

[7] T. Roshavelov, "Prediction of fresh concrete behavior based on analytical model for mixture proportioning", Cement and Concrete Research, 35, 831-835, 2005.

[8] Y. Tanigawa, H. Mori, K. Tsutsui, Y. Kurokawa, "Simple methods for estimating rheological constants of fresh concrete", Transactions of the Japan Concrete Institute, vol. 9, 131-138, 1987.

[9] A.N. Brooks, T.J.R. Hughes, "Streamline Upwind/Petrov-Galerkin Formulations for Convection Dominated Flows with Particular Emphasis on the Incompressible Navier-Stokes Equations", Computer Methods in App. Mechanics and Engng.,32, 199-259, 1982.

[10] T.E. Tezduyar, "Stabilized Finite Element Formulations for Incompressible Flow Computations", Advances in Applied Mechanics, 28:1-44, 1992.
[11] F. de Larrard, C.F. Ferraris, T. Sedran, "Fresh concrete: a Herschel-Bulkley material", Material and Structures, 31(211), 494-498, 1998.
[12] L.E. Brower, C.F. Ferraris, "Comparison of Concrete Rheometers", Concrete International, Vol. 25(8), 41-47, 2003.
[13] Y. Kurokawa, Y. Tanigawa, H. Mori, Y. Nishinosono, "Analytical study on effect of volume fraction of coarse aggregate on Bingham's constants of fresh concrete", Transactions of the Japan Concrete Institute 18, 3744, 1996.
[14] L.N. Thrane, P. Szabo, M. Geiker, M. Glavind, H. Stang, "Simulation and verification of flow in SCC test methods", Proc. of the 4th Int. RILEM Symp. on SCC, Hanley Wood, Chicago, 2005.
[15] N. Roussel, "Correlation between yield stress and slump: comparison between numerical simulations and concrete rheometers results", RILEM Materials and Structures, 39 (4), 501-509, 2006.
[16] F. Dufour, G. Pijaudier-Cabot, "Numerical modelling of concrete flow: homogenous approach", Int. J. Numer. Anal. Meth. Geomech., 29:395-416, 2005.
[17] M.A. Noor, T. Uomoto, "Three-dimensional discrete element simulation of rheology tests of self-compacting concrete", In Proceedings of the 1st International RILEM Symposium on Self-Compacting Concrete, Skarendahl A, Petersson O (eds). RILEM: Cachan, 3546, 1999.
[18] O. Petersson, "Simulation of self-compacting concrete - laboratory experiments and numerical modelling of testing methods, jring and l-box tests", In Proceedings of the 3rd International RILEM Symposium, Pro 33, Wallevik O, Nielsson I (eds). RILEM: Cachan, 202207, 2003.
[19] N.S. Martys, "Study of a dissipative particle dynamics based approach for modeling suspensions", Journal of Rheology, Vol. 49, No. 2, 401424, March/April 2005.
[20] N. Roussel, M.R. Geiker, F. Dufour, L.N. Thrane, P. Szabo, "Computational modeling of concrete flow: General overview", Cement and concrete research, 37, 1298-1307, 2007.
[21] J.M. Floryan, H. Rasmussen, "Numerical Methods for Viscous Flows with Moving Boundaries", Appl Mech Rev, 42, 323-341, 1989.
[22] C.W. Hirt, B.D. Nichols, "Volume of Fluid (VOF) Method for the Dynamics of Free Boundaries", Journal of Comput Physics, 39, 201-225, 1981.
[23] T.E. Tezduyar, "Interface-Tracking and Interface-Capturing Techniques for Finite Element Computation of Moving Boundaries and Interfaces", Computer Methods in App. Mechanics and Engng, 195, 2983-3000, 2006.
[24] T.E. Tezduyar, "Finite Element Methods for Flow Problems with Moving Boundaries and Interfaces", Archives of of Computational Methods in Engineering, 8, 83-130, 2001.
[25] W.J. Rider, D.B. Kothe, "Reconstructing Volume Tracking", J. Comput. Phys, 141, 112-152, 1998.
[26] J.A. Sethian, "Level Set Methods and Fast Marching Methods", Cambridge Uni-

versity Press, 1999.

[27] S. Osher, J.A. Sethian, "Fronts propagating with curvature-dependent speed: Algorithms based on Hamilton-Jacobi formulations", J. Comput. Phys., 79, 12-49, 1988.

[28] T.E. Tezduyar, "Computation of Moving Boundaries and Interfaces and Stabilization Parameters", Int. J. Numer. Meth. Fluids, 43, 555-575, 2003.

[29] T.E. Tezduyar TE, S. Sathe, "Stabilization Parameters in SUPG and PSPG Formulations", Journal of Comput. and Appl. Mechanics, 4(1), 71-88, 2003.

[30] T.C. Papanastasiou, "Flows of materials with yield", Journal of Rheology, 31, 385-404, 1987.

[31] A.J. Chorin, "Numerical solution of the Navier-Stokes equations", Mathematical Computation, 22, 745-762, 1968.

[32] M. Kawahara, K. Ohmiya, "Finite element solution of density flow using the velocity correction method", Int. J. Numer. Meth. Fluids, 5(11), 981-993, 1985.

[33] P.M. Gresho, "On the theory of semi-implicit projection methods for viscous incompressible flow and its implementation via a finite element method that also introduces a nearly consistent mass matrix. Part 1: Theory", Int. J. Numer. Meth. Fluids, 11, 587-620, 1990.

[34] K. Shahbazi, M. Paraschivoiu, J. Mostaghimi, "Second order accurate volume tracking based on remapping for triangular meshes", Journal of Comput. Physics, 188, 100-122, 2003.

[35] B. Patzák and Z. Bittnar, "Simulation of fresh concrete flow", In B. H. V. Topping, G. Montero, and R. Montenegro, editors, Proceedings of the Fifth International Conference on Engineering Computational Technology, Civil-Comp Press, Stirlingshire, UK, Paper 204, 2006.

[36] B. Patzák and Z. Bittnar, "Modeling of Fresh Concrete Flow", Computers and Structures, to appear, 2008.

[37] M.A. Cruchaga, D.J. Celentano, T.E. Tezduyar, "Moving-Interface Computations with the Edge-Tracked Interface Locator Technique (ETILT)", Int. J. Numer. Meth. Fluids, 47, 451-469, 2005.

[38] M.A. Cruchaga, D.J. Celentano, T.E. Tezduyar, "Collapse of a Liquid Column: Numerical Simulation and Experimental Validation", Computational Mechanics, published online, 2006.

[39] T.J. Barth, J.A. Sethian, "Numerical Schemes for the Hamilton-Jacobi and Level Set Equations on Triangulated Domains", J. Comp. Phys., 145, 1, 1–40, 1998.

[40] D. Enright, R. Fedkiw, J. Ferziger, I. Mitchell, "A Hybrid Particle Level Set Method for Improved Interface Capturing", J. Comput. Phys. 183, 83-116, 2002.

[41] P. Frolkovic, K. Mikula, "High-resolution flux-based level set method", SIAM J. Sci. Comp., to appear.

[42] M. Sussman, P. Smereka, S. Osher, "A Level Set Approach for Computing Solutions to Incompressible 2-Phase Flow", Journal of Computational Physics, 114(1): p. 146-159, 1994.

[43] B. Patzák, Z. Bittnar, "Design of object oriented finite element code", Advances in Engineering Software, 32, 759-767, 2001.

[44] B. Patzák, "OOFEM home page", http://www.oofem.org, 2008.
[45] F. Larrard, D. Fau, "Bétonlab: logiciel d'aide à la formulation des bétons", Presses de l'ENPC, Paris, 1992.

©Saxe-Coburg Publications, 2008.
Trends in Engineering Computational Technology
B.H.V. Topping and M. Papadrakakis, (Editors)
Saxe-Coburg Publications, Stirlingshire, Scotland, 81-101.

# Chapter 5

# Finite Element Modelling of Two- and Three-Dimensional Viscoelastic Polymer Flows

**R. Tenchev[1,3], O. Harlen[2], P.K. Jimack[1] and M.A. Walkley[1]**
**[1] School of Computing**
**University of Leeds, United Kingdom**
**[2] School of Mathematics**
**University of Leeds, United Kingdom**
**[3] LUSAS**
**Kingston upon Thames, United Kingdom**

## Abstract

We present an overview of a recently-developed finite element software tool for the simulation of non-Newtonian fluid flows arising in viscoelastic polymer melts. The software is based upon the use of quantitative macroscopic constitutive laws, defining the evolution of polymer stress, which contributes to the total fluid stress in the flow. A wide range of melt rheologies may be considered in this manner, and three specific examples are considered in this paper. In addition to providing a detailed overview of the software, a number of sample numerical simulations are presented to illustrate the capabilities of the code.

**Keywords:** polymer melts, viscoelastic flow, arbitrary Lagrangian-Eulerian meshes, finite element methods.

# 1 Introduction

Understanding and controlling the flow of polymer melts is of great importance in polymer processing industries. Both the qualitative and the quantitative features of such flows are determined by the rheology of the melt, which in turn depends upon the underlying molecular structures. Consequently, there are a number of different length scales at which simulations may be undertaken in order to obtain macro-scale flow descriptions. In this research we work only at the continuum scale by making use of quantitative macroscopic constitutive laws which are based upon theory involving the physics of molecular alignment and entanglement (see, for example, [1–3]).

In the remainder of this section some further background is provided to the problems of interest. Specifically, this includes a description of the underlying mathematical models that are used and a brief discussion of relevant numerical methods, applied elsewhere in the literature. Section 2 then describes the algorithmic approach that is used in this work: this includes a discussion of Eulerian versus Lagrangian meshes and an overview of the discretisation which follows. Section 3 goes into more detail of the software implementation, ranging from the problem specification, through the numerical methods, on to the presentation of results. The paper concludes with sections that illustrate the execution of the software via a selection of computational results, and a final section that draws brief conclusions and suggests areas for further research.

## 1.1 Governing Equations

For completeness, this section provides a full mathematical description of the viscoelastic models considered by our software. Although the equations are provided in full we do not attempt to justify them here: instead we provide sample references which go more deeply into the model derivations.

### 1.1.1 Stokes' equations

The flows considered in this work comprise of solutions of high molecular weight polymers at relatively low velocities. This leads to a model of a high viscosity fluid with relatively small transient and inertial effects. Hence the flow can be mathematically modelled by Stokes' equations:

$$\nabla\cdot\sigma + \mathbf{b} = \mathbf{0}; \tag{1}$$

$$\nabla\cdot\mathbf{u} = 0; \tag{2}$$

where $\mathbf{u} = (u, v, w)$ represents the velocity field, $\mathbf{b}$ represents the body forces present and $\sigma$ is the viscoelastic stress tensor. This stress tensor may be decomposed into Newtonian and non-Newtonian parts, $\sigma_\mathrm{N}$ and $\mathbf{T}$ respectively, with $\sigma_\mathrm{N}$ defined as

$$\sigma_\mathrm{N} = -p\mathbf{I} + \mu\left(\nabla\mathbf{u} + (\nabla\mathbf{u})^T\right) \tag{3}$$

where $p$ is the pressure field and $\mu$ is the Newtonian viscosity.

The additional polymer stress $\mathbf{T}$ is determined by the constitutive equation for the polymer and is considered in this work as an additional body force term in the momentum equation (1), which therefore re-arranges as:

$$-\nabla\cdot\sigma_\mathrm{N} = \mathbf{b} + \nabla\cdot\mathbf{T}. \tag{4}$$

The three viscoelastic constitutive models considered in this work are described in the following sub-sections.

### 1.1.2 An Oldroyd-B fluid

The Oldroyd-B model is a standard, single-mode, viscoelastic fluid model [4] for which the polymeric stress tensor $\mathbf{T}$ is given by

$$\mathbf{T} = G(\mathbf{A} - \mathbf{I}), \tag{5}$$

where $G$ is the elastic modulus of the fluid and the polymer deformation tensor $\mathbf{A}$ is assumed to obey the equation

$$\frac{\partial \mathbf{A}}{\partial t} + \mathbf{u} \cdot \nabla \mathbf{A} - \mathbf{A} \cdot \nabla \mathbf{u} - (\nabla \mathbf{u})^T \cdot \mathbf{A} = \frac{-1}{\tau} (\mathbf{A} - \mathbf{I}), \tag{6}$$

with $\tau$ the relaxation time for the polymer. For a given problem the parameters $G$ and $\tau$ must be specified to completely define the polymeric behaviour.

### 1.1.3 A Rolie-Poly fluid

The Rolie-Poly model [1] is a multi-mode constitutive law for linear (non-branched) polymers derived from a simplification of the tube theory for entangled polymers [2]. The stress tensor $\mathbf{T}$ for an $n$-mode model is given by

$$\mathbf{T} = \sum_{i=1}^{n} G_i (\mathbf{A}_i - \mathbf{I}), \tag{7}$$

where $G_i$ is the elastic modulus for mode $i$. The polymer deformation tensor for each mode, $\mathbf{A}_i$, satisfies

$$\frac{\partial \mathbf{A}_i}{\partial t} + \mathbf{u} \cdot \nabla \mathbf{A}_i - \mathbf{A}_i \cdot \nabla \mathbf{u} - (\nabla \mathbf{u})^T \cdot \mathbf{A}_i = \frac{-1}{\tau_{A,i}} \mathbf{A}_i + \frac{1}{\tau_{d,i}^{eff}} \mathbf{I}. \tag{8}$$

This model includes two relaxation times for each mode, which satisfy the following equations:

$$\frac{1}{\tau_{R,i}^*} = 2 \frac{\left(1 - \sqrt{3/tr\mathbf{A}_i}\right)}{\tau_{R,i}}; \tag{9}$$

$$\frac{1}{\tau_{d,i}^{eff}} = \frac{1}{\tau_{d,i}} + \frac{1}{\tau_{R,i}^*} \beta_i \left(\frac{tr\mathbf{A}_i}{3}\right)^{\delta_i}; \tag{10}$$

$$\frac{1}{\tau_{A,i}} = \frac{1}{\tau_{d,i}^{eff}} + \frac{1}{\tau_{R,i}^*}. \tag{11}$$

The reptation time $\tau_{d,i}$, Rouse time $\tau_{R,i}$, constraint release parameter $\beta_i$, tube deformation parameter $\delta_i$ and modulus $G_i$ must be specified for each mode to completely define the behaviour of the fluid for a given problem.

### 1.1.4 A pom-pom fluid

The pom-pom model is a multi-mode constitutive law for branched polymers derived from the tube theory applied to star and H-shaped polymers [3]. The stress tensor $\mathbf{T}$ for an $n$-mode model is given by

$$\mathbf{T} = \sum_{i=1}^{n} \frac{G_i \lambda_i^2}{(tr\mathbf{A}_i/3)} (\mathbf{A}_i - \mathbf{I}), \tag{12}$$

with elastic modulus $G_i$ and stretch parameter $\lambda_i$ given for each mode. The polymer deformation tensor for each mode, $\mathbf{A}_i$ evolves according to a similar equation to Oldroyd-B,

$$\frac{\partial \mathbf{A}_i}{\partial t} + \mathbf{u} \cdot \nabla \mathbf{A}_i - \mathbf{A}_i \cdot \nabla \mathbf{u} - (\nabla \mathbf{u})^T \cdot \mathbf{A}_i = \frac{-1}{\tau_{b,i}} (\mathbf{A}_i - \mathbf{I}), \tag{13}$$

but the stretch parameters $\lambda_i$ in the stress tensor (12) are governed by additional differential equations:

$$\frac{\partial \lambda_i}{\partial t} + \mathbf{u} \cdot \nabla \lambda_i = \frac{\lambda_i}{tr\mathbf{A}_i} (\mathbf{A}_i : \nabla \mathbf{u}) - \frac{\lambda_i - 1}{\tau_{s,i}} e^{\left(2\frac{\lambda_i - 1}{q_i - 1}\right)}, \tag{14}$$

where $q_i$ represents the maximum stretch, with $\lambda_i$ satisfying $0 < \lambda_i \leq q_i$, and $\tau_{b,i}$ and $\tau_{s,i}$ are, respectively, the alignment and stretch relaxation times. For a given problem the parameters $G_i$, $\tau_{b,i}$, $\tau_{s,i}$ and $q_i$ must be specified for each mode to completely define the polymeric behaviour.

## 1.2 Numerical Methods

There is a very substantial body of previous work that has developed numerical methods for the simulation of viscoelastic, especially Oldroyd-B, fluids. Indeed, the literature is far too vast for us to attempt any sort of review here. Instead, we refer to a selection of such work, based upon finite volume methods in two dimensions [5] or three dimensions [6], spectral methods [7], stabilised finite element methods [8] or mixed finite element methods [9]. These papers, and the many references cited therein, provide an illustration of the richness of this field.

The Rolie-Poly and pom-pom fluids, introduced above, have been less widely studied numerically. Consequently, we confine this discussion to one such implementation, [10], and refer the interested reader to references cited therein for a broader selection of numerical approaches. The *FlowSolve* code described in [10] is based upon a two-dimensional Lagrangian formulation of the problem, using mixed finite elements on an adapting mesh of triangular elements. The mixed formulation for the Stokes flow is inherently stable [11], and the fact that the mesh evolves with the flow ensures that the polymer transport equations are also trivial to integrate in time in a stable manner. The primary drawbacks of this approach are the need to frequently re-connect the mesh to avoid tangling (this is relatively cheap and easy in two dimensions but would be much more problematic in three), and the need to ensure that the initial mesh has sufficient nodes "upstream" to ensure a satisfactory mesh resolution throughout the simulation (or else a mechanism is required to introduce new nodes at the inflow boundary and remove nodes from the outflow boundary automatically).

As described in the next section, in this work we have chosen an Eulerian approach ahead of the Lagrangian technique. This is primarily due to the technical mesh-quality issues that arise with the latter in three dimensions. In order to allow domains that evolve with time, this is then extended to an arbitrary Lagrangian-Eulerian (ALE) formulation.

# 2 Software Overview

This section provides a very brief description of the software that is the subject of this paper. More details concerning the implementation are then provided in Section 3. Note that since the discretisation used in this work is based upon finite elements, the use of finite element meshes and spaces will be assumed throughout.

## 2.1 Eulerian and Lagrangian Meshes

As described in the previous section, there are two basic classes of approach that may be taken. In the Lagrangian approach the points of the finite element mesh move at a velocity that is equal to the instantaneous local fluid velocity. This has significant advantages when simulating the transport of the polymer stress, as in equations (6), (8) or (13) for example. By selecting an Eulerian approach there is a need to consider carefully the stability of any discretisation of these equations, and this matter is discussed further in Section 3.3 below. Clearly, a significant advantage of the Eulerian approach is that, once an initial high-quality mesh has been obtained for the domain of interest, this may be re-used for all time. This leads to the possibility of significant savings if it can be exploited in the solution procedure. In particular, in Section 3 we describe a penalty formulation that allows the pressure to be eliminated from the Stokes equations and the resulting discrete system to be factorised using a sparse or a banded LU factorisation [12]. This permits subsequent Stokes solves (required at each time step) to be completed at the cost of just a forward and a backward substitution, which is extremely efficient.

In the case where the fluid domain is time-dependent, due to the presence of a moving boundary or a free surface for example, the Eulerian approach requires some form of modification. In this work we have opted to exploit an ALE formulation. The principle behind this approach is that the mesh at the boundary of the domain should move with the correct normal velocity, and the velocity of the mesh elsewhere is selected primarily in order to maintain the local quality. In order to achieve this we make use of a simple Laplacian smoothing procedure which attempts to keep the interior nodes from coming together or the mesh itself from becoming tangled [13]. Having defined an appropriate velocity field for the mesh, $\mathbf{V}$ say, it is then necessary to modify the second terms appearing in the equations (6), (8) (13) and (14): replacing the advective velocity $\mathbf{u}$ by $(\mathbf{u} - \mathbf{V})$.

## 2.2 Structure of the Software

The structure of each time step within the code may be described as follows. Given an initial stress field $\mathbf{T}$, solve equations (4) and (2) for the resulting velocity field $\mathbf{u}$. Then use this instantaneous velocity field in equations (6), (8) or (13) in order to take a time step. The value of the stress tensor at the end of the time step may then be used to update the right-hand side of (4) at the start of the next time step.

Note that in three dimensions (4) and (2) represent 4 scalar partial differential equations (PDEs) whilst (8), for example, represents a further $6n$ PDEs (since each tensor $\mathbf{A}_i$ is symmetric). Fortunately, each of the $n$ sets of 6 PDEs given by (8) is independent of the others in the models that are considered here, and may therefore be solved independently.

As well as the time-stepping routines that lie at the core of the software, a number of pre- and post-processing routines are also provided. The former are described in Section 3.1 and include the ability to generate geometries, meshes and boundary conditions, as well as specifying fluid properties. The latter are described in Section 3.4, which provides an overview of the visualisation capabilities. The software itself is embedded in a user-friendly run-time environment that allows real-time visualisation of results and parameters to be modified during a computation. Figure 1 provides a sample snapshot of this environment.

# 3 Implementation Details

Having provided an overview of the software in the previous section, we now go on to describe a number of the key features in a little more detail. Space does not permit complete description of all aspects of the implementation and so the primary aim here is to present a discussion of selected features.

## 3.1 Problem Specification

The particular problem that is to be solved in any given run is specified by a number of components, such as its geometry, boundary conditions and fluid model parameters. In order to undertake a computation, additional numerical properties must also be specified, for example: the finite element mesh, the types of element to be used and the time-step size to take. Each of these attributes, and many more optional ones that will not be discussed here, must be defined through a single data file which is read by the program upon execution. Rather than describe the precise format of the data files here we focus on their key components.

The geometry is specified by the use of one or more quadrilateral or octahedral "super-elements" in 2-d or 3-d respectively (see Figure 1a for example). These super-elements may have curved as well as planar edges and faces. They are also used as an integral part of the built-in mesh generator, which is based upon a block-structured mesh (as illustrated in Figure 2a). By specifying the mesh dimensions within each block the user is able to build a mesh covering the entire geometry. It is also possible to grade these structured meshes towards, or away from, selected edges or faces: tools are provided to assist with ensuring that the mesh remains conforming between neighbouring blocks.

Having specified the geometry and the mesh for a given flow problem it is also necessary to prescribe the boundary conditions that should be used to solve the flow.

Different regions of the boundary may be identified in a variety of possible ways: however the simplest is to use the faces of the super-elements. For a given portion of the boundary one of a number of conditions may be selected. The most widely used of these are briefly summarised in the following list.

- No slip boundary at a solid wall. This simply involves prescribing the tangential (no slip) and normal (solid wall) components of the velocity to be zero.

- Inflow boundary. Rather similar to the above, this permits a specific fluid velocity profile to be prescribed. In this case however the normal flow is non-zero and into the domain.

- Developed inflow boundary. Frequently, especially for non-Newtonian fluids, the precise velocity profile of a fully developed flow is not known *a priori* however it is desirable for such a flow to be the input to the region of interest. This condition is imposed by requiring the user to define an estimate to the velocity profile and a plane that is parallel to the inflow boundary but a little way downstream of it. Then, at each time step the inflow velocities imposed are those that were recorded at the parallel plane for the previous time step. After a number of such time steps a fully developed flow is usually obtained.

- Outflow boundary. Here we allow use of the "no boundary condition" outflow boundary as described in [14]. This is designed to reduce disturbances in the solution at the outflow boundary and is very successful at doing so.

- Symmetry boundary condition. This is extremely important, especially in three dimensions, for reducing the size of a computational problem whose solution contains known symmetries. This condition applies to all of the unknowns and may be applied around a line in two dimensions or a plane in three dimensions.

- Free-surface condition. In the case where there is a free surface then it is necessary that the geometry is updated in a manner that is consistent with the fluid velocity at the free surface. This may be undertaken in a Lagrangian manner (*i.e.* the nodes on the free surface move with their fluid velocity), based upon the normal velocity only (*i.e.* the nodes on the free surface move perpendicular to the free surface with a consistent velocity), or based upon the velocity in a prescribed direction (*e.g.* through the use of predetermined "spines", [15]). In addition, there is a stress boundary condition which modifies the surfaces forces according to the local curvature and the surface tension (see, for example, [16]).

Other properties of a given problem that must be specified include the size and number of the time steps to be taken, the frequency at which outputs of the simulation should be written to file, and the properties of the fluid that is to be considered. For the latter, the user is able to select between each of the models described in Subsection 1.1, as well as a Newtonian fluid, and then to specify the number of modes to be used (for the Rolie-Poly and the pom-pom fluids) and the parameters that must be prescribed to uniquely define each of the governing equations.

## 3.2 Finite Element Solution

As described in Subsection 2.2, the equations for the evolution of the polymer deformation tensor (equations (6), (8) or (13)), as well as equation (14) in the pom-pom case, are solved using implicit time-stepping but with the velocity field frozen at the value from the beginning of the time step. This requires the use of a stabilised finite element method (see the following subsection) and yields sparse non-symmetric systems of linear algebraic equations that must be solved at each time step. In all of the calculations included in this paper piecewise bilinear or trilinear finite elements have been used for the trial spaces in two or three dimensions respectively. The resulting algebraic equations have been solved using the SPARSKIT library, [17], typically selecting the BiCGStab solver with an incomplete ILU(0) preconditioner.

Once the update to the polymer deformation field has been computed over the time step, a modified forcing term on the right-hand side of (4) may be computed and the Stokes' equations resolved. Two different approaches to solving these equations have been implemented in this work: one based upon a direct solver and one based upon an iterative method. The direct approach has significant advantages when a problem on a fixed domain, and therefore a fixed mesh, is solved since it is able to exploit the fact that only the right-hand side of (4) changes at each time step. The relative merits of the two approaches in the case of an evolving ALE mesh are less clear however, depending primarily on the quality of the preconditioner that can be obtained for the iterative solver. Both approaches are briefly outlined, although only the former is used for the results shown here (in fact the results obtained by the two methods are identical to a number of significant digits: the precise number of digits depending primarily on the choice of the penalty and convergence parameters that are used in the respective methods).

In order to develop a direct solver for the finite element discretisation of (4) it is desirable that the system be as small as possible, preferably with a small band-width. In order to reduce the size of the discrete problem we first apply a penalty approach in order to eliminate the pressure from the Stokes' system. This is achieved by replacing the incompressibility condition (2) with a new relation of the form

$$\nabla\cdot\mathbf{u} \;=\; -\frac{1}{\chi}p\,, \tag{15}$$

where the penalty parameter $\chi$ is suitably large (in particular $\chi \gg \mu$). Substituting (15) into (4) then yields

$$-\nabla\cdot\left(\mu\nabla\mathbf{u}+\chi\,(\nabla\mathbf{u})^{T}\right) \;=\; \mathbf{b}+\nabla\cdot\mathbf{T}\,, \tag{16}$$

which is the form used for the finite element discretisation. Note that the discrete system, obtained using a standard Galerkin approach, involves only the velocity unknowns. Due to the large penalty parameter it is not suitable for iterative solution (it is poorly conditioned) but, so long as the choice of penalty value is not too large for the precision of floating point arithmetic that is used, it is well suited to a banded or a sparse direct solution method. In this work we use a banded solver from the IMSL library but any other such solver is suitable (*e.g.* [12]). The key requirement is that the LU factorisation of the Stokes' matrix should be stored, either in primary memory or on disk (depending upon the size of the problem), so that subsequent Stokes' solvers require only forward and backward substitutions.

An iterative solver based upon a mixed finite element formulation of the Stokes problem has also been implemented. This follows the standard approach that is developed in detail in [11], for example. In its simplest form a stable pair of finite element spaces must be selected for the velocities and pressure respectively: we use biquadratic/bilinear and triquadratic/trilinear in two and three dimensions respectively. This leads to a coupled discrete system for the velocity and the pressure unknowns that is symmetric but indefinite. Suitable choices of iterative solver are described in [11] however the effectiveness of this solver depends critically upon the efficiency of the preconditioner that is used. Highly efficient preconditioners are possible; however implementation of an optimal iterative solver is not a task that we have so far undertaken.

## 3.3 Stabilisation

A number of stability issues need to be addressed in this work, some of which have already been introduced in the sections above. These include the stability of the finite element discretisation, both for hyperbolic equations or for mixed systems (such as the Stokes' equations), and the requirement to ensure that certain quantities remain within predetermined bounds. Each of these issues is discussed in turn.

The Galerkin method is not appropriate for the discretisation of purely hyperbolic equations such as (6), (8), (13) or (14). In this work we make use of a Petrov-Galerkin approach based upon streamline upwinding [18]. This has a stabilising effect through the introduction of a minimal amount of diffusion along the streamlines. We also permit the mixed finite element solution of the Stokes' equations to be undertaken using unstable pairings of finite element spaces for the velocities and the pressures, typically piecewise bilinear (or trilinear in three dimensions) spaces for all unknowns. Here the stabilisation may be introduced at the solution stage, as described in [11] for example.

Certain components of the solutions that are computed have specific physical constraints whose violation can lead to further instabilities in the simulations. Examples include the requirement that all components of the deformation tensors remain positive, or the upper and lower bounds on the stretch parameters $\lambda_i$ in the pom-pom model. There are a number of possible ways of ensuring that these bounds are respected by the numerical solution, two of which are used within our code. The first of these is to put a large penalty term on the right-hand side of each evolution equation, which only becomes active when a constraint for the degree of freedom in question is violated: this acts as a crude barrier and is effective when a low order time integration scheme is being used, as is typically the case here. The second approach is to prevent the constraint from being violated in the first place by ramping up the penalty term gradually as the constraint is being approached from within the set of allowed values. This is more appropriate when it is essential that the constraint is never violated, even for a single time step, or when a higher degree time integration scheme is being used. It does tend to over-penalise however, meaning that the limiting value is never reached.

### 3.4 User Interface

We conclude this section with a brief description of the user interface that is provided with our software. This is an optional Windows interface and the solver may also be run without it, under Linux for example. In the latter case a choice of output formats are available for compatibility with the visualisation tools GMV [19] and Tecplot [20]. Indeed, some of the results presented in the following sections exploit this option.

The user interface that is provided comes with substantially more functionality than can be described here. We provide a brief flavour of this however in the following paragraphs on the pre-processing and the solution phases respectively.

The preprocessing phase allows the user to undertake a number of tasks that are designed to assist with the setting up of the problem. These include options to: plot the super elements and their associated numberings, as illustrated in Figure 1(a), plot the mesh and its dimensions, plot boundary faces of different types, plot and display node numbers, show different types of boundary condition, *etc.* All of these facilities allow the user to check that the problem has been properly defined, and all objects may be viewed from different positions and with varying levels of magnification. Once the user is satisfied that the problem is correctly defined they may then begin the execution.

As a run is executing the graphical output is updated after each time step, regardless of the frequency with which data is being saved to file. It is possible to pause the execution after any time step and then to resume, possibly having changed certain parameters (*e.g.* the time step size). There is a large choice of solution fields that may be displayed: any of the velocity components, illustrated in Figure 1(b), any of the deformation components (either for individual modes or summed over all modes), the parameters $\lambda_i$, pressure, *etc.* The default is to use colouring for contours; however discrete contour lines may be selected instead. In addition, streamlines, velocity vectors and birefringence patterns may be displayed. As with the geometry and the mesh, each of the above displays may be rotated or magnified as required.

## 4 Computational Results

The following sections describe numerical experiments that illustrate the functionality of the software. Three rheologies, given in Tables 1-3, are used as examples of the Oldroyd-B, Rolie-Poly and pom-pom fluid models.

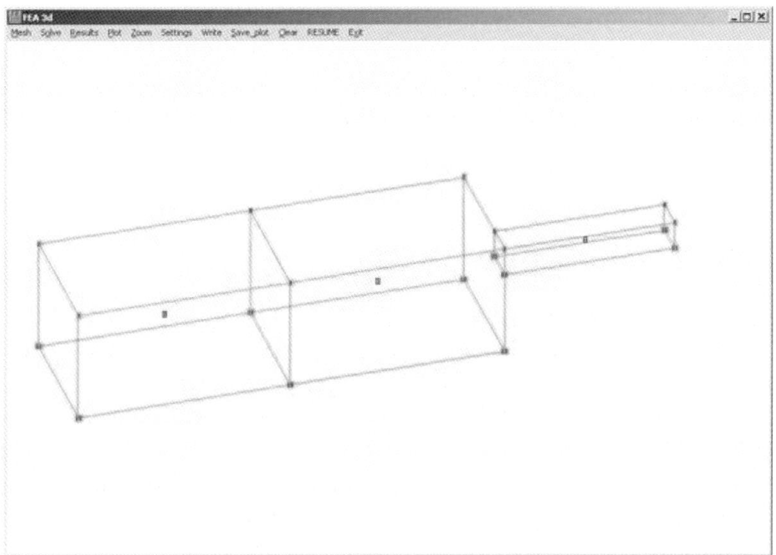

(a) Input super-element mesh

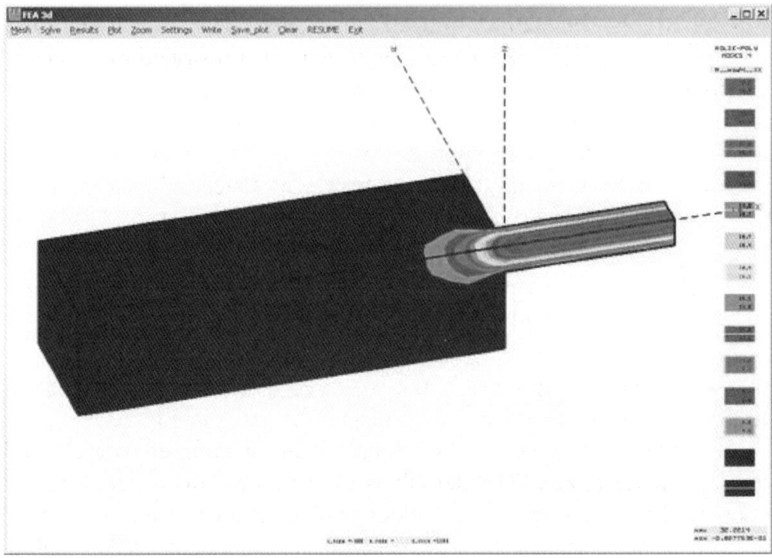

(b) Computed velocity profile

Figure 1: The Windows program interface

## 4.1 Two-Dimensional Flow Around a Cylinder

Flow past a solid cylinder is computed in 2d for the Rolie-Poly fluid with a unit radius cylinder and a solid wall at $y = \pm 2$. A symmetric half of the domain is solved and the inflow and outflow boundaries are placed at $x = -16$ and $x = 10$ respectively.

| $G$ | $\tau$ |
|---|---|
| 1.36667 | 0.3 |

Table 1: Oldroyd-B fluid model: $\mu = 0.59$

| $i$ | $G_i$ | $\tau_{d,i}$ | $\tau_{R,i}$ | $\beta_i$ | $\delta_i$ |
|---|---|---|---|---|---|
| 1 | 1286.0e-6 | 2.3780 | 0.05263 | 0.0 | -0.5 |
| 2 | 4407.0e-6 | 0.7519 | 0.03008 | 0.0 | -0.5 |
| 3 | 11639.0e-6 | 0.2378 | 0.01530 | 0.0 | -0.5 |
| 4 | 21043.0e-6 | 0.0752 | 0.00752 | 0.0 | -0.5 |

Table 2: Rolie-Poly fluid model: $\mu = 1123.0e - 6$, $n = 4$

| $i$ | $G_i$ | $\tau_{b,i}$ | $\tau_{b,i}/\tau_{s,i}$ | $q_i$ |
|---|---|---|---|---|
| 1 | 8314.05e-6 | 0.393901 | 5.0 | 3.0 |
| 2 | 4590.42e-6 | 1.18237 | 5.0 | 4.0 |
| 3 | 1907.74e-6 | 3.73898 | 5.0 | 4.0 |
| 4 | 612.395e-6 | 11.8237 | 4.0 | 7.0 |
| 5 | 159.889e-6 | 37.3898 | 3.0 | 8.0 |
| 6 | 26.2254e-6 | 118.237 | 2.0 | 9.0 |

Table 3: Pom-pom fluid model: $\mu = 5122.9157e - 6$, $n = 6$

Figure 2 shows the super-elements employed and the bilinear quadrilateral mesh that is produced from the super-element description. The computed steady-state solution shows the expected acceleration due to the contraction around the cylinder and also the increase in polymer stress on the cylinder surface and on the exterior wall.

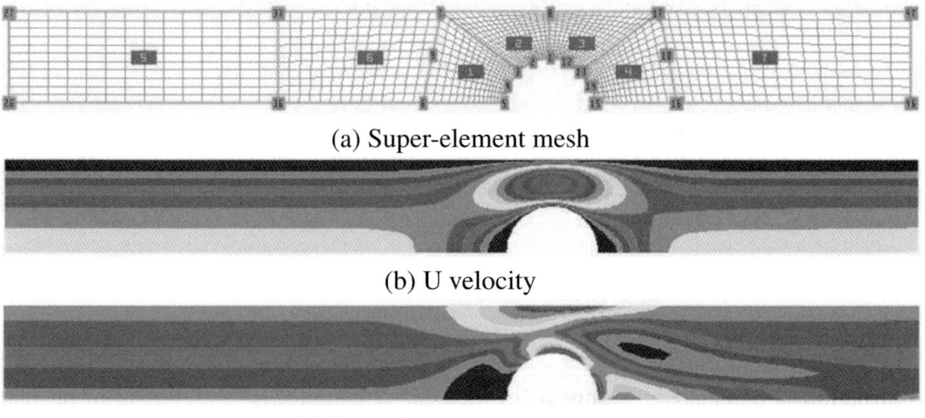

(a) Super-element mesh

(b) U velocity

(c) Total $A_{xx}$ stress component

Figure 2: Flow around a cylinder

## 4.2 Two-Dimensional Contraction Flow

The contraction geometry, shown in Figure 3, is modelled with the Rolie-Poly fluid. A symmetric half of the domain is solved. The re-entrant corner is rounded to prevent the formation of a stress-singularity at, or just after, this point. The streamlines clearly show the presence of a recirculation region prior to the contraction. The size and shape of this region is strongly determined by the fluid rheology. The stress birefringence pattern represents visually regions where the stress gradient is high. It can be seen here that there is stress build up on the centreline before the contraction and a subsequent release farther downstream. At the wall there is a high stress gradient at the corner as expected.

## 4.3 Three-Dimensional Contraction-Expansion Flow

The contraction-expansion flow shown in Figure 4 is modelled with the Oldroyd-B fluid. The mesh for a symmetric quarter of the domain is visualised with GMV [19] in Figure 4(a). Contours of velocity magnitude are shown in Figure 4(b) with velocity vectors superimposed. A small recirculation region is visible in the corners above the contraction region.

## 4.4 Three-Dimensional Y-shaped Channel Flow

This Rolie-Poly flow comprises of a three-dimensional domain with two converging channels meeting at an angled junction. The re-entrant corner is rounded to prevent a stress singularity at that point. A symmetric half of the domain is visualised through GMV [19] in Figure 5(a) coloured with contours of velocity magnitude. The acceleration caused by the convergence into one channel is clearly seen. Figure 5(b) shows a component of the total fluid stress, $\sigma$ in Eq. (1), illustrating the peak in stress that is produced at the re-entrant corner.

## 4.5 Three-Dimensional Flow Past a Cylinder

The flow of a typical pom-pom fluid through a thin channel with a cylindrical obstruction is depicted in Figure 6 (in fact this calculation is undertaken using just a single pom-pom mode, rather than the 6-mode model given in Table 3). Qualitative differences in the stress pattern for this fluid are obvious in comparison with the two-dimensional simulation shown in Figure 2. It is also possible to visualize the additional $\lambda$ field for each mode, satisfying the equations (14).

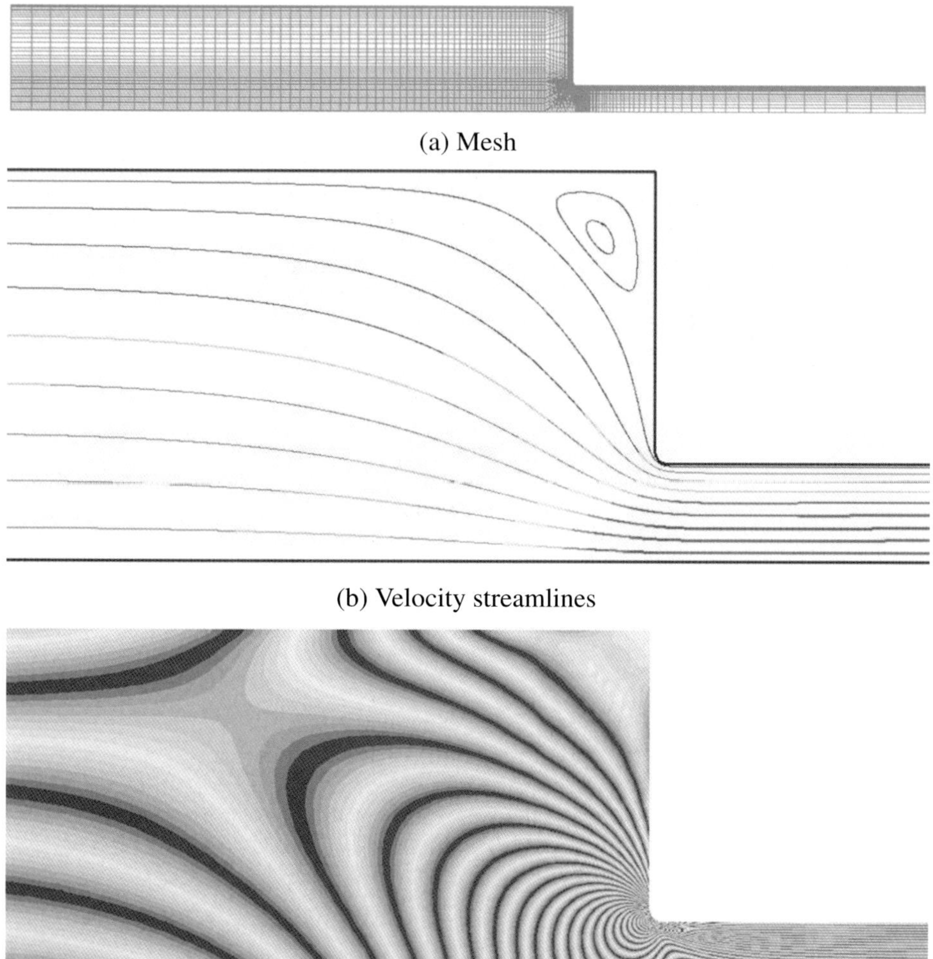

(a) Mesh

(b) Velocity streamlines

(c) Stress birefringence pattern

Figure 3: Contraction flow

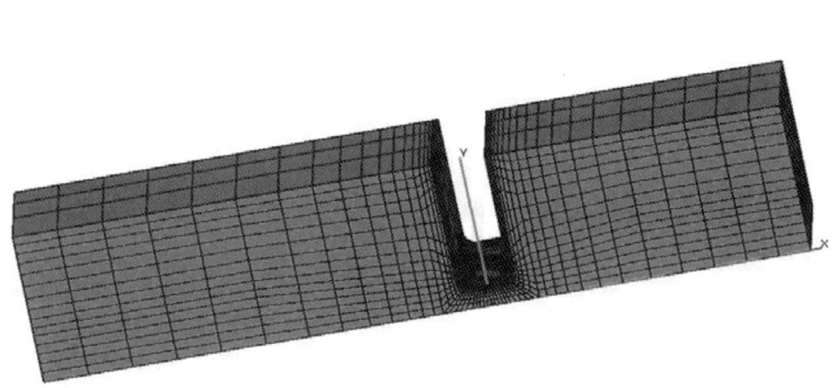


(a) Mesh

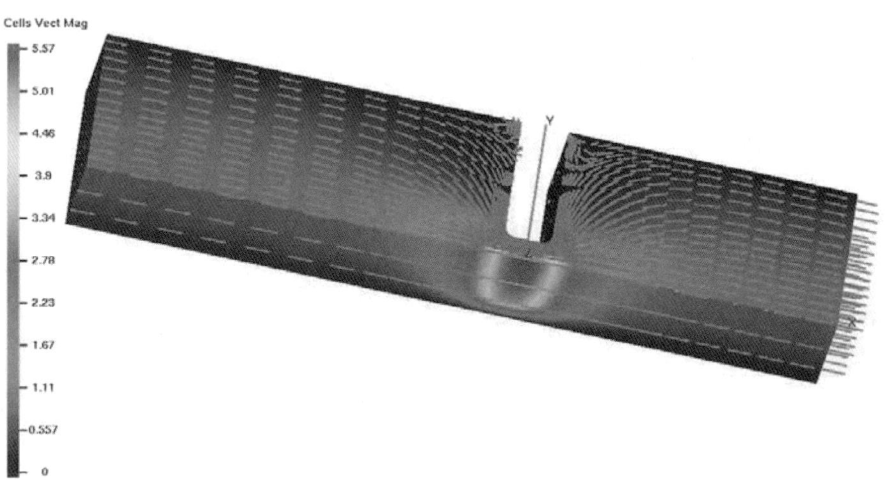


(b) Velocity vectors

Figure 4: Contraction-expansion flow

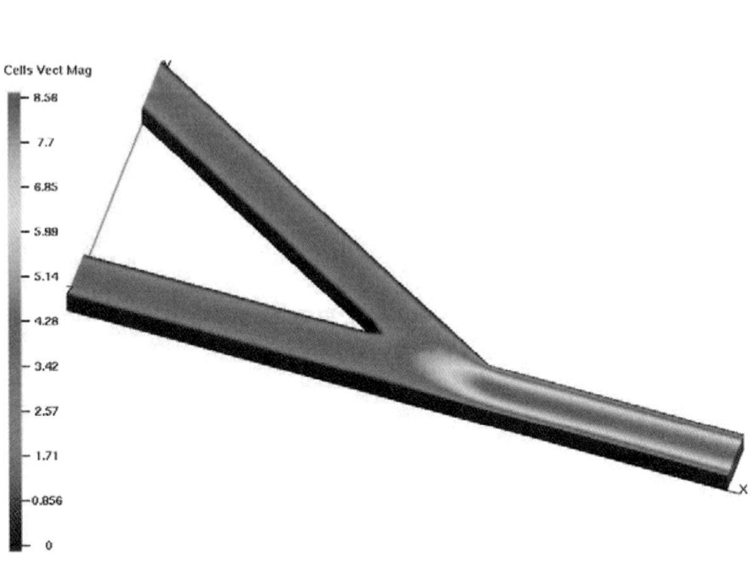


(a) Contours of velocity magnitude

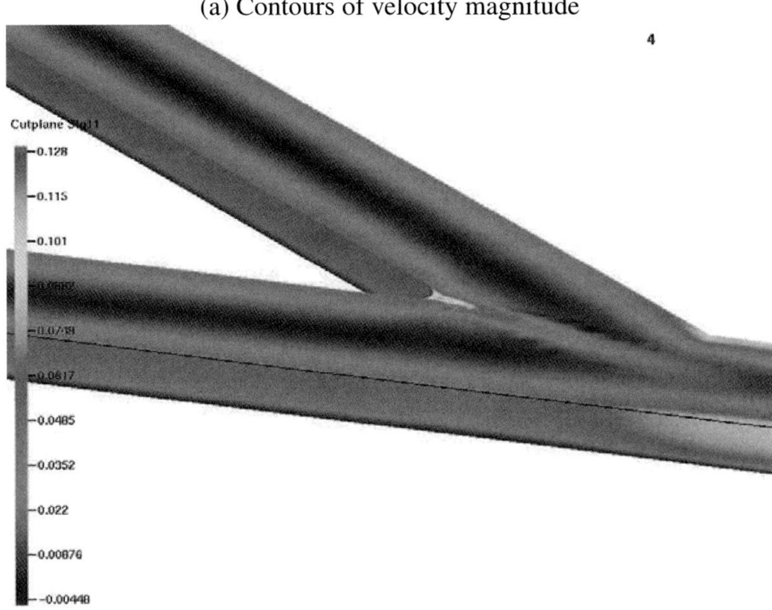


(b) Contours of total stress $\sigma_{xx}$

Figure 5: Y-shaped channel

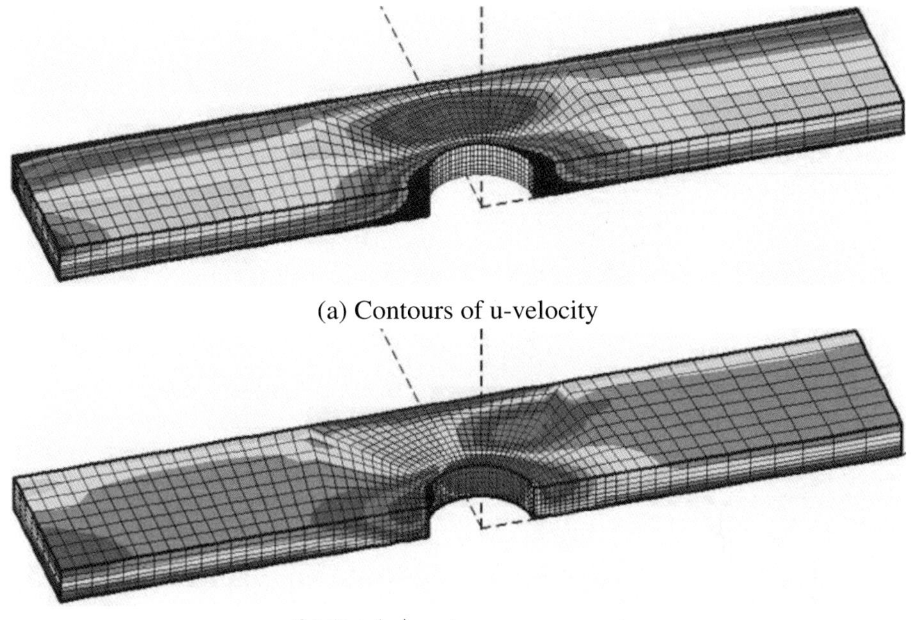

(a) Contours of u-velocity

(b) Total $A_{xx}$ stress component

Figure 6: Channel flow past a cylinder

# 5 Further Enhancements

The previous sections present some representative results for flows in relatively simple, fixed, geometries in two and three dimensions. We now illustrate some of the enhancements of the software, to allow more complex geometries which may be time-dependent.

## 5.1 Flow With Rigid Particles

The direct simulation of particle suspensions represents a very important class of problem in the area of polymer processing [21]. We consider a unit-cell from an infinite domain with the particle size varied to produce different volume-fractions. The particle undergoes shear and a rotation is induced, which can be simulated without moving the mesh. Two-dimensional results for the Rolie-Poly fluid are shown in Fig 7(a) and the extension of these ideas to the Oldroyd-B fluid in three-dimensions in Figure 7(b). In each case the rotation rates obtained converge to the theoretical results as volume-fraction is reduced.

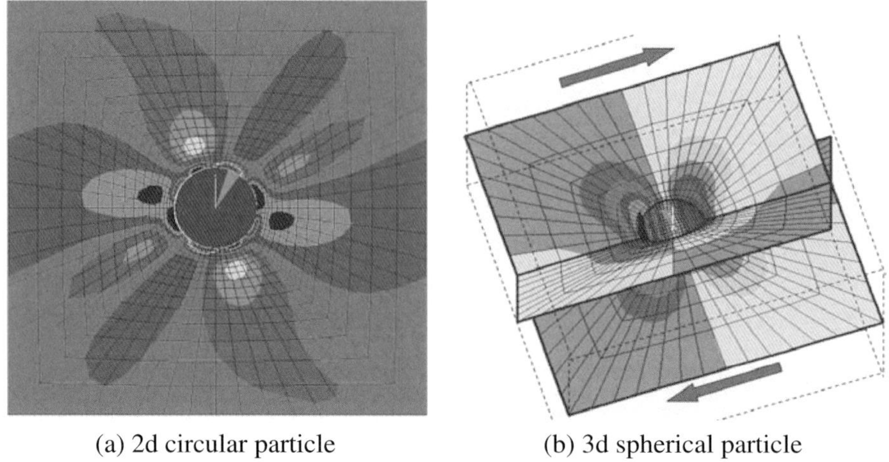

(a) 2d circular particle (b) 3d spherical particle

Figure 7: Shear flow around a rigid particle

## 5.2 Arbitrary Lagrangian Eulerian Implementation

Shear flow around a solid elliptical particle is used to illustrate the ALE remeshing possible within the software. The shear flow drives a periodic rotation of the particle that requires mesh movement due to the variation in the radius. Results in Figure 8 illustrate the mesh at two different times in the simulation. It can be seen clearly how the ALE adaptivity copes with the particle rotation. The mesh is continuously deformed through a Laplacian smoothing approach in order to maintain the quality of the quadrilateral elements. Contours of the vertical velocity component illustrate the variation in the flow.

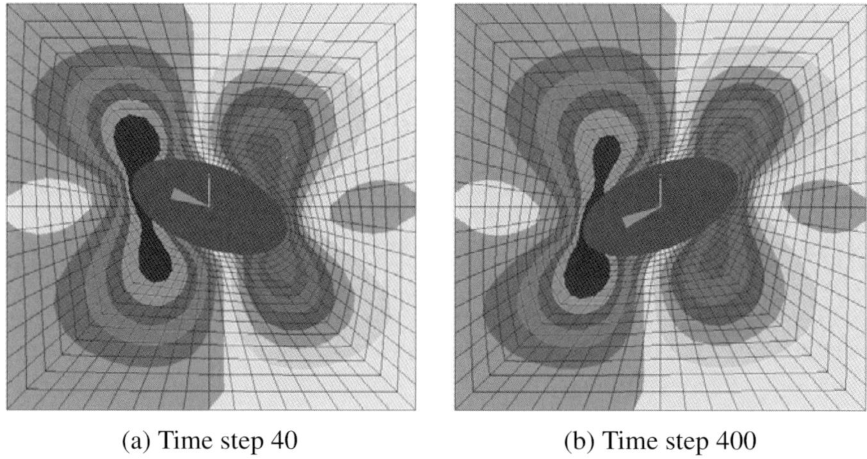

(a) Time step 40 (b) Time step 400

Figure 8: ALE grid for shear flow around an elliptical particle

## 5.3 Free-Surface Flows

Die-swell represents an important industrial process and, although the current software implementation is designed only for problems with a relatively small boundary deformation, it is possible to model such cases. A Rolie-Poly fluid is used and initially the fluid surface is horizontal. A symmetric half of the domain is used and the mesh is refined near the outlet. The spine method is used, as described in Section 3.1, [15], where the mesh motion is limited to vertical displacements, driven by the free surface motion. Results in Figure 9 show contours of u-velocity at three times in the simulation. Initially the free surface bulges at the outlet but the magnitude of this expansion decreases slightly as a steady state is reached.

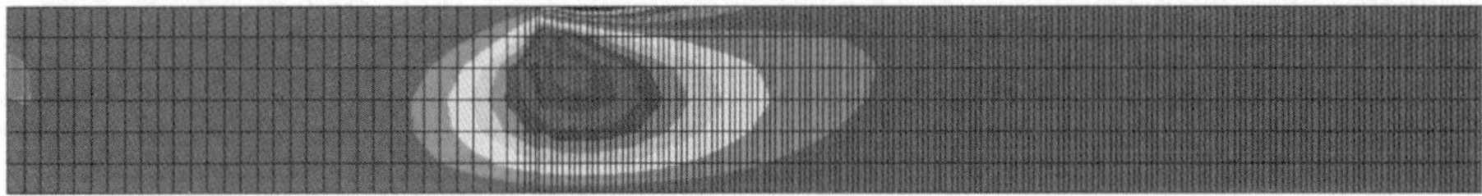

(a) Time step 20, max u = 0.6

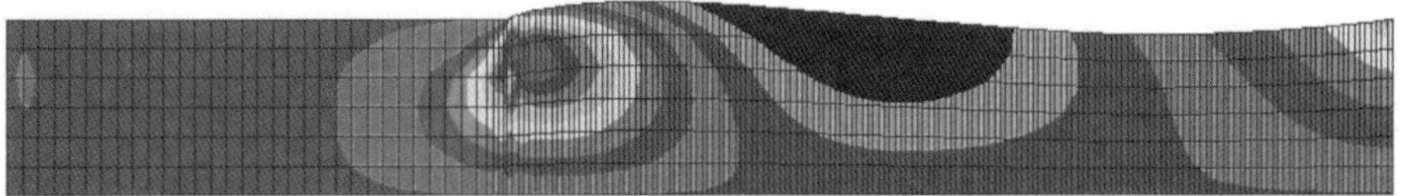

(b) Time step 600, max u = 1.16

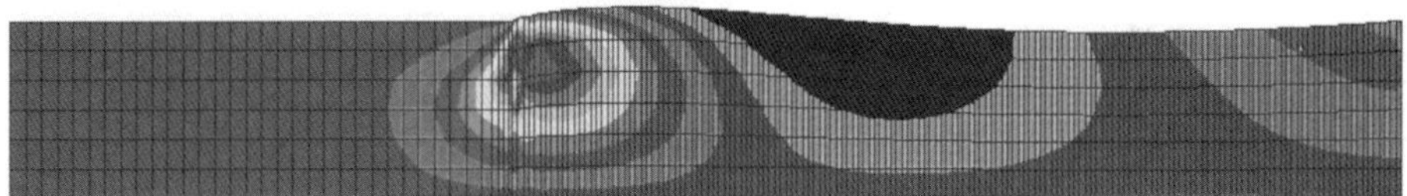

(c) Time step 1500, max u = 1.12

Figure 9: 2d die swell problem, u-velocity contours

# 6 Conclusions

We have presented an introduction to a new software tool that has been developed to allow the numerical solution of two- and three-dimensional viscoelastic flow problems. The tool allows a variety of constitutive laws to be considered for fluids within a wide range of geometries, including those with moving or free boundaries. A selection of numerical results has been presented in order to convey the versatility of the software.

Current work is ongoing to benchmark the results produced by this code against those obtained using other computational tools and from experimental observation and measurement. Preliminary results suggest that the robustness of the software is matched by its accuracy and computational efficiency.

The next stage of the work will involve applying the code to the simulation of a variety of real polymers in flow regimes of practical interest and importance. This is likely to involve further development, for example to allow more substantial evolution of free surfaces in both two and three dimensions. For the ALE formulation in three dimensions it is also likely to be beneficial to improve the efficiency of the iterative Stokes' solver that has been implemented, through the use of optimal preconditioning for example [11].

## Acknowledgements

This research was undertaken as part of EPSRC grant GR/T11807/01.

## References

[1] A.E. Likhtman, R. Graham, "A simple constitutive equation for linear polymer melts derived from molecular theory: the Rolie-Poly model", *J. Non-Newtonian Fluid Mech.*, 114(1):1–12, 2003.

[2] A.E. Likhtman, T.C.B. McLeish, "Quantitative theory for linear dynamics of linear entangled polymers", *Macromol.*, 35:6332–6343, 2002.

[3] T.C.B. McLeish, R.G. Larson, "Molecular constitutive equations for a class of branched polymers: the pom-pom polymer", *J. Rheol.*, 42(1):81–110, 1998.

[4] H.A. Barnes, J.F. Hutton, K. Walters, *An Introduction to Rheology*, Elsevier, 4th edn, 1996.

[5] M.A. Alves, P.J. Oliviera, F.T. Pinho, "The flow of viscoelastic fluids past a cylinder: finite-volume high-resolution methods", *J. Non-Newtonian Fluid Mech.*, 97:207–232, 2001.

[6] M. Sahin, H.J. Wilson, "A semi-staggered dilation-free finite volume method for the numerical solution of viscoelastic fluid flows on all-hexahedral elements", *J. Non-Newtonian Fluid Mech.*, 147:79-91, 2007.

[7] R.G. Owens, C. Chauvière, T.N. Philips, "A locally-upwinded spectral technique (LUST) for viscoelastic flows", *J. Non-Newtonian Fluid Mech.*, 108:49–71, 2002.

[8] Y.R. Fan, R.I. Tanner, N. Phan-Thien, "Galerkin/least-squares finite-element methods for steady viscoelastic flows", *J. Non-Newtonian Fluid Mech.*, 84:233–256, 1999.

[9] R. Guenette, M. Fortin, "A new mixed finite element method for computing viscoelastic flows", *J. Non-Newtonian Fluid Mech.*, 60:27–52, 1995.

[10] G.B. Bishko, O.G. Harlen, T.C.B. McLeish, T.M. Nicholson, "Numerical simulation of the transient flow of branched polymer melts through a planar contraction using the 'pom-pom' model", *J. Non-Newtonian Fluid Mech.*, 82:255–273, 1999.

[11] H.C. Elman, D.J. Silvester, A.J. Wathen, *Finite Elements and Fast Iterative Solvers with applications in incompressible fluid dynamics*, OUP, 2005.

[12] X.Y.S. Li, "An overview of SuperLU: Algorithms, implementation, and user interface", *ACM Trans. on Math. Software*, 31:302–325, 2005.

[13] R.C. Peterson, P.K. Jimack, M.A. Kelmanson, "The solution of two-dimensional free-surface problems using automatic mesh generation", *Int. J. Numer. Meths. Fluids*, 31:937–960, 1999.

[14] D.F. Griffiths, "The "no boundary condition" outflow boundary condition", *Int. J. Numer. Meths. Fluids*, 24:393–411, 1997.

[15] S.F. Kistler, L.E. Scriven, "Coating flows", In *Computational Analysis of Polymer Processing*, London Applied Science Publishers (eds. J.R.A. Perason & S.M. Richardson), 243–299, 1983.

[16] M.A. Walkley, P.H. Gaskell, P.K. Jimack, M.A. Kelmanson, J.L. Summers, "Finite element simulation of three-dimensional free-surface flow problems", *J. Sci. Comput.*, 24:147-162, 2005.

[17] Y. Saad, *Iterative Methods for Sparse Linear Systems*, PWS, 1996.

[18] X.-L. Luo, R.I. Tanner, "A decoupled finite element streamline-upwind scheme for viscoelastic flow problems", *J. Non-Newtonian Fluid Mech.*, 31:143-162, 1989.

[19] GMV - The General Mesh Viewer, `http://laws.lanl.gov/XCM/gmv/`.

[20] Tecplot - CFD Post Processing, Plotting, Graphing and Data Visualization Software, `http://www.tecplot.com/`.

[21] A. Malidi, O. Harlen, "A Lagrangian Simulation Method for Suspensions in Viscoelastic Fluids", *Proceedings of ECCOMAS CFD*, 2006.

©Saxe-Coburg Publications, 2008.
Trends in Engineering Computational Technology
B.H.V. Topping and M. Papadrakakis, (Editors)
Saxe-Coburg Publications, Stirlingshire, Scotland, 103-126.

Chapter 6

# Particle Swarm Optimization: Fundamental Study and its Application to Optimization and to Jetty Scheduling Problems

**J. Sienz and M.S. Innocente**
**ADOPT Research Group, School of Engineering**
**Swansea University, United Kingdom**

## Abstract

The advantages of evolutionary algorithms with respect to traditional methods have been much discussed in the literature. While particle swarm optimizers share such advantages, they outperform evolutionary algorithms in that they require lower computational cost and easier implementation, involving no operator design and few coefficients to be tuned. However, even marginal variations in the settings of these coefficients greatly influence the dynamics of the swarm. Since this paper does not intend to study their tuning, general-purpose settings are taken from previous studies, and virtually the same algorithm is used to optimize a variety of notably different problems. Thus, following a review of the paradigm, the algorithm is tested on a set of benchmark functions and engineering problems taken from the literature. Later, complementary lines of code are incorporated to adapt the method to combinatorial optimization as it occurs in scheduling problems, and a real case is solved using the same optimizer with the same settings. The aim is to show the flexibility and robustness of the approach, which can handle a wide variety of problems.

**Keywords:** particle swarms, artificial intelligence, optimization, scheduling.

## 1 Introduction

The characteristics of the objective variables, the function to be optimized and the constraint functions severely restrict the applicability of traditional optimization algorithms. The variables and both the objective and constraint functions must comply with a number of requirements for a given traditional method to be applicable. Furthermore, traditional methods are typically prone to converge towards local optima. By contrast, population-based methods such as evolutionary algorithms (EAs) and particle swarm optimization (PSO) are general-purpose optimizers, which are able to handle different types of variables and functions with few or no adaptations. Besides, although finding the global optimum is not guaranteed, they are able to escape poor local optima by evolving a population of interacting individuals which profit from information acquired through experience, and use stochastic weights or operators to introduce new responses. The lack of limitations to the features of the variables and functions that model the problem enable these methods to handle models whose high complexity does not allow traditional, deterministic, analytical approaches. While the advantages of PSO and EAs with respect to traditional methods are roughly the same, the main advantages of PSO when compared to EAs are its lower computational cost and easier implementation. Regarding their drawbacks, both these methods require higher computational effort, some constraint-handling technique incorporated, and find it hard to handle equality constraints.

Population-based methods like EAs and PSO are considered *modern heuristics* because they are not designed to optimize a given problem deterministically but to carry out some procedures that are not directly related to the optimization problem. Optimization occurs without evident links between the implemented technique and the resulting optimization process. They are also viewed as *Artificial Intelligence* (AI) *techniques* because their ability to optimize is an emergent property that is not specifically intended, and therefore not implemented in the code. Thus, the problem *per se* is not analytically solved, but artificial-intelligent entities are implemented, which are expected to find a solution themselves. In particular, *Swarm Intelligence* (SI) is the branch of AI concerned with the study of the collective behaviour that emerges from decentralized and self-organized systems. It is the property of a system whose individual parts interact locally with one another and with their environment, inducing the emergence of coherent global patterns that the individual parts are unaware of. PSO is viewed as one of the most prominent SI-based methods. Either modern heuristics or AI-based optimizers, these methods are not deterministically designed to optimize. EAs perform some kind of artificial evolution, where individuals in a population undergo simulated evolutionary processes which results in the maximization of a fitness function, resembling biological evolution. Likewise, PSO consists of a sort of simulation of a social milieu, where the ability of the population to optimize its performance emerges from the cooperation among individuals.

Although the basic particle swarm optimizer requires the tuning of a few coefficients only, even marginal variations in their values have a strong impact on the dynamics of the swarm. The general-purpose settings used throughout this paper were taken from previous studies (see Innocente [1]) because the objective is not to study their tuning but to demonstrate that virtually the same algorithm is able to cope with a variety of problems that would require different traditional methods to be solved. Thus, this paper intends to introduce the PSO method, pose the optimization and scheduling problems, and solve them with essentially the same algorithm. Additional code is required to turn the continuous search algorithm into a scheduler.

## 2 Mathematical optimization

For problems where the quality of a solution can be quantified in a numerical value, *optimization* is the process of seeking the permitted combination of variables that optimizes that value. Thus, different combinations of *variables* allow trying different candidate solutions, the *constraints* limit the valid combinations, and the *optimality criterion* allows differentiating better from worse. The suitability of traditional methods is limited by the nature of the variables, and by requirements that the objective and constraint functions must comply with.

The process of problem-solving consists of two major steps: developing a model of the problem, and finding a solution to it. The behaviour of an existing system sometimes can be analyzed by observation, but the number of situations analyzed is necessarily limited. When different alternatives need to be considered – for instance for optimization processes –, the development of a model becomes essential. Either mathematical or physical (*prototype*), any model is just an interpretation of a problem, which always implies simplifications. This work is concerned with mathematical models, so that the plain words in which real problems are posed must be turned into mathematical language (*formulation of the problem*). The question is whether to introduce numerous simplifications so that the available methods are able to solve the model, or to develop a model with higher fidelity and approximate the solving techniques. Traditional methods require specific characteristics of the equations involved, forcing the model to take the corresponding simplifying assumptions to suit the solver. This results in an exact solution of an approximate model. The alternative is to develop more precise models, and then attempt to solve them using a set of toolboxes that do not severely limit the model. This results in an approximate solution of a more precise model (*e.g.* finite element method, EAs and PSO). The second approach usually outperforms the first for complex, real-world problems.

While the *objective function* relates the real problem to the model, the *cost function* (also *evaluation function*) is the function to be minimized. The *problem variables* are called *object variables*, or just *variables*. Since *parameters* might stand for *variables*, the *parameters* of an algorithm are referred to as *coefficients* in this work.

## 2.1 Types of optimization problems

Optimization problems can be differentiated according to the models that represent them. Thus, they can be classified as *linear* or *non-linear*; *continuous*, *discrete* or *mixed-integer*; *convex* or *non-convex*; *differentiable* or *non-differentiable*, *smooth* or *non-smooth*, *etc.* It is interesting to classify them here into *continuous* and *discrete* because the PSO method was originally designed for continuous search-spaces, regardless of whether the problem is or is not *linear*, *convex*, *differentiable*, or *smooth*.

### 2.1.1 Continuous optimization

The variables in continuous problems can take real values. Since there are infinite real numbers, trying every possible solution is not an option, and the problem has no solution when the landscape that represents it is random. However, there is almost always a function that models the landscape – not necessarily continuous –, which returns information about the systematic relationships between its points.

### 2.1.2 Discrete optimization

A discrete problem has a finite number of possible solutions, although such number becomes intractable for real-world problems. Again, trying them all is out of question. Some branches of discrete optimization are binary optimization, linear programming, integer programming and combinatorial optimization. Due to the application of the PSO method to scheduling problems, the latter is of special interest here. The aim in combinatorial optimization is to find the optimal arrangement of the elements that optimizes a result. The simplest object variable is a list, where the different solutions consist of its permutations. A classic example is the travelling salesman problem. Traditional methods for combinatorial problems are sequential.

## 2.2 General optimization problem

Let $\mathcal{S}$ be the search-space, and $\mathcal{F} \subseteq \mathcal{S}$ its feasible part. A minimization problem consists of finding $\hat{\mathbf{x}} \in \mathcal{F}$ such that $f(\hat{\mathbf{x}}) \leq f(\mathbf{x})\ \forall \mathbf{x} \in \mathcal{F}$, where $f(\hat{\mathbf{x}})$ is a global minimum and $\hat{\mathbf{x}}$ its location. The problem can be formulated as in Equation (1):

$$\text{Minimize } f(\mathbf{x}), \quad \text{subject to} \begin{cases} g_j(\mathbf{x}) \geq 0 & ; \quad j = 1, \ldots, q \\ g_j(\mathbf{x}) = 0 & ; \quad j = q+1, \ldots, m \end{cases} \tag{1}$$

where $\mathbf{x} \in \mathcal{S}$ is the vector of object variables; $f(\cdot): \mathcal{S} \to \mathcal{E}$ is the cost function; $g_j(\cdot)$ is the $j^{\text{th}}$ constraint function; and $\mathcal{S} \subseteq \mathcal{R}^n \wedge \mathcal{E} \subseteq \mathcal{R}$ for continuous problems ($\mathcal{R}$ is the set of real numbers).

### 2.2.1 The objective and the cost functions

The *objective* of an optimization problem is given in plain words. Multi-objective problems seek to achieve several – typically conflicting – objectives. The *formulation* of the objective is called the *objective function*, whose output is a measure of the fulfilment of the objective(s). The *cost function* is the function to be minimized, and maps the output of the *objective function* to the real numbers so that the minimum matches the best fulfilment. The *cost* and the *objective* functions may coincide; be proportional; or there can be an arbitrarily complex mapping between them.

### 2.2.2 Constraints

Although the constraints allow concentrating the search into limited areas, each potential solution must be verified to comply with them. Hence the problem usually becomes harder to solve than its unconstrained counterpart. The most appropriate technique to handle a constraint depends on the type of constraint. Some common types of constraints are as follows:

- *Inequality constraint*: function of the object variables that must be greater than or equal to a constant, as shown in Equation (1).
- *Equality constraint*: function of the object variables that must be equal to a constant, as shown in Equation (1).
- *Boundary constraint*: instance of inequality constraints, consisting of functions that define boundaries that contain the feasible space (if the boundary constraints are given by a hyper-rectangle, they are also called *interval* or *side constraints*).

While a constrained optimization problem was defined as the problem of finding the combination of variables that minimizes the cost function and satisfies all constraints, real-world problems sometimes do not lend themselves to such strict conditions. Frequently, all the constraints cannot be strictly satisfied simultaneously, and the problem turns into finding a trade-off between minimizing the cost function and minimizing the constraints' violations. Thus, a constraint is said to be *hard* if it does not admit any degree of violation, and *soft* is there is some given tolerance.

### 2.2.3 Constraint-handling techniques

Constraint-handling techniques are usually classified into groups, although such classifications are not unified. Besides, while some methods are sometimes regarded as belonging to one class, the boundaries are frequently blurry, and the memberships not unique. The focus here is on those techniques suitable for PSO.

Rejection of infeasible solutions

This is a *penalization* method, sometimes referred to as the *death penalty* due to the elimination of infeasible solutions. The algorithm does not need to evaluate infeasible solutions, thus saving computational effort. The method is suitable when $\mathcal{F}$ is convex, and when it constitutes an important part of $\mathcal{S}$. It presents some problems like the fact that the solution(s) must be initialized within $\mathcal{F}$, and cannot be placed within infeasible regions on the way towards other disjointed regions of $\mathcal{F}$.

Penalization of infeasible solutions

The evaluation of infeasible solutions might be useful to guide the search towards more promising areas. Thus, the value of the *cost function* is increased for infeasible solutions as in Equation (2), and the problem is treated as if it was unconstrained.

$$fp(\mathbf{x}) = f(\mathbf{x}) + Q(\mathbf{x}) \tag{2}$$

where $f(\mathbf{x})$ and $fp(\mathbf{x})$ are the cost and penalized cost functions, respectively, while $Q(\mathbf{x})$ is the overall penalization due to the infeasibility of solution $\mathbf{x}$.

Penalties are usually not fixed but linked to the amount of infeasibility. A classical penalization for the $j^{th}$ constraint violation is shown in Equation (3):

$$f_j(\mathbf{x}) = \begin{cases} \max\{0, g_j(\mathbf{x})\} & ; \quad 1 \le j \le q \\ g_j(\mathbf{x}) & ; \quad q < j \le m \end{cases} \tag{3}$$

Thus, the function to be optimized could be as follows (additive penalization):

$$fp(\mathbf{x}) = f(\mathbf{x}) + \sum_{j=1}^{m} \lambda_j^{(t)} \cdot \left(f_j(\mathbf{x})\right)^{\alpha} \tag{4}$$

where the penalization coefficients $\lambda_j^{(t)}$ and $\alpha$ may be constant, dynamic, or adaptive. Typically, $\lambda_j^{(t)}$ is set to high and $\alpha$ to small values. A high penalization might lead to infeasible regions of $\mathcal{S}$ not being explored, thus converging to a non-optimal but feasible solution. A low penalization might lead to the system evolving solutions that are violating constraints but present themselves as having lower costs than feasible solutions. The definition of the penalty functions is not trivial, and it plays a critical role in the performance of the algorithm. A wide variety of penalization methods can be found in the literature, including the multiplicative ones.

Preserving feasibility

Infeasible solutions are ignored, and candidate solutions quickly pulled back towards $\mathcal{F}$. While it *rejects infeasible solutions*, it is considered here a *preserving feasibility* method in agreement with Hu *et al.* [2]. Its drawbacks are that the solution(s) must be initialized within $\mathcal{F}$; it is inefficient in handling small and disjointed feasible spaces; and cannot handle equality constraints. Its advantage is that it only needs a slight adaptation from the unconstrained algorithm, without coefficients to be tuned.

Cut off at the boundary

When a solution is moved to an infeasible location, its displacement vector is simply cut off. Either the solution is relocated on the boundary the nearest possible to the attempted new location, or the direction of the displacement vector is kept unmodified. Typically, the first alternative is preferred.

Bisection method

The *cut off at the boundary* technique works well when the boundaries are limited by interval constraints, and when the solution is located on the boundary. However, solutions might get trapped in the boundaries when the optimum is near them. Foryś *et al.* [3] proposed a *reflection technique* instead, while Innocente [1] proposed the *bisection method*: if the new position is infeasible, the displacement vector is split in halves, and the constraints verified. If the new position is still infeasible, the vector is split again, and so on. A feasible position will be found in time, unless the particle is already on the boundary. Hence an upper limit of iterations is set, and the particle keeps its position if no new feasible location can be found. Its drawbacks are that it cannot pass through infeasible space, and that the solution(s) must be initialized within $\mathcal{F}$. These drawbacks are shared with the *cut off at the boundary*, and to some extent with the *preserving feasibility* methods. Its main advantages are that it needs no adaptations to handle different inequality constraints, and its fast convergence.

Repair infeasible solutions

Engelbrecht [4] defines them as methods that "...allow particles to move into infeasible space, but special operators (methods) are then applied to either change the particle into a feasible one or to direct the particle to move towards feasible space. These methods usually start with initial feasible particles". Note that *particle* stands for *candidate solution.*

According to Engelbrecht [4], the *preserving feasibility* methods are those which ensure that adjustments to the particles do not violate any constraint. Hence the *cut off at the boundary* technique and the *bisection method* could be considered as *preserving feasibility* methods, while the *preserving feasibility* method discussed above could be considered as a method that *rejects infeasible solutions.*

## 2.3 Optimization methods

The important differences among the different types of optimization problems make it necessary to handle each type by means of completely different approaches. Optimization methods can be classified according to the **type of problems they are able to handle** (*i.e. linear*, *integer*, *mixed-integer*, *binary*, *quadratic*, or *non-linear*; *differentiable* or *non-differentiable*; *smooth* or *non-smooth*; *continuous* or *discrete*), or according to **the features of the algorithms** (*i.e. gradient-based* or *gradient-free*; *exact* or *approximate*; *deterministic* or *probabilistic*; *analytical* or *heuristic*, *single-based* or *population-based*). There are numerous different and often conflicting classifications. In addition, words like *heuristics* do not have a precise definition in this context. In general, it refers to techniques that do not guarantee to find anything, and are usually based on common sense, natural metaphors, or even methods whose behaviour is not fully understood. The classification offered in Figure 1 adheres to this definition. Since the membership of some paradigms to one or another class is sometimes unclear and conflicting, the criterion used is that if a *heuristic* qualifies as a *traditional method*, the latter prevails (*e.g. Tabu Search*). Purely deterministic methods are always viewed as traditional here, while probabilistic methods as heuristics.

*Modern heuristics* are general-purpose methods that will do well on most problems, although a problem-specific algorithm probably would be more efficient. They use techniques that are not directly related to optimization processes, heavily relying on stochastic operators. Their status of general-purpose algorithms together with their ability to escape local optima, ridges, and plateaus are their major advantages.

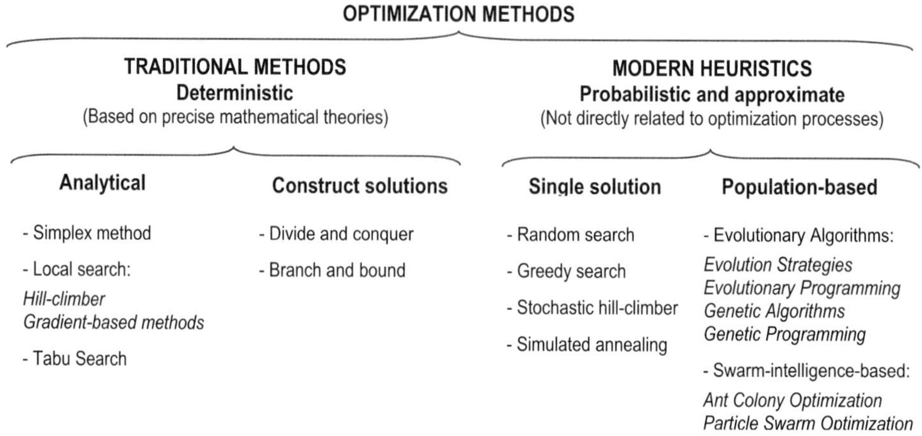

OPTIMIZATION METHODS

| TRADITIONAL METHODS<br>Deterministic<br>(Based on precise mathematical theories) | | MODERN HEURISTICS<br>Probabilistic and approximate<br>(Not directly related to optimization processes) | |
|---|---|---|---|
| **Analytical** | **Construct solutions** | **Single solution** | **Population-based** |
| - Simplex method<br>- Local search:<br>*Hill-climber*<br>*Gradient-based methods*<br>- Tabu Search | - Divide and conquer<br>- Branch and bound | - Random search<br>- Greedy search<br>- Stochastic hill-climber<br>- Simulated annealing | - Evolutionary Algorithms:<br>*Evolution Strategies*<br>*Evolutionary Programming*<br>*Genetic Algorithms*<br>*Genetic Programming*<br>- Swarm-intelligence-based:<br>*Ant Colony Optimization*<br>*Particle Swarm Optimization* |

Figure 1: Classification of different types of optimization methods

## 2.4 Population-based methods

Population-based methods have proved to be applicable to a variety of optimization problems, data mining, pattern recognition, classification, machine learning, scheduling, supply-chain management, *etc*. These outstandingly robust methods outperform traditional methods when dealing with problems of high complexity. They successively update a population of candidate solutions, thus performing a parallel exploration instead of the sequential exploration carried out by traditional methods, whose point-to-point search is usually unable to overcome local pathologies. In addition, these methods only require the cost function information to guide the search, without the need for auxiliary information such as gradients and Hessian matrices.

# 3 Particle Swarm Optimization

The original PSO method was designed by social-psychologist James Kennedy and electrical-engineer Russell C. Eberhart in 1995 [5]. The main algorithm was inspired by earlier bird flock simulations framed within the field of social psychology. Therefore, the method is closely related to other simulations of social processes and experimental studies in social psychology, while also having strong roots in both optimization and AI. To be more precise, it can be placed within the subfields *Computational Intelligence* (CI)[1] and/or *Artificial Life* (AL)[2]. Some of the experimental stud-

---

[1] AI divides into different schools of thought with vague boundaries. Some suggest two mainstreams: **classical AI** refers to the *symbolic paradigm* and its derivatives, and **CI** to all other paradigms such as *artificial neural networks* (ANNs), *fuzzy systems*, EAs and PSO. Others include both the *symbolic* and the *connectionist* (*i.e.* ANNs) *paradigms* into **classical AI**, while **modern AI** comprises the rest.

[2] *Artificial life* (AL) encompasses paradigms based upon metaphors of the behaviour of biological organisms, such as *cellular automaton*, EAs and PSO. The aim is not to model biological life but to create artificial beings that exhibit degrees of aliveness and intelligence by exploiting principles underlying biological organisms, where evolution and the ability to learn are critical for their survival.

ies in social psychology that influenced the method are Lewin's field theory; Gestalt theory; Sherif's and Asch's experiments; Latané's social impact theory; Bandura's no-trial learning; simulations of the spread of culture in a population, and of the behaviour of social animals such as bird flocks and social insects (refer to Kennedy *et al.* [6] or Innocente [1] for further reading). From an optimization point of view, it is a search method suitable for optimization problems whose solutions can be represented as points in an $n$-dimensional space. While the variables needed to be real-valued in the original version, binary and other discrete versions have also been developed (e.g. Kennedy *et al.* [6,7], Clerc [8], Mohan *et al.* [9]).

The PSO method relies on random coefficients to introduce creativity. It is a bottom-up approach in the sense that the system's intelligent behaviour emerges in a higher level than the individuals', evolving intelligent solutions without using the programmers' expertise on the subject matter. While this feature makes it difficult to understand the way optimization is actually performed, it shows astonishing robustness in handling many kinds of complex problems that it was not specifically designed for. However, it presents the disadvantage that its theoretical bases are extremely difficult to understand deterministically. Nevertheless, considerable theoretical work has been carried out on simplified versions of the algorithm, extrapolated to the full version, and finally supported by experimental results (*e.g.* Trelea [10], Clerc *et al.* [11], Ozcan *et al.* [12]). Refer to Kennedy *et al.* [6], Engelbrecht [4] and Clerc [13] for a general, comprehensive review of the paradigm.

## 3.1 Continuous, unconstrained PSO

The PSO method was originated on the simulation of a simplified social milieu, where individuals (*i.e. particles*) were thought of as collision-proof birds. The population is usually referred to as the *swarm*, while the function to be minimized is called here *conflict* function due to the social-psychology metaphor that inspired the method. That is, each individual searches the space of beliefs, seeking the minimization of the *conflicts* among the beliefs it holds by using the information gathered by both its own experience and those of others. Individuals indirectly seek agreement by clustering in the space of beliefs, which is – broadly speaking – the result of all the individuals imitating their most successful peer(s), thus becoming more similar to one another as the search progresses. However, the clustering is delayed by their own previous successful experiences, which each individual is reluctant to disregard.

To summarize, while the emergent properties of the PSO algorithm result from local interactions among the particles in the swarm, the behaviour of a single individual particle can be summarized in three sequential processes:

- **Evaluation**. The particle evaluates its position in the environment, which is given by the associated value of the conflict function. Following the social psychology metaphor, this would stand for the conflict among its current set of beliefs.
- **Comparison**. Once the particle's position in the environment is evaluated, it is not straightforward to tell how good it is. Experiments and theories in social psy-

chology suggest that humans judge themselves by comparing themselves to others, thus telling better from worse rather than good from bad. Therefore, the particle compares the conflict among its current set of beliefs to those of its neighbours.
- **Imitation**. The particle imitates those whose performances are superior in some sense. In the basic PSO (B-PSO), only the most successful neighbour is imitated.

These three processes are implemented within PSO, where the only sign of individual intelligence is a small memory. The basic update equations are as follows:

$$v_{ij}^{(t)} = w \cdot v_{ij}^{(t-1)} + iw \cdot U_{(0,1)} \cdot \left(pbest_{ij}^{(t-1)} - x_{ij}^{(t-1)}\right) + sw \cdot U_{(0,1)} \cdot \left(gbest_{j}^{(t-1)} - x_{ij}^{(t-1)}\right) \quad (5)$$

$$x_{ij}^{(t)} = x_{ij}^{(t-1)} + v_{ij}^{(t)} \quad (6)$$

where $x_{ij}^{(t)}$ and $v_{ij}^{(t)}$ are the $j^{th}$ coordinate of the position and velocity, respectively, of particle $i$ at time-step $t$; $U_{(0,1)}$ is a random number from a uniform distribution in the range [0,1] (resampled anew every time it is referenced); $w$, $iw$ and $sw$ are the inertia, individuality, and sociality weights, respectively; $pbest_{ij}^{(t-1)}$ and $gbest_{j}^{(t-1)}$ are the $j^{th}$ coordinate of the best position found by particle $i$ and by any particle in the swarm, respectively, by time-step $(t-1)$.

As shown in Equations (5) and (6), there are three coefficients in the basic algorithm which rule the dynamics of the swarm: the *inertia* ($w$), the *individuality* ($iw$), and the *sociality* ($sw$) *weights*. The $iw$ and the $sw$ are sometimes referred to as the *learning weights*, while their aggregation is called here the *acceleration weight* ($aw$).

Thus, the performance of a particle in its current position is **evaluated** in terms of the *conflict* function. In order to decide upon its next position, the particle **compares** its current conflict to those associated to both its own and its neighbours' best previous experiences (*pbest* and *gbest*, respectively). Finally, the particle **imitates** the best experience of its most successful neighbour, but without disregarding its own.

The relative importance given to $iw$ and $sw$ results in the particles exhibiting more self-confident or more conformist behaviour, while the random weights introduce creativity into the system: since they are resampled anew for each time-step, for each particle, for each component, and for each term in Equation (5), the particles display uneven trajectories that allow better exploration of the search-space. In addition, resampling them anew for the individuality and the sociality terms − together with setting $iw = sw$ − leads to the particles alternating self-confident and conformist behaviour. Typically, $iw = sw = 2$ (*i.e.* $aw = 4$). Every particle also tends to keep its current velocity, where the strength of this tendency is governed by $w$. The relative importance between $w$ and $aw$ results in more explorative or more exploitative behaviour of the swarm.

In the original PSO, Kennedy *et al.* [5] did not consider $w$ (*i.e.* $w = 1$), and suggested setting $iw = sw = 2$. However, the particles tended to diverge rather than cluster, and the swarm appeared to perform a so-called *explosion*. It was found that if the components of the particles' velocities were bounded, the explosion could be controlled and the particles ended up clustering around a solution. Later, aiming to control the explosion, Shi *et al.* [14] proposed the incorporation of the inertia weight ($w$).

Based on the B-PSO, countless variations can be thought of. For instance, Clerc *et al.* [13] proposed the use of a constriction factor – which can be viewed as an instance of the $w$ – to control the explosion of the original PSO and guarantee convergence; different settings of the coefficients can be thought of (which can be constant, deterministically dynamic, or adaptive); the coefficients and velocity constraint can be scalars or impose different behaviour for different dimensions; the neighbourhoods can be *global* or *local*; different neighbourhoods' architecture can be designed; the best experiences can be updated after all the particles' positions in the neighbourhood have been updated or every time a particle's position is updated, *etc.* As to the discrete version (D-PSO), alternatives like rounding off the particles' positions to match the discrete values and the binary PSO (b-PSO) can be considered.

### 3.1.1 About PSO as AI-based optimization method

AI is a polemical field due to controversies in the very definition of intelligence. Countless definitions can be found, most of them involving concepts like exhibiting verbal, mathematical and problem-solving abilities; capacity to think and reason; and/or ability to acquire, store and apply knowledge. While these definitions enumerate qualities, it can be argued that those are symptoms of intelligence. If intelligence is to be defined by enumerating its symptoms, all of them should be specified. If intelligence could be unquestionably defined, it would be reasonable to assert that AI is the intelligence exhibited by any artificial entity. However, definitions for biological intelligence and AI frequently differ. Fogel *et al.* [15] claim that a proper definition should apply to humans and machines equally well, and suggested that "intelligence may be defined as the capability of a system to adapt its behaviour to meet its goals in a range of environments" (quoted in [15]). Thus, it makes sense to speak of biological and artificial entities exhibiting degrees of intelligence, as different features of intelligence – and to different extents – are displayed.

The swarm in the PSO method displays problem-solving ability; each particle learns and profits from its own and others' previous experiences (stored in a small memory); and most importantly, the swarm *adapts its behaviour to meet its goals in a range of environments*, which gives the algorithm the status of a *general-purpose* problem-solver. Therefore, it can be claimed to exhibit some degree of intelligence.

## 3.2 Binary, unconstrained PSO

The b-PSO was originally proposed by Kennedy *et al.* [7] as a variation of the continuous version. The search-space consists of an $n$-dimensional binary hyper-cube;

the particles are represented by bit-strings; and the conflict function $c : \{0,1\}^n \rightarrow \mathcal{R}$. The basic update equations are now as follows:

$$v_{ij}^{(t)} = v_{ij}^{(t-1)} + iw \cdot U_{(0,1)} \cdot \left(pbest_{ij}^{(t-1)} - x_{ij}^{(t-1)}\right) + sw \cdot U_{(0,1)} \cdot \left(gbest_{j}^{(t-1)} - x_{ij}^{(t-1)}\right) \quad (7)$$

$$p_{ij}^{(t)} = \frac{1}{1+e^{-v_{ij}^{(t)}}} \in [0,1] \subset \mathcal{R} \quad (8)$$

$$x_{ij}^{(t)} = \begin{cases} 1 & \text{if} \quad U_{(0,1)} < p_{ij}^{(t)} \\ 0 & \text{otherwise} \end{cases} \quad (9)$$

where $p_{ij}^{(t)}$ stands for the probability of a bit adopting state 1. That is, the probability that individual $i$ has of holding belief $j$ at time-step $t$. The other parameters in Equation (7) are the same as in Equation (1). The velocities can still take real values.

### 3.3 Constrained PSO

The *additive penalization* as posed in Equation (4), the *preserving feasibility* and the *bisection* methods were implemented. The penalization coefficients were not tuned but intuitively set as follows: $\lambda = 10^6$ and $\alpha = 2$. The study of their tuning and self-adaptation is beyond the scope of this paper, as well as the adaptations of the *preserving feasibility* and *bisection* methods to handle equality constraints.

After extensive study of the impact of the three basic coefficients in Equations (5) on the behaviour of the swarm (see Innocente [1]), it was concluded that there is no general-purpose setting that together provides the abilities to fine-cluster and to escape poor sub-optimal solutions. The strategy adopted was to sub-divide the swarm into three sub-swarms whose set of coefficients grant them complementary abilities.

## 4 Benchmark optimization problems

It is important to keep in mind that the objective here is not fast convergence but to show that the same algorithm with the same settings can handle different problems.

### 4.1 First suite: unconstrained benchmark functions

The proposed general-purpose settings have to be tested on a suite of unconstrained benchmark problems, so that their evaluation is independent from the constraint-handling method. The suite is that used by Eberhart *et al.* [16] and Trelea [10] for comparison. Refer to their work for the equations and stopping criteria.

Typically, termination conditions in the literature consist of a maximum number of iterations and a given function target, only applicable for problems whose solution is known. The ones implemented in our solver are based on those studied by Innocente [1], suitable for real-world problems. However, in order to compare our results to those obtained by Eberhart *et al.* [16] and Trelea [10], the attainment of function targets terminates the search, the maximum number of time-steps is set to 10000, and the swarm size equals 30. A global (GP-GPSO) and a local version with 3-particle neighbourhood (GP-LPSO) are tested, and the results shown in Table 1.

| Function | Optimizer | Time-steps to meet function goal | | Success rate | Expected function evaluations to meet function goal | Average solution found | Best solution found | Time-steps | Function Evaluations |
|---|---|---|---|---|---|---|---|---|---|
| | | Average | Minimum | | | | | | |
| Sphere | Trelea Set 1 [10] | 344.00 | 266 | 1.00 | 1.03E+04 | - | - | - | - |
| | Trelea Set 2 [10] | 395.00 | 330 | 1.00 | 1.19E+04 | - | - | - | - |
| | Eberhart *et al.* [16] | 529.65 | 495 | 1.00 | 1.59E+04 | - | - | - | - |
| | GP-GPSO | **462.40** | **383** | **1.00** | **1.39E+04** | **1.42E-120** | **9.14E-134** | **1.00E+04** | **3.00E+05** |
| | GP-LPSO | **5977.75** | **4177** | **1.00** | **1.79E+05** | **2.36E-04** | **1.54E-07** | **1.00E+04** | **3.00E+05** |
| Rosenbrock | Trelea Set 1 [10] | 614.00 | 239 | 1.00 | 1.84E+04 | - | - | - | - |
| | Trelea Set 2 [10] | 900.00 | 298 | 1.00 | 2.70E+04 | - | - | - | - |
| | Eberhart *et al.* [16] | 668.75 | 402 | 1.00 | 2.01E+04 | - | - | - | - |
| | GP-GPSO | **679.35** | **336** | **1.00** | **2.04E+04** | **7.58E+00** | **4.34E-04** | **1.00E+04** | **3.00E+05** |
| | GP-LPSO | **6262.00** | **4840** | **0.30** | **6.26E+05** | **1.69E+02** | **4.02E+01** | **1.00E+04** | **3.00E+05** |
| Rastrigrin | Trelea Set 1 [10] | 140.00 | 104 | 0.90 | 4.67E+03 | - | - | - | - |
| | Trelea Set 2 [10] | 182.00 | 123 | 0.95 | 5.75E+03 | - | - | - | - |
| | This paper Set 1 | **102.53** | **67** | **0.95** | **3.24E+03** | **6.78E+01** | **3.48E+01** | **1.00E+04** | **3.00E+05** |
| | This paper Set 2 | **156.40** | **106** | **1.00** | **4.69E+03** | **6.17E+01** | **3.58E+01** | **1.00E+04** | **3.00E+05** |
| | Eberhart *et al.* [16] | 213.45 | 161 | 1.00 | 6.40E+03 | - | - | - | - |
| | GP-GPSO | **211.50** | **117** | **1.00** | **6.35E+03** | **2.61E+01** | **9.95E+00** | **1.00E+04** | **3.00E+05** |
| | GP-LPSO | **2802.05** | **202** | **0.90** | **9.34E+04** | **6.92E+01** | **5.02E+01** | **1.00E+04** | **3.00E+05** |
| Griewank | Trelea Set 1 [10] | 313.00 | 257 | 0.90 | 1.04E+04 | - | - | - | - |
| | Trelea Set 2 [10] | 365.00 | 319 | 0.90 | 1.22E+04 | - | - | - | - |
| | Eberhart *et al.* [16] | 312.60 | 282 | 1.00 | 9.38E+03 | - | - | - | - |
| | GP-GPSO | **601.53** | **362** | **0.95** | **1.90E+04** | **4.00E-02** | **0.00E+00** | **1.00E+04** | **3.00E+05** |
| | GP-LPSO | **5860.00** | **3724** | **0.80** | **2.20E+05** | **4.67E-02** | **1.60E-04** | **1.00E+04** | **3.00E+05** |
| Schaffer f6 | Trelea Set 1 [10] | 161.00 | 74 | 0.75 | 6.44E+03 | - | - | - | - |
| | Trelea Set 2 [10] | 350.00 | 102 | 0.60 | 1.75E+04 | - | - | - | - |
| | Eberhart *et al.* [16] | 532.40 | 94 | 1.00 | 1.60E+04 | - | - | - | - |
| | GP-GPSO | **694.44** | **142** | **0.90** | **2.31E+04** | **3.70E-04** | **0.00E+00** | **1.00E+04** | **3.00E+05** |
| | GP-LPSO | **779.95** | **101** | **1.00** | **2.34E+04** | **0.00E+00** | **0.00E+00** | **1.00E+04** | **3.00E+05** |

Table 1: Summary of the results obtained for the suite of unconstrained benchmark functions. The settings in Trelea Set 2 [10] and Eberhart *et al.* [16] are the same, while those in Trelea [10] were replicated in "This paper" for the Rastrigrin function.

Eberhart *et al.* [16] and Trelea [10] implemented settings that favour fine-clustering and convergence, so that their optimizers quickly achieve the conveniently set function targets. However, premature clustering takes place for time-extended searches −especially for multi-modal problems such as the Rastrigrin function−, as studied by Innocente [1]. In contrast, further improvement is possible for our optimizers because a third of the swarm has the ability to escape local optima. In order to illustrate this, the function targets were removed and our optimizers also tested for the whole 10000 time-steps. The settings in Eberhart *et al.* [16] and Trelea [10] were replicated and also tested along 10000 time-steps (on the Rastrigrin function only), showing their premature clustering. The results are shown in Table 1, where the white background indicates those corresponding to the function targets (see Trelea [10]), and the shaded background the results corresponding to the whole 10000 time-steps. The coefficients in Trelea *Set 2* [10], Eberhart *et al. [16]* and *This paper Set 2* coincide. The same is true for Trelea *Set 1* [10] and *This paper Set 1*.

Although its coefficients were not aimed at fast convergence, the number of time-steps that the GP-PSO require to meet the targets for the first three functions is comparable to those of Eberhart *et al.* [16] and Trelea *Set 2* [10], while its success rate outperforms those in Trelea [10] overall. As to the neighbourhood topology, the objective of the local versions is to help avoid premature clustering. Therefore, it is not surprising that the GP-LPSO – whose both coefficients and neighbourhood topology delay convergence – require many more time-steps to meet the targets. It is important to remark that the failures to achieve such goals along the permitted 10000 time-steps is due to the delay in the particles' clustering rather than to premature convergence as in the cases of Eberhart *et al.* [16] and Trelea [10]. This can be inferred from the fact that the improvement of solutions does not stagnate, and by comparing the evolution of the measures of clustering between the GP-GPSO and the GP-LPSO (not included here). Thus, the local version delays the loss of diversity and hence convergence, keeping the ability to escape local optima for longer. This is especially convenient for multi-modal problems such as the Rastrigrin and Schaffer f6 functions. The former is 30-dimensional in the test suite, which results in the permitted 10000 time-steps being insufficient for the particles to fine-cluster, whereas the latter is only 2-dimensional. Thus, the poor solutions found by the GP-LPSO for the first four 30-dimensional functions notably improve for searches extended beyond the permitted 10000 time-steps, while it finds the exact solutions for the 2-dimensional Schaffer f6 function in every run (see last row in Table 1).

The settings in Trelea *Set 1* [10] were replicated in *This paper Set 1*; those corresponding to both Trelea *Set 1* [10] and Eberhart *et al. [16]* in *This paper Set 2*; and both tested on the Rastrigrin function for the whole 10000 time-steps. It can be seen that they meet the target in less than 160 time-steps on average, their best solution out of 20 runs is around 35, and their average solution over 60. In contrast, the GP-GPSO meets the target in over 210 time-steps on average, but its best solution after 10000 time-steps is notably better (9.95), while its average solution in 20 runs (26) is notably smaller even than the best found by the other optimizers! The GP-LPSO is far from converging, and improvement still occurs even after 30000 time-steps.

In summary, settings that favour convergence and a global version are convenient for expensive function evaluations (*fes*) and for convex functions, while settings that do not favour convergence and local versions are convenient when exploration needs to be greatly improved (*e.g.* multi-modal functions, heavily constrained problems and disjointed feasible spaces). The GP-GPSO is a robust, general-purpose trade-off.

## 4.2 Second suite: constrained benchmark functions

The optimizer also needs to be tested on a suite of constrained functions because such constraints affect the normal behaviour of the system.

### 4.2.1 Pressure vessel problem

The problem – as posed by Hu *et al.* [17] – is mixed-discrete (m-d), where $x_1$ and $x_2$ are integer multiples of 0.0625, while $x_3$ and $x_4$ are continuous. Thus, a swarm composed of 20 particles, 10000 time-steps and 11 runs per test are implemented, as in Hu et al. [17]. The results are also compared to those obtained by Coello Coello [18] in Table 2 and Table 3. Since Hu *et al.* [17] and Coello Coello [18] reported their results as the best among other authors', such comparisons are omitted here.

| PRESSURE VESSEL | | 1 neighbour | | 2 neighbours | | Global | |
|---|---|---|---|---|---|---|---|
| | | Real | Penalized | Real | Penalized | Real | Penalized |
| Penalization | Best | 6033.618818 | 6050.947591 | 6040.076927 | 6049.857978 | **6040.078181** | **6049.857964** |
| | Mean | 6173.760461 | 6179.567556 | 6127.535794 | 6135.206712 | **6398.899925** | **6411.395265** |
| Preserving feasibility | Best | **6066.259215** | - | 6090.526202 | - | 6090.526202 | - |
| | Mean | **6175.946584** | - | 6328.738977 | - | 6547.361378 | - |
| Bisection | Best | 6090.526202 | - | **6059.714335** | - | 6090.526202 | - |
| | Mean | 6503.193272 | - | **6476.062984** | - | 6539.554416 | - |
| Hu *et al.* [17] | Best | - | - | **6059.131296** | - | - | |
| Coello Coello [18] | Best | **6288.744500** | | | | | |
| | Mean | **6293.843232** | | | | | |

Table 2: Minima found by our optimizer with three constraint-handling methods and three neighbourhood sizes, and by two other authors for the m-d pressure vessel problem

| Pressure Vessel | $x_1$ | $x_2$ | $x_3$ | $x_4$ |
|---|---|---|---|---|
| Penalization | 0.812500 | 0.437500 | 42.260491 | 174.638894 |
| Preserving feasibility | 0.812500 | 0.437500 | 42.044793 | 177.302575 |
| Bisection | 0.812500 | 0.437500 | 42.098446 | 176.636596 |
| Hu *et al.* [17] | 0.812500 | 0.437500 | 42.098450 | 176.636600 |
| Coello Coello [18] | 0.812500 | 0.437500 | 40.323900 | 200.000000 |

Table 3: Coordinates of the minima found by our optimizer and by two other authors for the m-d pressure vessel problem

He *et al.* [19] proposed a hybrid PSO (simulating annealing incorporated), and tested it on this problem using 250 particles, 81000 *fes*, and 30 runs for the statistics. They reported having outperformed several other authors. Hence our optimizer is tested again using the same number of particles and *fes*, and the results compared only to theirs (see Table 4). It must be noted that the general-purpose coefficients used here were designed for a swarm of 20-50 particles, 1000-30000 time-steps, and 1 to 30-dimensional problems. While the appropriate swarm-size, degree of locality, and number of object variables are clearly related, the study of such relationships is beyond the scope of this paper. It is fair to mention, however, that fast convergence results not only from appropriate coefficients' settings but also from small swarm sizes, highly connected neighbourhoods, and constraint-handling methods that quickly decrease the particles' momentum. Notice that the swarm size and *fes* used by He *et al.* [19] result in a more parallel search and shorter evolution. Therefore, coefficients' settings that favour convergence would be probably more appropriate.

| PRESSURE VESSEL | | 2 neighbours | | Global | |
|---|---|---|---|---|---|
| | | Real | Penalized | Real | Penalized |
| Penalization | Best | **6036.019786** | **6050.289209** | 6073.148086 | 6081.860801 |
| | Mean | **6054.780217** | **6064.294442** | 6309.524952 | 6317.137956 |
| Preserving feasibility | Best | 6061.308346 | - | **6059.714335** | - |
| | Mean | 6092.492880 | - | **6422.404101** | - |
| Bisection | Best | **6059.714335** | - | 6059.714335 | - |
| | Mean | **6073.456533** | - | 6360.990084 | - |
| He *et al.* [19] | Best | - | - | **6059.714300** | - |
| | Mean | - | - | **6099.932300** | - |

Table 4: Minima found by our optimizer with three constraint-handling methods and two neighbourhood sizes, and by He *et al.* [19] for the m-d pressure vessel problem

Although de Freitas Vaz *et al.* [20] claimed that their optimizer returned better results than those reported by Hu *et al.* [17], the comparison is invalid because they posed the problem as continuous. The same is true in Foryś *et al.* [3] and Innocente [1]. In order to compare our results to those of de Freitas Vaz *et al.* [20], 30 analyses are performed using their settings: continuous variables and 879000 *fes*. The results are also compared to those of Foryś *et al.* [3] for reference (see Table 5 and Table 6).

| PRESSURE VESSEL | | 2 neighbours | | Global | |
|---|---|---|---|---|---|
| | | Real | Penalized | Real | Penalized |
| Penalization | Best | 5851.887101 | 5879.828359 | **5854.933947** | **5870.123975** |
| | Mean | 5872.459619 | 5886.238461 | **5854.933953** | **5870.123975** |
| Preserving feasibility | Best | 5888.171310 | - | **5886.483117** | - |
| | Mean | **5944.336034** | - | 6130.527590 | - |
| Bisection | Best | **5887.586796** | - | 5918.329579 | - |
| | Mean | **6046.356830** | - | 6362.007338 | - |
| de Freitas Vaz *et al.* [20] | Best | - | - | **5885.330000** | - |
| Forys et al. [3] | Best | **5885.490000** | | | |

Table 5: Minima found by our optimizer with three constraint-handling methods and two neighbourhood sizes, and by two other authors for the pressure vessel problem

| Pressure Vessel | $x_1$ | $x_2$ | $x_3$ | $x_4$ |
|---|---|---|---|---|
| Penalization | 0.774549 | 0.383204 | 40.319619 | 200.000001 |
| Preserving feasibility | 0.778841 | 0.384982 | 40.354458 | 199.515579 |
| Bisection | 0.779484 | 0.385300 | 40.387763 | 199.053548 |
| de Freitas Vaz *et al.* [20] | 0.778169 | 0.384649 | 40.319600 | 200.000000 |
| Forys [3] | 0.778300 | 0.384700 | 40.324400 | 199.933000 |

Table 6: Coordinates of the minima found by our optimizer and by two other authors for the pressure vessel problem

Given that a swarm of 30 particles is implemented here, the search is carried out throughout 29300 time-steps, so as to match the number of *fes* used in de Freitas Vaz *et al.* [20]. This long search combined with this small swarm and a low-dimensional problem gives time for the local versions to achieve a reasonable degree of clustering, although improvement still does not stagnate for local versions with penalization and preserving feasibility methods by the time the search is ended.

### 4.2.2 Welded beam problem

For the formulation of this problem, refer to Hu *et al.* [17] or Coello Coello [18]. In order to compare our results to those obtained by Hu *et al.* [17], their settings are replicated, and also compared to those of other authors for reference despite the algorithmic details being different. Some of the results are shown in Table 7.

| **WELDED BEAM** | | 1 neighbour | | 2 neighbours | | Global | |
|---|---|---|---|---|---|---|---|
| | | Real | Penalized | Real | Penalized | Real | Penalized |
| Penalization | Best | 1.72485136 | 1.72485185 | **1.72485135** | **1.72485185** | 1.72485136 | 1.72485185 |
| | Mean | 1.74076752 | 1.74076893 | **1.72485139** | **1.72485185** | 1.73616402 | 1.73616456 |
| Preserving feasibility | Best | 1.72485231 | 1.72485231 | **1.72485231** | - | 1.72485231 | 1.72485231 |
| | Mean | 1.72487563 | 1.72487563 | **1.72485234** | - | 1.73528547 | 1.73528547 |
| Bisection | Best | 1.72485231 | 1.72485231 | **1.72485231** | - | 1.72485231 | 1.72485231 |
| | Mean | 1.72548316 | 1.72548316 | **1.72485807** | - | 1.74321178 | 1.74321178 |
| Hu *et al.* [17] | Best | - | - | **1.72485084** | - | - | - |
| Coello Coello [18] | Best | **1.74830941** | | | | | |
| | Mean | **1.77197269** | | | | | |
| He *et al.* [19] | Best | - | - | - | - | **1.724852** | - |
| | Mean | - | - | - | - | **1.749040** | - |
| Foryś *et al.* [3] | Best | **1.7248** | | | | | |
| de Freitas Vaz *et al.* [20] | Best | - | - | - | - | **1.814290** | - |

Table 7: Minima found by our optimizer with three constraint-handling methods and three neighbourhood sizes, and by five other authors for the welded beam problem

### 4.2.3 Tension/compression spring design

For the formulation of this problem, refer to Hu *et al.* [17] and Coello Coello [18]. The test conditions used in Hu *et al.* [17] are replicated, and the results are also compared to those of other authors for reference (see Table 8).

### 4.2.4 Himmelblau's nonlinear optimization problem

For the formulation of the problem, refer to Hu *et al.* [17], Coello Coello [18] and Toscano Pulido *et al.* [21]. Invalid comparisons can be found in the literature, as the problem formulations differ! Thus, the third terms in the conflict and the constraint functions in Hu *et al.* [17], Coello Coello [18] and Toscano Pulido *et al.* [21] are different. It must be noted that not all the formulations of all the problems in all the papers referenced in this work were verified in detail. The settings in Hu *et al.* [17]

were replicated for comparison, and it was assumed that de Freitas Vaz *et al.* [20] used the same formulation. The results are shown in Table 9 and Table 10.

| TENSION / COMPRESSION SPRING DESIGN | | 1 neighbour | | 2 neighbours | | Global | |
|---|---|---|---|---|---|---|---|
| | | Real | Penalized | Real | Penalized | Real | Penalized |
| Penalization | Best | 0.0126703 | 0.0126703 | 0.0126747 | 0.0126747 | **0.0126652** | **0.0126652** |
| | Mean | 0.0128350 | 0.0128350 | 0.0128103 | 0.0128103 | **0.0127956** | **0.0127956** |
| Preserving feasibility | Best | **0.0126654** | **0.0126654** | 0.0126672 | 0.0126672 | 0.0126700 | 0.0126700 |
| | Mean | **0.0127360** | **0.0127360** | 0.0127507 | 0.0127507 | 0.0128180 | 0.0128180 |
| Bisection | Best | **0.0126652** | **0.0126652** | 0.0126652 | 0.0126652 | 0.0126670 | 0.0126670 |
| | Mean | **0.0126860** | **0.0126860** | 0.0127252 | 0.0127252 | 0.0129549 | 0.0129549 |
| Hu *et al.* [17] | Best | - | - | **0.0126661** | - | - | - |
| Coello Coello [18] | Best | **0.0127048** | | | | | |
| | Mean | **0.0127692** | | | | | |
| He *et al.* [19] | Best | - | - | - | - | **0.0126652** | - |
| | Mean | - | - | - | - | **0.0127072** | - |
| de Freitas Vaz *et al.* [20] | Best | - | - | - | - | **0.0131926** | - |

Table 8: Minima found by our optimizer with three constraint-handling methods and three neighbourhood sizes, and by four other authors for the tension/compression minimum weight string design problem

| HIMMELBLAU'S PROBLEM | | 1 neighbour | | 2 neighbours | | Global | |
|---|---|---|---|---|---|---|---|
| | | Real | Penalized | Real | Penalized | Real | Penalized |
| Penalization | Best | **-31025.8178** | **-31025.6896** | -31025.8178 | -31025.6896 | -31025.8178 | -31025.6896 |
| | Mean | **-31025.8178** | **-31025.6896** | -31025.8178 | -31025.6896 | -31025.8178 | -31025.6896 |
| Preserving feasibility | Best | -31025.5474 | -31025.5474 | -31025.5614 | -31025.5614 | **-31025.5614** | **-31025.5614** |
| | Mean | -31025.0798 | -31025.0798 | -31025.5548 | -31025.5548 | **-31025.5614** | **-31025.5614** |
| Bisection | Best | **-31025.5614** | **-31025.5614** | -31025.5614 | -31025.5614 | -31025.5614 | -31025.5614 |
| | Mean | **-31025.5614** | **-31025.5614** | -31025.5614 | -31025.5614 | -31022.7249 | -31022.7249 |
| Hu *et al.* [17] | Best | - | - | **-31025.5614** | - | - | - |
| de Freitas Vaz *et al.* [20] | Best | - | - | - | - | **-31012.1000** | - |

Table 9: Minima found by our optimizer with three constraint-handling methods and three neighbourhood sizes, and by two other authors for the Himmelblau's problem as posed in Hu *et al.* [17]

| Himmelblau's Problem | $x_1$ | $x_2$ | $x_3$ | $x_4$ | $x_5$ |
|---|---|---|---|---|---|
| Penalization | 77.99997324 | 32.99997741 | 27.07010953 | 45.00001602 | 44.96929014 |
| Preserving feasibility | 78.00000000 | 33.00000000 | 27.07099711 | 45.00000000 | 44.96924255 |
| Bisection | 78.00000000 | 33.00000000 | 27.07099711 | 45.00000000 | 44.96924255 |
| Hu *et al.* [17] | 78.00000000 | 33.00000000 | 27.07009700 | 45.00000000 | 44.96924255 |
| de Freitas Vaz *et al.* [20] | 78.0000 | 33.0000 | 27.1106 | 45.0000 | 45.0000 |

Table 10: Coordinates of the minima found by our optimizer and by two other authors for the Himmelblau's problem as posed in Hu *et al.* [17]

In order to compare our results to those of other authors, the problem is formulated as in Toscano Pulido *et al.* [21], and the results also compared to those of He *et al.* [19] and Hu *et al.* [2] for reference (see Table 11 and Table 12).

| HIMMELBLAU'S PROBLEM | | 1 neighbour | | 2 neighbours | | Global | |
|---|---|---|---|---|---|---|---|
| | | Real | Penalized | Real | Penalized | Real | Penalized |
| Penalization | Best | **-30665.953** | **-30665.746** | **-30665.953** | **-30665.746** | **-30665.953** | **-30665.746** |
| | Mean | **-30665.953** | **-30665.746** | **-30665.953** | **-30665.746** | **-30665.953** | **-30665.746** |
| Preserving feasibility | Best | -30665.538 | -30665.538 | **-30665.539** | **-30665.539** | **-30665.539** | **-30665.539** |
| | Mean | -30665.488 | -30665.488 | **-30665.539** | **-30665.539** | **-30665.539** | **-30665.539** |
| Bisection | Best | **-30665.539** | **-30665.539** | **-30665.539** | **-30665.539** | -30665.539 | -30665.539 |
| | Mean | **-30665.539** | **-30665.539** | **-30665.539** | **-30665.539** | -30624.748 | -30624.748 |
| Toscano Pulido *et al.* [21] | Best | - | - | - | - | **-30665.500** | - |
| | Mean | - | - | - | - | **-30665.500** | - |
| He *et al.* [19] | Best | - | - | - | - | **-30665.539** | - |
| | Mean | - | - | - | - | **-30665.539** | - |
| Hu *et al.* [2] | Best | | | | | **-30665.500** | |
| | Mean | - | - | - | - | **-30665.500** | - |

Table 11: Minima found by our optimizer with three constraint-handling methods and three neighbourhood sizes, and by three other authors for the Himmelblau's problem as posed in Toscano Pulido *et al.* [21].

| Himmelblau's Problem | $x_1$ | $x_2$ | $x_3$ | $x_4$ | $x_5$ |
|---|---|---|---|---|---|
| Penalization | 78.0000 | 33.0000 | 29.9938 | 45.0000 | 36.7766 |
| Preserving feasibility | 78.0000 | 33.0000 | 29.9953 | 45.0000 | 36.7758 |
| Bisection | 78.0000 | 33.0000 | 29.9953 | 45.0000 | 36.7758 |

Table 12: Coordinates of the minima found by our optimizer for the Himmelblau's problem as posed in Toscano Pulido *et al.* [21].

## 4.3 General comments

The PSO algorithm is a robust, general-purpose optimization method that can cope with very different continuous problems regardless of their features. It can also cope with problems where the variables can take a set of discrete values either by rounding-off the particles' trajectories or by the binary version of the algorithm. However, different settings of the coefficients in the particles' velocity update equation considerably affect the dynamics of the swarm. The general-purpose settings used here together with a fully connected neighbourhood topology result in a general-purpose optimizer that performs well on most problems (see Table 1 to Table 12). In some cases, using some local version might conveniently keep diversity for a longer period of time. Our optimizer is flexible, and all the settings can be introduced by the user, including any number of neighbours influencing every particle. As to the constraint-handling techniques, the penalization method results in the best exploration of the search-space because it turns the problem into unconstrained, PSO being an inherently unconstrained optimization method. The drawbacks are the need to tune additional coefficients, and solutions often moderately violating constraints. While self-adaptive penalizations are an active, ongoing line of research, some tolerance is typically acceptable in real-world applications for marginally activated constraints.

# 5 The jetty scheduling problem

Every scheduling problem can be represented by a set of $n$ jobs to be carried out on a set of $m$ machines, where each job consists of a set of operations to be performed on one or more machines. In particular, the job-shop scheduling problem (JSSP) is an NP-hard combinatorial problem consisting of $n$ jobs to be processed through $m$ machines, where each job must be processed on each machine only once, in a particular order, and with no precedence constraints among different job operations. Each machine can handle one job at a time – which cannot be interrupted – and the processing time is fixed and known. The problem consists of finding an optimum schedule, typically associated to the minimum *makespan* (time required to complete all jobs).

If the space of schedules is relatively small, all combinations might be tried and the best one selected (so-called *exhaustive search*). However this becomes infeasible even for small real-world problems, and more intelligent methods that allow cutting off regions of the space of schedules are required (*branch and bound methods*). *Hill-climbing methods* have also been used to handle these problems, where a schedule is randomly generated and successively improved by altering it and keeping the best between the original and the altered schedules. Since these traditional methods get trapped in local optima, modern methods that can escape local optima such as *Tabu Search* and *Simulated Annealing* have been successfully used. More recently, *Genetic Algorithms* (GA) have been reported to find better solutions for scheduling problems. For applications of GAs to scheduling problems, refer to Yamada *et al.* [22]. In the last few years, the PSO method – originally suitable for continuous problems only – has also been successfully applied to scheduling problems. In most publications, the PSO method undergoes numerous adaptations to be made suitable for such problems (*e.g.* Xia *et al.* [23] and Sha *et al.* [24]). In contrast, the aim here is to keep the method as close to the original version as possible, so as to show its robustness in handling different types of problems with few or no adaptations.

The jetty scheduling problem can be seen as an instance of the general JSSP, which involves the allocation of vessels to the berths at a port along a given period of time in order to minimize the costs incurred through demurrage. The *makespan* minimization might also be desirable as a secondary objective. Our case study consists of a jetty scheduling problem in an oil refinery, where there are numerous constraints that prevent a given vessel mooring up to a given berth (*e.g.* unfitness of the vessel entering the berth due to vessel length or insufficient depth of water; products' unavailability; berth out of order for a period of time; *etc.*). Thus, the problem is posed as a set of $n$ shipments to be served on a set of $m$ jetties, where a set of berthing operations are to be performed for each shipment served on a given jetty. For the case study considered here, it is assumed that every jetty has the ability to handle all the operations associated to any shipment feasible to be served.

# 6 PSO-based jetty scheduler

Although the objective is to use basically the same optimizer as for the previous benchmark problems, additional code is required to turn the search algorithm into a scheduler. Thus, the jetty scheduling problem is viewed as a search problem on a discrete space, and the PSO equations are transformed in the simplest way possible: the particles' continuous trajectories are rounded off to match the discrete values.

The search-space is limited to its discrete part, and the continuous *Estimated Time of Berthing* (ETB) values are handled deterministically by additional implementations (so far) unrelated to the PSO code. Hence a search-space composed of as many dimensions as shipments are to be scheduled is constructed, where these variables can take any of a set of integer values representing the berths in the port at the refinery. A point in such search-space allocates the shipments that are to be served on each berth, without giving any information about their ETBs, or even about the order in which they are to be served. Therefore, a solution to the problem is given at present by a point in the search-space plus some deterministic rules attached. A future stage in our research consists of considering each berth and its allocated shipments as a travelling salesman problem, so that the order in which they are to be served can be treated as an independent, embedded optimization problem.

Thus, the information regarding shipments and berths is input into the scheduler. This includes fitness information relating vessels and berths; residency time for each vessel on each berth; each shipment's estimated time of arrival (ETA), *laycan window*[3], and demurrage rate. The precedence must be clearly defined at this stage, as the order of the shipments in a berth is deterministically defined. Thus, earlier ETAs award higher precedence to a job. This will be done by an embedded optimization algorithm in a future stage, which would also consider other aspects like the demurrage rate. While the function to be minimized is the total demurrage at present, the *makespan* minimization will be incorporated as a secondary objective in the future.

Our optimizer also allows fixing some vessels to some berths, fixing some vessels' ETBs, and fixing both, while the remaining shipments and berths are scheduled around those constraints. The activation of multi-berth serving is also possible in our optimizer, although this eliminates the chance to embed the optimization algorithm to decide upon the order of the shipments at each berth, since each berth is not independent from the others for multi-berth visits activated. The constraint-handling technique consists of a deterministic repair algorithm, which searches for the closest berth in which the infeasible shipment can be served.

In order to make the comparison fair, a real-world schedule is fed to our PSO-based scheduler, returning a demurrage cost equal to $2.63835\times10^9$. Next, the PSO-based scheduler described in the previous section is run using the same coefficients as when optimizing the benchmark problems in previous sections. Since the problem

[3] It is the time-span in which the vessel has to be served through its completion since its arrival.

is high-dimensional, a bigger swarm would probably be more convenient. However, the aim is to find a quick solution to this 142-dimensional NP-hard problem. Therefore, the swarm size is kept to 30, a fully connected topology is chosen to favour fast convergence, and the search is performed along 200 time-steps only. The minimum demurrage found equals $1.36016\times10^{8}$, while the average among the demurrages of all 30 particles equals $3.79679\times10^{8}$. Notice that, since a repairing algorithm is used, every particle represents a feasible solution. The search was carried along 8.3 minutes only, using the software *matlab* and a 1.73 GHz Intel Pentium M processor. The evolution of the average and minimum demurrage is shown in Figure 2.

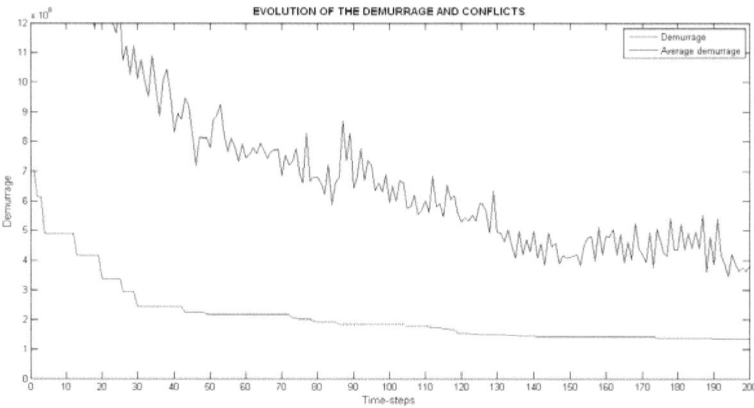


Figure 2: Evolution of the best and average demurrage for a single run of the PSO-based scheduler on a problem with 142 shipments and 10 berths

## 7 Concluding remarks and future work

The objective to show the robustness of the PSO approach to cope with notably different problems with the same basic algorithm was successfully achieved. In order to strengthen the algorithm's general-purpose characteristics, it is important to use general-purpose settings for the coefficients in the velocity update rule, such as those studied by Innocente [1]. However, not all settings can be kept problem-independent because the conditions of the different problems define different computer power requirements as acceptable. Such settings should keep a balance between exploration and exploitation in order to avoid premature clustering on the one hand, and the lack of fine-tuning or unnecessarily long searches on the other. It appears convenient to use general-purpose settings for the coefficients, and control the exploration/exploitation ratio by setting convenient swarm sizes, maximum number of time-steps permitted, and neighbourhoods' topology. An acceptable number of *fes* always needs to be set in any algorithm, while the topology of the neighbourhood can be set in our optimizer by simply entering the number of neighbours influencing every particle (ring topology). Finally, given an acceptable number of *fes*, the greater the swarm size the more parallel the search – useful in some problems – in detriment of the time-steps assigned for the search to evolve. A swarm of around 30 particles is enough for most problems, but the use of greater swarms clearly should not be disregarded.

Three different constraint-handling methods were implemented: the *penalization*, the *preserving feasibility*, and the *bisection* methods. As much as very good results were obtained, there is room for improvement in each of these techniques. For instance, the self-adaptation of penalization coefficients and the development of techniques to delay the loss of particles' momentum and to generate initial feasible solutions other than brute force are promising lines of research for further improvement.

While binary versions of the method to handle discrete variables comprise promising lines for future research, the algorithm was adapted here to discrete problems by rounding-off the particles' trajectories. Thus, the continuous algorithm was kept virtually unmodified. In turn, the jetty scheduling problem was set as a search problem by generating a discrete space where the variables represent shipments and the discrete values represent the berths where they are to be served. Every solution returned by the PSO algorithm *per se* consists of a list of shipments to be served on each berth. The order is deterministically decided at present according to the ETAs.

The next steps in our research consist of designing more developed priority rules, incorporating the demurrage rate in addition to the ETAs; testing binary versions of the PSO algorithm; considering other constraint-handling techniques such as the penalization method; including the minimization of the *makespan* as a secondary objective; and considering embedding an independent optimization algorithm on every berth to optimize the order in which the shipments allocated are to be served.

## Acknowledgements

The authors would like to acknowledge the financial and technical support of Maron Systems.

## References

[1] M.S. Innocente, "Population-Based Methods: Particle Swarm Optimization - Development of a General-Purpose Optimizer and Applications", Master's thesis, Universitat Politècnica de Catalunya - CIMNE, Barcelona, Spain, 2006.

[2] X. Hu, R.C. Eberhart, "Solving Constrained Nonlinear Optimization Problems with Particle Swarm Optimization", In Proceedings of the 6th World Multiconference on Systemics, Cybernetics and Informatics, 2002.

[3] P. Foryś, B. Bochenek, "A New Particle Swarm Optimizer and its Application to Structural Optimization", In Proceedings of 5th ASMO UK / ISSMO Conference on Engineering Design Optimization, Stratford upon Avon, 2004.

[4] A.P. Engelbrecht, "Fundamentals of Computational Swarm Intelligence", John Wiley & Sons Ltd, England, 2005.

[5] J. Kennedy, R.C. Eberhart, "Particle Swarm optimization", In Proceedings of IEEE Int. Conference on Neural Networks, Piscataway, 1942-1948, 1995.

[6] J. Kennedy, R.C. Eberhart, "Swarm Intelligence", Morgan Kaufmann Publishers, 2001.

[7] J. Kennedy, R.C. Eberhart, "A Discrete Binary Version of the Particle Swarm Algorithm", 0-7803-4053-1/97 – IEEE, 4104-4108, 1997.

[8] M. Clerc, "Discrete Particle Swarm Optimization", New Optimization Techniques in Engineering: Springer-Verlag, 2004.

[9] C. Mohan, B. Al-kazemi, "Discrete particle swarm optimization", In Proceedings of the Workshop on Particle Swarm Optimiz.", Indianapolis, 2001.

[10] I. Trelea, "The particle swarm optimization algorithm: convergence analysis and parameter selection", Elsevier Science, 2002 – Information Processing Letters 85, 317-325, 2003.

[11] M. Clerc, J. Kennedy, "The Particle Swarm—Explosion, Stability, and Convergence in a Multidimensional Complex Space", IEEE Transactions on Evolutionary Computation, Vol. 6, No. 1, 58-73, February 2002.

[12] E. Ozcan, C. Mohan, "Particle Swarm Optimization: Surfing the Waves", In Proceedings of IEEE Congress on Evolutionary Computation, Piscataway, New Jersey, 1999.

[13] M. Clerc, "Particle Swarm Optimization", ISTE, 2006.

[14] Y. Shi, R.C. Eberhart, "A Modified Particle Swarm Optimizer", IEEE Int. Conference on Evolutionary Computation, Alaska, 0-7803-4869-9/98, 1998.

[15] D. Fogel, L. Fogel, "Evolution and Computational Intelligence", IEEE Spectrum, 1995.

[16] R.C. Eberhart, Y. Shi, "Comparing Inertia Weights and Constriction Factors in Particle Swarm Optimization", IEEE 0-7803-6375-2/00, 2000.

[17] X. Hu, R.C. Eberhart, Y. Shi, "Engineering Optimization with Particle Swarm", in Swarm Intelligence Symposium, Proceedings of the IEEE, 2003.

[18] C.A. Coello Coello, "Use of a self-adaptive penalty approach for engineering optimization problems", Computers in Industry, vol. 41, no. 2, 113-127, 2000.

[19] Q. He, L. Wang, "A hybrid particle swarm optimization with a feasibility-based rule for constrained optimization", Applied Mathematics and Computation 186, 1406-1422, 2007.

[20] A.I. de Freitas Vaz, E.M. da Graça Pinto Fernandes, "Optimization of Nonlinear Constrained Particle Swarm", in Technological and Economic Development of Economy, Vol XII, No 1, 30-36, 2006.

[21] G. Toscano Pulido, C.A. Coello Coello, "A Constraint-Handling Mechanism for Particle Swarm Optimiz.", IEEE 0-7803-8515-2, 1396-1403, 2004.

[22] T. Yamada, R. Nakano, "Genetic Algorithms for Job-Shop Scheduling Problems", in Proceedings of Modern Heuristic for Decision Support, 67-81, UNICOM seminar, 18-19, London, March 1997.

[23] W. Xia, Z. Wu, "An effective hybrid optimization approach for multi-objective flexible job-shop scheduling problems", Computers & Industrial Engineering 48, 409-425, 2005.

[24] D.Y. Sha, C.-Y. Hsu, "A hybrid particle swarm optimization for job-shop scheduling problem", Computers & Industrial Engineering 51, 791-808, 2006.

©Saxe-Coburg Publications, 2008.
Trends in Engineering Computational Technology
B.H.V. Topping and M. Papadrakakis, (Editors)
Saxe-Coburg Publications, Stirlingshire, Scotland, 127-150.

# Chapter 7

# Optimization Techniques in Human Movement Analysis

**A. Eriksson**
**Department of Mechanics, KTH Engineering Sciences**
**Royal Institute of Technology, Stockholm, Sweden**

## Abstract

This paper discusses a few aspects of human movements when seen as an optimal control problem for a mechanism. In particular, the redundancy of the muscle force actuators in simplified models of a body or body part is a main concern. The criteria governing the control of human movement planning are far from fully known, but a general optimization setting allows investigation of alternatives. The numerical description of muscular force production is complex, and its history dependence only partially known. The paper discusses a setting of the dynamic optimization problem, which is reliable and efficient compared to shooting-type methods, and which allows several criteria for optimal performance. The method is demonstrated by a few aspects of a weight-lifting problem. The need for muscular models compatible with the basic formulation, and accurately representing the history dependence in force production is emphasized.

**Keywords:** human movements, optimization, mechanisms, muscular forces.

# 1 Introduction

Computational modeling of load-carrying structures is a conventional tool in mechanical, structural and robotic engineering. These modeling possibilities are more or less general in their representation of the behavior of the structures and components, when subjected to load. Many such tools have also been used for bio-mechanical problems, in the form of special simulation packages, such as the Danish project AnyBody (http://anybody.auc.dk), as specialized additions to general simulation programs for rigid mechanisms, *e.g.* ADAMS (MSC Software, Santa Ana, CA, USA), or as toolboxes developed close to motion capture systems used for the registration and analysis of movement patterns, such as OpenSim (http://simtk.org/home/opensim).

The numerical modeling and simulations of human movements are essentially of two types. The inverse dynamics approach starts by measuring body segment kinematics, external forces and segment inertial characteristics; these are used to compute the variations of joint moments and forces during a movement sequence. The forward dynamics approach takes the opposite route: measured forces are used to find the movement of the system, [1]. Inverse dynamics calculations are less expensive computationally, but require accurate measurements of external forces and body motions [2]. Recent advances in computing resources have made forward dynamics calculations reasonably efficient, and permit more advanced mechanical knowledge in the simulations.

For both classes of simulations, significant simplifications of the real behavior of the biological systems are necessary. Typically, simulations see the human body parts as rigid, un-deformable links, connected at perfect hinges and affected by very simplified muscle models. The bio-mechanical models created can be subjected to active or passive forces, and allow prediction with some accuracy of the expected response of the modeled object, at least for common situations of repetitive movements at reasonable loading levels.

The musculoskeletal system is from the computational viewpoint normally a highly redundant force system for most loading cases, [3]. The indeterminacy in force distribution is related to the high number of alternative components over most joints. It is an interesting fact that a major part of this redundancy problem is related to the modeling of the human body with simplified, e.g., plane models, as the differences in actions between different muscles vanish, when the kinematics and kinetics are not fully resolved.

The redundancy means that a load on the skeletal system can be carried in several ways, by activating different sets of muscles. It also allows very complex movements under neural control, but causes computational difficulties when attempts are made to simulate the behavior. The control system must be modeled in simulations of any will-controlled movements, [4], but is not easily represented, as it consists of several levels of learnt and instant control, [5]. The exact functioning of the control system is not fully known, but a common approach is to assume that the muscular forces are chosen from some optimization rule: minimum forces, nominal efforts, activation, energy consumption, or maximum smoothness, *etc.* In simulations, these can be formulated as mathematical rules. A special aspect of this is the distribution of forces between synergistic muscles, [6], which is often seen as a static or dynamic optimization problem, [7]. Several methods and criteria for static optimization can easily be set in a common algorithm, [8], by using a more systematic mathematical approach, separating the cost function formulation from the numerical solution algorithm.

In an inverse simulation of human movement, where a recorded sequence is used as basis for the simulation, the resultant forces needed can be rather easily calculated from basic mechanical equations. Assuming a quasi-static situation at one particular time instance recorded, a static optimization can be used to give some information on the force distribution at this instant. If inertial forces are considered in the formulation of the resultants from the kinematic recordings, the static optimization can be assumed to give reasonable distributions of forces in static postures and in slow movements, [9]. This does, however, imply a rather static view of muscular force production, where the force is computed from, at best, the current length and lengthening velocity, disregarding the history in muscular force production, [10–14].

Forward simulations assume that the forces acting on the mechanical system are known. From these known forces, the resulting movement is easily calculated within the framework of the chosen mechanical model. With the forces accurately measured for a specific case, the result is thereby easily obtained. The forces can, however, be *a priori* unknown, and iteratively sought to produce a certain movement, in which case the forward simulation is conducted based on these assumed forces. The movement to match can then be known as either a recorded real movement, or as just an initial and a final configuration, with the movement between these as the answer. Depending on the setting of the problem, there can exist a number of possible movements, and an optimal one can be sought. The term 'optimal' is, however, far from obvious.

One of the key issues in this context is the numerical description of the muscular actuators in the mechanical model. The simplest view is that muscles can always deliver an arbitrarily large mechanical force immediately. With this assumption, the mechanical problem is easily solved. More realistic descriptions of muscle activity have to consider that the muscles have limited and configuration-dependent capacities, that they are not immediately recruitable, and that the force production is governed by a neural control system.

This paper discusses numerical simulation of musculoskeletal systems from two viewpoints. The first important aspect is related to the dynamic simulations of the body, or parts of it, when seen as a mechanical system. This modeling is in itself not very complicated, although the human anatomical geometry is complex. The approximation of the skeleton as rigid links, and the muscles as localized force actuators is seemingly straight-forward. The more complicated part is the evaluation of optimal movement strategies, as the criteria for optimality are not obvious, and, as pointed out in e.g. [15, 16], probably not universal for all situations: common and well-practiced movements under limited loadings are probably optimized in another way than the more extreme situations. An interesting question is related to the similarities and differences obtained with different criteria for optimality, giving a basis for discussions on neural control. Systematic simulations can also show the effects on movement patterns from kinematical and strength restrictions in the search for optimality. Although the results can be questioned as the answer to the sensorimotor control of human movement, [17], they can give important information concerning the envelope of possible movements. It must also be noted that the simulation, at best, can reproduce a planned motion, but completely lacks consideration of the feed-back error-control mechanisms, essential for real human movement, [18].

The second aspect of musculoskeletal movement is related to the general and detailed numerical representation of muscle behavior. Although physiological knowledge about muscle action is extensive, it is not always immediately suited for the numerical simulations. To be useful in this context, the muscular model must describe the relation between force and muscle lengths for arbitrary movement histories. The effects from an interaction between the muscular forces and the partly externally induced movements are only to some degree understood and implemented in numerical models. These effects are of critical importance for the results, in particular for fast impact situations.

The paper is presented as follows. First, a discussion is given on the treatment of redundant force systems, where the choice of alternative force patterns can be based on some optimality criterion: both static and dynamic viewpoints are considered. Thereafter, a few aspects of the numerical modeling of muscles are considered. The focus is set on aspects of muscle physiology which are important for general simulations of musculoskeletal systems, and where relevant physiological experiments are still scarce. A brief numerical example gives a few aspects of human movement simulations.

# 2 Redundant systems

Most of the systems found in nature and in man-made structures are redundant: a specified load may be equilibrated in an infinite number of ways. Indeed, most structures allow different equilibrium paths for any loading, but it is noted that a chosen modeling can affect the redundancy situation. Often a numerical model is a simplification of reality which does not allow all movements, and thereby leads to a seemingly unique solution. The opposite is, however, often true for models of the musculo-skeletal system, when small differences in paths of action for muscles disappear when a simplified model is created. Various formulations have been used for the treatment of redundant force systems, [3].

The following will discuss optimization aspects of redundant system in two sections. First, situations where redundancy affects the static equilibrium will be described. Thereafter, dynamic situations will be considered, where the definition of a motion from just its initial and final states will create a dynamic form of redundancy. In both forms, references will be given to human movement problems.

## 2.1 Static optimization

### 2.1.1 Basic variables

The configuration of a general mechanical system is given by a set of $N_d$ discrete coordinates:

$$\boldsymbol{q} = \begin{pmatrix} q_1 & q_2 & \dots & q_{N_d} \end{pmatrix}^{\mathrm{T}} \tag{1}$$

In a musculoskeletal context, these often represent orientation angles of limb segments, or joint flexion angles. The applied loads acting on the system are defined by a set of force components, attached to the displacement coordinates:

$$\boldsymbol{p} = \begin{pmatrix} p_1 & p_2 & \dots & p_{N_d} \end{pmatrix}^{\mathrm{T}} \tag{2}$$

Typically, $\boldsymbol{p}$ contains both gravity loads, $\boldsymbol{p}_g$, and external loads, $\boldsymbol{p}_e$, both of which may be dependent on the current configuration of the system, $\boldsymbol{p} = \boldsymbol{p}(\boldsymbol{q})$. Many engineering problems also see the load as parameterized by a set of parameters, [19, 20]:

$$\boldsymbol{p} = \boldsymbol{p}(\boldsymbol{\lambda}) \tag{3}$$

The loads $\boldsymbol{p}$ are carried by a set of $N_m$ unknown internal forces, $\boldsymbol{t}$, corresponding to nominal stresses:

$$\boldsymbol{t} = \begin{pmatrix} t_1 & t_2 & \dots & t_{N_m} \end{pmatrix}^{\mathrm{T}} = \mathbf{A}\boldsymbol{\sigma} \tag{4}$$

where $\mathbf{A}$ is a diagonal matrix with cross-sectional areas of the system members, and the stresses are

$$\boldsymbol{\sigma} = \begin{pmatrix} \sigma_1 & \sigma_2 & \dots & \sigma_{N_m} \end{pmatrix}^{\mathrm{T}} \tag{5}$$

An internal force may be limited as:

$$t_i^- \leq t_i \leq t_i^+ \tag{6}$$

where $t_i^- = 0$ when considering only tensile forces. The force capacity of the member $t_i^+$ would in a musculo-skeletal context typically correspond to the maximum isometric muscular force, but could be refined to reflect a non-optimal muscle fibre length in a current configuration.

The above quantities allow formulation of a set of $N_d$ static equilibrium equations, which for fixed load parameters are written:

$$\mathbf{E}_t \boldsymbol{t} = \boldsymbol{p} \tag{7}$$

where the $N_d \times N_m$ equilibrium matrix $\mathbf{E}_t = \mathbf{E}_t(\boldsymbol{q})$ reflects the projections of the internal forces on the equilibrium equations for a given configuration $\boldsymbol{q}$. For a musculoskeletal problem, the matrix $\mathbf{E}_t$ often contains the moment arms of the muscles at the joints where moment equilibrium is formulated, but other force or moment projections are also possible, depending on model and chosen equations. It is noted that the equilibrium equations need be conjugated to the chosen displacement coordinates, but the internal forces can be of other types, just needing the correct projection in $\mathbf{E}_t$. For instance, when using stresses $\boldsymbol{\sigma}$ as unknowns, Equation (4), the equilibrium is, analogously, written as:

$$\mathbf{E}_\sigma \boldsymbol{\sigma} = \boldsymbol{p} \qquad \text{with} \qquad \mathbf{E}_\sigma = \mathbf{E}_t \mathbf{A} \tag{8}$$

Also other, often completely equivalent, forms have been proposed in literature.

### 2.1.2 Static equilibrium solution for redundant systems

For a redundant system with $N_d \leq N_m$, the equilibrium solution can be found from the minimization of a cost function:

$$\Pi = \Pi_t + \boldsymbol{\mu}^{\mathrm{T}}(\boldsymbol{p} - \mathbf{E}_t \boldsymbol{t}) \tag{9}$$

where $\boldsymbol{\mu}$ contains a set of Lagrange multipliers enforcing the equilibrium. Potentials based on the sum of squared forces, $\Pi_t = \frac{1}{2}\boldsymbol{t}^{\mathrm{T}}\boldsymbol{t}$, or stresses, $\Pi_t = \frac{1}{2}\boldsymbol{\sigma}^{\mathrm{T}}\boldsymbol{\sigma}$, are often used in redundant bio-mechanical problems [8, 21].

Based on the expression in Equation (9), an algorithm is discussed in [8] for the evaluation of optimal forces in a redundant static force system. The algorithm is developed as a function for Matlab (The MathWorks, Natick, MA, USA), and considers that the distribution of load on a redundant set of internal forces can be dependent on loading level. The formulation allows the equilibrium matrix to be dependent on the coordinates used, $\mathbf{E}_t = \mathbf{E}_t(\boldsymbol{q})$. Such a dependence can also be easily introduced into the force capacities $t_i^+ = t_i^+(\boldsymbol{q})$.

Allowing general cost functions, the minimization must be performed by demanding the differentials of Equation (9) to vanish, $\Pi_{,\boldsymbol{t}} = \mathbf{0}$, $\Pi_{,\boldsymbol{\mu}} = \mathbf{0}$, where subscripts following commas denote differentials. The system becomes non-linear, and is of the form:

$$\boldsymbol{g}(\boldsymbol{z}) = \begin{pmatrix} \Pi_{t,\boldsymbol{t}} + \mathbf{E}_t^T \boldsymbol{\mu} \\ \boldsymbol{p} - \mathbf{E}_t \boldsymbol{t} \end{pmatrix} = \mathbf{0} \quad , \quad \boldsymbol{z} = \begin{pmatrix} \boldsymbol{t}_q \\ \boldsymbol{\mu} \end{pmatrix} \tag{10}$$

for a chosen configuration $\boldsymbol{q}$, and if muscular forces are seen as unknowns. This can be solved by a Newton-type iteration method, [22]; the needed differentials are easily obtained.

Limits in force values introduce equality constraints for the variables, and lead to a formulation:

$$\mathbf{E}_{tb}\, \boldsymbol{t}_q = \begin{bmatrix} \mathbf{E}_t \\ \mathbf{E}_b \end{bmatrix} \boldsymbol{t}_q = \begin{pmatrix} \boldsymbol{p} \\ \boldsymbol{t}_b \end{pmatrix} = \boldsymbol{t}_{tb} \tag{11}$$

where the limits are $t_i = t_i^+$ or $t_i = t_i^-$, depending on context, and are described by $\mathbf{E}_b \boldsymbol{t}_q = \boldsymbol{t}_b$. The system can be solved by the same methods as discussed above, as long as the side conditions are of limited number, [8], and for loads below a certain magnitude. This non-linearity makes the solution be affected by the load level. The algorithm therefore includes a strategy for successive introduction of constraints, moving the solution along the bounding hypersurfaces of the feasible region in the search for the optimal solution.

When sufficiently many components have reached their capacities, the feasible region does not contain any solutions, and the system has passed its capacity. This is utilized in a step-wise solution, where the external forces are one-parametric, Equation (3), as:

$$\boldsymbol{p} = \lambda \cdot \boldsymbol{p}_{,\lambda} \tag{12}$$

and the muscular forces needed for equilibrium are variable, $\boldsymbol{t}_q = \boldsymbol{t}_q(\lambda)$. For $\boldsymbol{q}$ fixed, the algorithm solves $\boldsymbol{t}_q(\lambda)$ for increasing factors $\lambda \geq 0$, and introduces new constraints by $(\mathbf{E}_b, \boldsymbol{t}_b)$, until no equilibrium can be found, above a load multiplier $\lambda_{\max}$, [8].

## 2.2 Dynamic optimization

The dynamic optimization is here seen as related to a forward simulation, where *a priori* unknown forces are introduced as controls in the mechanical system. The simulation aims at an optimization of a whole movement sequence.

### 2.2.1 Basic quantities and equilibrium

The kinematic description of motion, in a dynamic simulation of human movement, assumes that the configuration of the studied system at time $t$ can be described by a set of $N_d$ coordinates $q_i(t)$, cf. Equation (1):

$$\boldsymbol{q}(t) = \begin{pmatrix} q_1(t) & q_2(t) & \ldots & q_{N_d}(t) \end{pmatrix}^{\mathrm{T}} \tag{13}$$

where the coordinates can be chosen freely, as long as they uniquely define any configuration within the assumptions stated for the movement.

For the kinetic description, gravity and configuration-independent applied forces are defined as $\boldsymbol{p}(t)$. The algorithm also considers a set of unknown control forces, $c_i(t)$ $(1 \leq i \leq N_c)$. At time $t$, the acting control forces are collected as:

$$\boldsymbol{c}(t) = \begin{pmatrix} c_1(t) & c_2(t) & \ldots & c_{N_c}(t) \end{pmatrix}^{\mathrm{T}} \tag{14}$$

Including inertia and configuration-dependent forces, the dynamic problem can be seen as governed by $N_e \equiv N_d$ equilibrium equations. Allowing very general formulations, these are for a specific time instance $t$ collected as, [23]:

$$\mathbf{M}\boldsymbol{a}(t) + \boldsymbol{f}(\boldsymbol{q}(t), \boldsymbol{v}(t)) - \boldsymbol{p}(t) - \mathbf{E}_c\boldsymbol{c}(t) = \mathbf{0} \tag{15}$$

using the mass matrix $\mathbf{M}$ to form inertia forces from accelerations $\boldsymbol{a}(t)$, and introducing velocities $\boldsymbol{v}(t) = \dot{q}(t)$, where a superposed dot represents time differentials. The vector $\boldsymbol{f}$ describes all internal forces and configuration-dependent loads. The effects of control forces $\boldsymbol{c}(t)$ are described by an action description matrix $\mathbf{E}_c$, of size $N_d \times N_c$. With these assumptions, a time instance residual form is written:

$$\boldsymbol{e}(t) \equiv \boldsymbol{e}(\boldsymbol{q}(t), \boldsymbol{v}(t), \boldsymbol{a}(t), \boldsymbol{p}(t), \boldsymbol{c}(t); t) = \mathbf{0} \tag{16}$$

This equation must be formally derived for a specific mechanical system, using any method which is general enough for the problem context, [24].

The explicit derivation of the instantaneous equilibrium equations is not necessary. When discretizing the movement into a number of time nodes, the necessary interpolation can be introduced in different ways in relation to the Lagrangian functional, found from the kinematic description through potential and kinetic energies. The different forms will tend to emphasize different interpolations, although they are formally equivalent. With the above formulation of the instantaneous equilibrium equations from the Euler-Lagrange equations, the algorithm developed has used the discretization and interpolation below. As, however, the choices of functional form, discretization and optimization criterion will affect the results obtained in a complex inter-relation, these choices have led to further studies.

### 2.2.2 Time of movement

When studying a human movement, the timing of the movement is of major importance for the relation between the resulting movement and the forces actuating it. In particular, this is true when quick movements are desired, close to the capacity of the musculo-skeletal system under study. In the algorithm developed, a pre-defined time interval $0 \leq t \leq T$ is considered, but a generalization of the present form where the time is seen as a free variable is possible, *cf.* [25].

The time interval studied is represented by the elements between $N_t + 1$ equally spaced time nodes:

$$t^j = j \cdot \frac{T}{N_t} \tag{17}$$

for $(0 \leq j \leq N_t)$. These time nodes are used for the kinematic description, and discretize the displacement coordinates.

### 2.2.3 Description of motion

In a discretization of a movement, a locally defined Hermitian third-order approximation was used to interpolate all kinematic coordinates. Defining a time interval as a temporal finite element, both the coordinates and the corresponding time differentials were chosen as nodal quantities. A configuration is thereby described by a vector of double size:

$$\boldsymbol{Q}(t) = \begin{pmatrix} q_1(t) & \dot{q}_1(t) & q_2(t) & \dots & \dot{q}_{N_d}(t) \end{pmatrix}^{\mathrm{T}} \tag{18}$$

The initial state is specified by the configuration $\boldsymbol{Q}(0)$: position and velocities. For a target movement, which is a movement from one pre-defined configuration to another, a final configuration can be specified by $\boldsymbol{Q}(\theta)$, where $\theta$ is often equal to the considered time $T$. A target configuration does not necessarily need to specify all displacement and velocity components.

At each discrete time node, the configuration is thus specified by the degrees of freedom:

$$\boldsymbol{Q}^j = \boldsymbol{Q}(t^j) \tag{19}$$

The whole movement is described by the collection of all the time nodes:

$$\boldsymbol{Q} = \left[\boldsymbol{Q}^{0\,\mathrm{T}}, \boldsymbol{Q}^{1\,\mathrm{T}}, \boldsymbol{Q}^{2\,\mathrm{T}}, \dots, \boldsymbol{Q}^{N_t\,\mathrm{T}}\right]^{\mathrm{T}} \tag{20}$$

which is a vector of length $2\,N_d(N_t+1)$.

The Hermitian interpolation of each kinematic variable gives $C^1$ continuity of the coordinates over element borders, according to, [26]:

$$q_i(t) = N_1(t)q_i^j + N_2(t)\dot{q}_i^j + N_3(t)q_i^{j+1} + N_4(t)\dot{q}_i^{j+1} \qquad \text{for} \qquad t^j \leq t \leq t^{j+1} \quad (21)$$

When collected for all coordinates, the interpolation can formally be seen as:

$$\boldsymbol{q}(t) = \mathbb{N}\boldsymbol{Q} \quad (22)$$

with a very sparse matrix $\mathbb{N}$ of size $N_d \times 2N_d(N_t+1)$.

From this interpolation, velocity, acceleration and jerk components are consistently obtained as:

$$\boldsymbol{v}(t) \equiv \dot{\boldsymbol{q}}(t) = \dot{\mathbb{N}}\,\boldsymbol{Q} \quad (23)$$

$$\boldsymbol{a}(t) \equiv \ddot{\boldsymbol{q}}(t) = \ddot{\mathbb{N}}\,\boldsymbol{Q} \quad (24)$$

$$\boldsymbol{j}(t) \equiv \dddot{\boldsymbol{q}}(t) = \dddot{\mathbb{N}}\,\boldsymbol{Q} \quad (25)$$

where the latter are constants within each interval, with the interpolation presented.

### 2.2.4 Description of control forces

The external forces are assumed to be known at every time instance $t$, as are the expressions for configuration-dependent loads as parts of $\boldsymbol{f}$. In the present algorithm, the acting control forces are discretized at $N_k$ time nodes $\tau^j$:

$$\boldsymbol{C}^j = \boldsymbol{c}(\tau^j) \quad (26)$$

with $0 \leq \tau^j \leq T$, and $1 \leq j \leq N_k$. A linear interpolation was chosen for the control forces, according to:

$$\boldsymbol{c}(t) = \mathbb{N}_c\boldsymbol{C} \quad (27)$$

where the matrix $\mathbb{N}_c$ is $N_c \times N_c N_k$, but very sparse: at most two values in each row are non-zero. This discretization was chosen as compatible with the kinematic description described above, but other forms are fully possible, and can be more natural when the simulation model is based on other fundamental expressions. The discretizations were chosen with the aim of having coinciding time nodes for both sets, but experience has shown that the control force nodes can be farther apart, but should preferably coincide with a subset of the displacement time nodes. Shorter time intervals in the discretization of the controls than for the coordinates lead to an ill-conditioned problem, when used with the overall formulation and optimization discussed below, but is allowable with other formulations.

With chosen time nodes for the controls, the whole set of unknown control force components is collected as:

$$\boldsymbol{C} = \left[\boldsymbol{C}^{1\,\mathrm{T}}, \boldsymbol{C}^{2\,\mathrm{T}}, \ldots, \boldsymbol{C}^{N_k\,\mathrm{T}}\right]^{\mathrm{T}} \quad (28)$$

The representation of the muscular controls through discrete values is considered as the least accurate part of the algorithm presented here, as it implicitly assumes that the forces can be infinitely quickly regulated. A more relevant representation would need to use an indirect formulation, where either the neural stimulation or the muscular activation of each muscle at each time node are seen as unknown variables. The control force $i$ at time $\tau^j$ is then seen as:

$$c_i^j = c_i^j(\xi_0^{\tau_j}) \tag{29}$$

with $\xi_0^{\tau_j}$ representing the history of a stimulation $\xi_i(t)$ for the time interval $0 \leq t < \tau_j$. The required modeling is further discussed below.

### 2.2.5 Algebraic equations

The main idea of the proposed method for solution of optimal movements between initial and target states is the establishment of a full set of equations for immediate solution of both displacement variables $\boldsymbol{Q}$ and controls $\boldsymbol{C}$. For this, the time instance form of dynamic equilibrium given by Equation (16) is used in the algorithm developed. A point collocation with two points in each time interval is used to establish an algebraic set of $2N_dN_t$ equations:

$$\boldsymbol{E}(\boldsymbol{Q}, \boldsymbol{C}, \boldsymbol{P}) = 0 \tag{30}$$

with $\boldsymbol{P}$ a consistent representation of the external forces $\boldsymbol{p}(t)$ for all collocation points. Based on the same equations, other algebraic representations can also be used, [27,28], or the transcription methods discussed in [29].

A Galerkin-type weak representation can also be used for establishing the set of equations needed for the solution of the full problem. When integrating over the time elements, another basic formulation than the time instance equilibrium equations is, however, preferable, as this can allow more general forms of interpolations for both displacements and controls.

### 2.2.6 Boundary values and restrictions

In addition to the equilibrium equations, a set of $N_b$ linear boundary conditions on the discrete coordinates are introduced by:

$$\boldsymbol{B}(\boldsymbol{Q}) \equiv \mathrm{B}_Q\,\boldsymbol{Q} - \boldsymbol{b}_Q = 0 \tag{31}$$

At least $2N_d$ conditions are needed to define an initial state at $t = 0$. Excessive boundary conditions define a target state, and imply the need for at least $N_cN_k = N_b - 2N_d$ free control force components.

The element based interpolation of displacement variables allows an introduction of zero accelerations in the variables at one or both ends of the time interval considered.

This will have rather small effects on the movement calculated, but will ensure that the controls are in static equilibrium for the posture defined by the coordinate values.

During the dynamic process, movement is mechanically or physiologically restricted. Kinematic restrictions can often be seen as linear inequalities in the coordinates. Similarly, the controls might be kinetically restricted. Typically, either type of restriction is valid in all time nodes, and the restrictions are thereby numerous. The restrictions can be seen as:

$$\mathbf{B}_z \boldsymbol{z} \leq \mathbf{0} \qquad \text{where} \qquad \boldsymbol{z} = \begin{pmatrix} \boldsymbol{Q} \\ \boldsymbol{C} \end{pmatrix} \tag{32}$$

### 2.2.7 Optimality of movement

With excessive control force components available, an optimal solution can be sought. Optimality can, in general, be seen in many different ways, but the focus here is on criteria related to the performance of the motion, *i.e.*, a minimum of used forces, a maximum of some movement aspect, *etc.* This optimality is thereby of another type than criteria expressing a best-fit between a simulated and a measured movement, [27,28].

The present formulation uses a 'cost' or 'performance' function, for which the minimum is sought. The criteria discussed are all based on the variables in $\boldsymbol{Q}$ and $\boldsymbol{C}$, although other indirect measures could well be introduced. One class of cost functions measure the control forces. An integrated sum of squared control forces over the considered time interval gives a cost function:

$$\Pi_{cc} \equiv \frac{1}{2} \int_0^{\mathrm{T}} \left( \sum_i (c_i(t))^2 \right) dt = \sum_i \sum_j \Pi_i^j = \boldsymbol{C}^{\mathrm{T}} \mathbb{C}_{cc} \boldsymbol{C} \tag{33}$$

with simple form for the matrix $\mathbb{C}_c$, utilizing that the control variations are systematically interpolated in the formulation, [23]. Similarly, it is also straight-forward to introduce a similar measure for the variations of the controls, $\Pi_{ct}$ with another matrix $\mathbb{C}_{ct}$, [30]; this criterion emphasizes a smoothness in forces used.

Other cost functions are related to the displacements. Seeking the smoothest possible movement, a jerkiness 'cost' can be formulated, based on the idea in [31]. An expression for the integrated sum of squared jerk components:

$$\Pi_{qj} \equiv \frac{1}{2} \int_0^{\mathrm{T}} \left( \sum_i (j_i(t))^2 \right) dt = \boldsymbol{Q}^{\mathrm{T}} \mathbb{C}_{qj} \boldsymbol{Q} \tag{34}$$

can be obtained through the temporal interpolation, Equation (25), giving a simple matrix $\mathbb{C}_{qj}$ from the time element length $\Delta t = \frac{T}{N_t}$, [32]. Similarly, costs for accelerations, $\Pi_{qa}$, and velocities, $\Pi_{qv}$, can be defined through similar matrices $\mathbb{C}_{qa}$, $\mathbb{C}_{qv}$.

The algorithm allows arbitrary combinations of the above cost terms, *i.e.*:

$$\Pi(\boldsymbol{z}) = \boldsymbol{Q}^{\mathrm{T}}(\alpha_{qj}\mathbb{C}_{qj} + \alpha_{qa}\mathbb{C}_{qa} + \alpha_{qv}\mathbb{C}_{qv})\boldsymbol{Q} + \boldsymbol{C}^{\mathrm{T}}(\alpha_{cc}\mathbb{C}_{cc} + \alpha_{ct}\mathbb{C}_{ct})\boldsymbol{C} \tag{35}$$

Different combinations of the $\alpha$ parameters are easily introduced into the algorithm, yielding different results; these can be compared to experimentally documented movement strategies. It has been noted that all choices can not handle statically redundant controls.

### 2.2.8 Optimization algorithm

Based on the formulations above, the studied constrained optimization problem can be classified as, [33]:

$$\begin{aligned} \text{mimimize} \quad & \Pi_z \equiv \Pi(\boldsymbol{z}) \\ \text{under equality constraints} \quad & \boldsymbol{b}_1(\boldsymbol{z}) = 0 \\ \text{and inequalities} \quad & \boldsymbol{b}_2(\boldsymbol{z}) \geq 0 \end{aligned} \tag{36}$$

where the unknown $\boldsymbol{z}$, as in Equation (32), contains both the displacements and the controls, and the cost function is based on these variables, Equation (35).

The equality constraints are of two types: a non-linear part corresponding to the set of equilibrium equations, and a linear part representing the boundary conditions, *cf.* Equations (30) and (31):

$$\begin{pmatrix} \boldsymbol{E}(\boldsymbol{z}, \boldsymbol{P}) \\ \mathrm{B}'_Q\, \boldsymbol{z} - \boldsymbol{b}_Q \end{pmatrix} = \boldsymbol{0} \tag{37}$$

with $\mathrm{B}'_Q$ a representation of $\mathrm{B}_Q$ from Equation (31).

Restrictions are linear, as in Equation (32), but can also be non-linear, representing, *e.g.*, contact conditions:

$$\boldsymbol{h}(\boldsymbol{z}) \leq 0 \tag{38}$$

which needs to be formulated in relation to a specific problem formulation.

Several optimization algorithms have been used for the problem stated. Initially, an in-house optimization function based on sequential linear programming was developed, [23, 34]. More general, although also more computationally demanding, approaches have also been tested. One is the function 'fmincon' included in the 'Optimization toolbox' of Matlab, which offers a suitable format for solving the stated problem; it uses a sequential quadratic programming approach. An estimate of the Hessian of the Lagrangian is updated in each iteration using a BFGS method, [33].

An even more robust, but also more time-consuming, algorithm is the function 'optnlin' included in the software 'Comsol Script' (version Aug 2006, Comsol AB, Stockholm, Sweden). This non-linear optimization function is based on the 'SNOPT' package (version 7.2–1), [35]. The examples presented in this paper were evaluated with the latter algorithm, with very good convergence properties.

Work is presently in progress to use an algorithm based on a Method of Moving Asymptotes ('MMA'), [36, 37], in particular its 'globally convergent' version. As this algorithm in itself does not consider equality constraints, the general problem needs some re-writing. This can be done through a part-inversion of the equilibrium equations, giving only controls $\boldsymbol{C}$ as design variables, and displacements $\boldsymbol{Q} = \boldsymbol{Q}(\boldsymbol{C})$.

Another possibility is to replace the equality constraints with $\varepsilon$-bounded inequalities. This potentially robust and efficient approach is currently studied.

### 2.2.9 Algorithmic implementation

The above sections provide the basis for a general algorithm for the solution of a targeted dynamic problem, including control forces to be optimized, and a choice of optimization criterion. In general, the basis of the problem formulation is the time instance equilibrium, Equation (15). From this, by introducing the interpolation for the involved quantities within the relevant time element, and differentiating with respect to the discrete variables, the operators are readily obtainable by symbolic manipulation with, *e.g.*, Mathematica, (Wolfram Research, Champaign, IL., USA). In the present implementation of the algorithm, problem specific procedures deliver the needed parts of the stated system and its differential matrices to the main algorithm, as functions of the current approximation to the solution. The size of the problems, normally considering thousands of variables, and often even higher numbers of equality and inequality constraints, are important problem aspects when choosing optimization algorithm. The formulation is, however, extremely sparse for large problems, a fact which can guide significant improvements in efficiency, [38].

# 3 Muscle modelling

Extensive knowledge about skeletal muscles is available in the physiological literature. Most experimental studies have focussed on the static, or at least step-wise static, force production at either isometric or isotonic conditions, or possibly the transition between two such states. The results are thereby suitable for static equilibrium analyses, where the muscular description only needs maximum values of forces and elongations of the muscles.

In order to represent dynamic situations, commonly used equations come from Hill, [39,40], and Fung [41]. As also discussed by Fung, the so-called Hill model has been modified over the years, and the term now seems to denote any muscular model, where rheological components are used to represent the behavior. The models aim at a description of the relation between force in a muscle and its length and shortening or stretch rate. Set in a displacement-based form, this means that the muscular force:

$$F(t) = F(l, \dot{l}, q; t) \tag{39}$$

is a function of current length $l$, its time differential $t$, and an activation measure $q$ at time $t$. Most Hill models also require that the length of the contractile part of the muscle is tracked over time. It is well established that a muscle acts very differently in different regions of its span of lengths, with maximum capability for force production under fully activated isometric conditions at an 'optimal length'. There is a significant difference in behavior between shorter ('ascending limb', sub-optimal) and longer ('descending limb', super-optimal) lengths. In simulations, the length and differential are dependent on the surrounding compliance of the whole system. With respect to the length ratio, eccentric contractions, during which the muscle is stretched while activated, are particularly interesting, as they are typical of landing and other impact situations, [10, 42], with maximal muscular forces produced. The force production in the muscles counteracting the externally induced stretching is time-dependent, and very important in the resulting movement pattern, [10].

One frequently noted shortcoming of the Hill models is that they do not connect the activation level and the length–velocity values correctly in common situations of movement, [43]. From the simulation viewpoint, however, the most severe shortcoming is the lack of memory. It is in [11] pointed out that the frequently adopted Hill-type models show an incapability to 'account for what is called memory-dependent phenomena, such as force depression following shortening or force enhancement following stretch'. The assumption that once a stretching is completed, the force in the model is described completely by the isometric force-length curve is strongly contradicted by experiments, [44]. A special instance of this is when activation is discontinued, [12].

The history effects in muscular force production are not completely understood, although studies of the macroscopic muscular force production have been performed, *e.g.*, [14,45,46] focussing on the force depression following active muscle shortening. The studies have proposed descriptive measures for the modifications of force production, but neither numerical models nor phenomenological descriptions are finally established.

The cause of force enhancement is largely unknown, but has been attributed to passive structures in the muscles, [47]. The idea comes from an observation that the passive force after deactivation is increased by active stretching [48], and verified with experiments on cat soleus muscles, [44]. It is pointed out that muscle lengths beyond the optimal one give significantly higher enhancement of forces. In addition, the passive forces in the muscle are non-zero only for super-optimal lengths, at which they grow rapidly for increasing length; they are believed to be damaging to the muscle, [49].

The force variation, important in daily walking as well as in athletic jumping, is from the application viewpoint seen as a result of a 'stretch shortening cycle', SSC, [50, 51], which is seen and utilized but not fully understood. Having shown that the isometric force in the muscle is increased by a preceding stretch, and decreased by a shortening, the question is what happens after an SSC?

In order to introduce a simple form of memory in a muscle, [11] develops a model, where to the common components of most muscle models is added an 'elastic rack': a parallel elastic element, which is engaged upon muscle activation. The force from the contractile element is chosen as the tetanized force at the isometric length during activation, without rise phase, and not considering the length change after the activation. The length change thereafter, positive or negative, not only has an effect on the series of damper and spring, but also on the parallel elastic rack.

A very important aspect of the muscular modeling in dynamic simulations is the activation dynamics. In brief, a neural stimulation, often denoted $u$, causes an activation of the muscle, $q$, which can be seen as related to the $Ca^{2+}$ level of the muscle. This in turn can be seen as a measure for how large a force the muscle can produce. The activation $q$, typically normalized as $(0 \leq q \leq 1)$, is governed by the stimulation $u$ as a first order differential equation $\dot{q} = \dot{q}(q, u)$. The delays caused by this indirect activation of muscular force is a fundamental aspect of the musculo-skeletal modeling.

One of the more sophisticated muscular models today, with a large number of parameters is described by Günther *et al.* [52]. This model handles one of the shortcoming of most Hill models, and connects the activation level and the length–velocity values, [43]. It does not, however, include any memory. This aspect is described by Figure 1, which qualitatively shows the effects from the contraction memory. Subfigure 1.a shows the force produced by a stimulated mouse muscle under different length regimes, similar to the experiments reported in [46]. The cases correspond to an isometric contraction, where the length is held constant during stimulation, a concentric contraction where a shortening takes place during stimulation, an eccentric contraction where the muscle is instead stretched, and a stretch-shortening cycle. All the length variations are ending with an isometric state at the optimal length of the muscle.

Subfigure 1.b shows simulation results obtained with the muscular model from [52]. The model is adopted in the experimental setting by an introduction of just five out of a total of 23 parameters: the optimal length of contractile component $l_{\text{CE,opt}} = 0.011\,\text{m}$, the maximum isometric force $F_{\text{max}} = 0.22\,\text{N}$, the force-free length of the series elastic component $l_{\text{SEE,0}} = 0.002\,\text{m}$, the transition force for the series elastic component $\Delta F_{\text{SEE,0}} = 2 \cdot F_{\text{max}}$, and the activation time constant $\tau_q = 0.05\,\text{s}$; the rest of the parameters and a force-dependent serial damping were chosen as in the reference. A full optimization of the model parameters could certainly give better correlation between simulations and experiments, but the main conclusion from the figure is obvious: the history aspects are needed in a reasonably accurate transient model for muscular force production.

The above-mentioned description of history from [11] seems to give a reasonably good basis for the creation of a phenomenologically based muscle model with some of the important memory aspects represented. The inclusion of such a memory model together with the muscle model from [52] in the temporal description is currently studied.

# 4 Numerical example

Only two aspects of one example are shown here. Further aspects of the same problem, and other human movement problems are presented in [3, 30, 34, 53]. These examples show the dependence of results on the discretizations of controls and displacements, the effects from different choices of cost function, and the differences in movement patterns when restrictions of different severities are introduced.

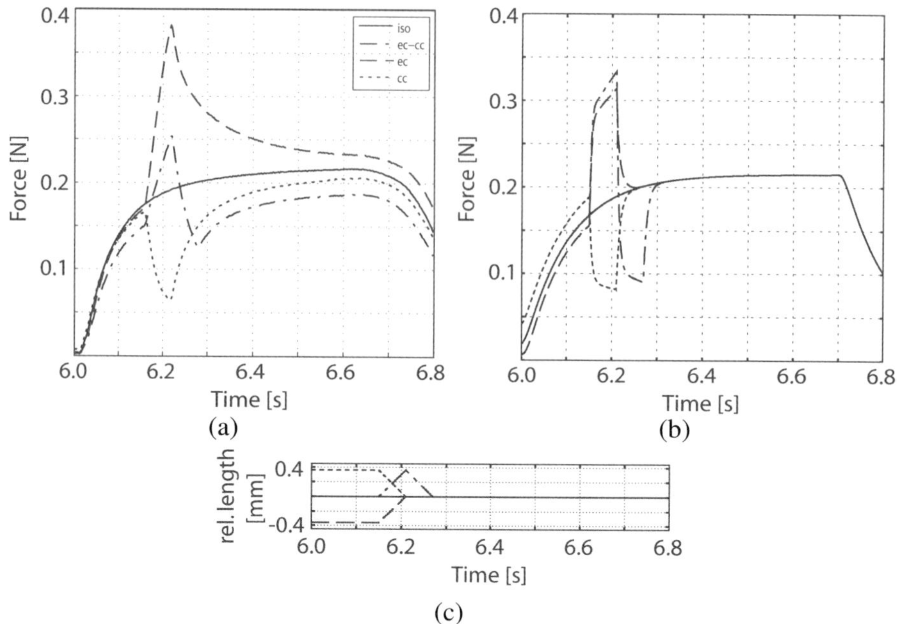


Figure 1: Comparison of experimental and simulated force production in mouse muscle under isometric, concentric, eccentric and stretch-shortening contractions. a) are experimental time-variations; b) are results simulated with a muscular model from [52], with basic data adopted to experimental data, but all dimension-less data kept as in the reference; c) is the length variation, measured as a difference to the muscle optimal length

## 4.1 Problem definition

A 'clean-and-jerk' lifting problem, shown in Figure 2 was studied as a schematic sagittal model, with the use of five coordinates and five control moments; dynamical equilibrium was formulated from kinetic and potential energies of body segments and barbell. Masses were chosen as $m_1 = 6.8\,\text{kg}$, $m_2 = 14.8\,\text{kg}$, $m_3 = 36.4\,\text{kg}$, $m_4 = 4.2\,\text{kg}$, $m_5 = 3.2\,\text{kg}$, $m_h = 7.0\,\text{kg}$, $g = 10\,\text{m/s}^2$, and lengths as $L_1 = L_2 = 0.47\,\text{m}$, $L_3 = 0.54\,\text{m}$, $L_4 = 0.35\,\text{m}$, $L_5 = 0.42\,\text{m}$, $L_h = 0.40\,\text{m}$; all segments were seen as prismatic (head and trunk as one unit). Joint angles were restricted, according to $40° \leq q_1 \leq 100°$, $0° \leq q_2 - q_1 \leq 160°$, $-135° \leq q_3 - q_2 \leq 15°$, $-225° \leq q_4 - q_3 \leq 0°$, $0° \leq q_5 - q_4 \leq 170°$. Initial and final positions were estimated from a set of photos, where a weight-lifter lifts a rather low weight. Accelerations were set to zero at the final state, representing a stable posture, but were free at the beginning. Non-linear inequalities at all time nodes restricted the barbell from going below $0.2\,\text{m}$ above the ankle joint, and closer than $0.10\,\text{m}$ in front of the most anterior part of the leg-trunk segment center line at the same height. The discretizations were chosen as $N_t = T/0.05\,\text{s}$, and $N_k = N_t/2 + 1$. The initial guess to the solution was

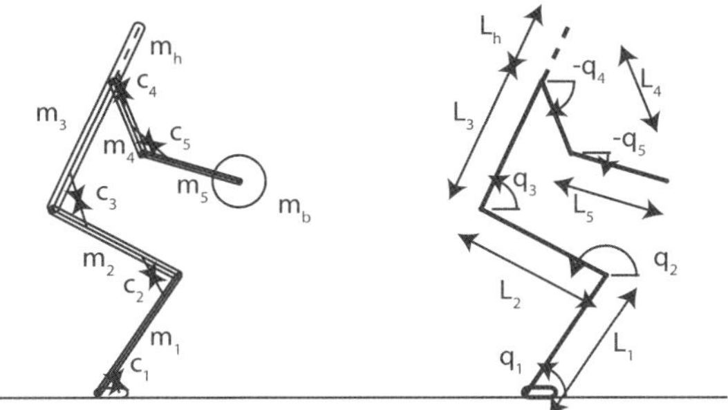


Figure 2: Complex lifting problem, with five displacement coordinates $q_i$ and five controls $c_i$, being joint moments (positive for flexion)

a linear interpolation of displacements between initial and final values, together with zero controls.

Simulations were optimized as $\alpha_{cc} = 1$, or $\alpha_{ct} = 1$. The effects from chosen fixed time intervals $T = 1.0, 1.5, \ldots, 3.5$ s were particularly studied, as was the relation between joint resultant capacities and weight lifted.

## 4.2 Results

Among the obtained results, Figure 3 shows the obtained angular variations over the time interval for the two optimization criteria, with a barbell mass of 50 kg, $T = 2.5$ s, and unrestricted joint moments. The figure demonstrates two features of the obtained solutions, which are also verified by other simulations. First, a movement obtained by minimizing any aspect of the control forces, will often show a pendulum type of motion with rapidly varying forces, when the time $T$ is long enough; in this case primarily the elbow joint will be 'pumping' the load upwards. Second, when an unnecessarily long time is specified, the movement will rest during some part of the interval; the resting will take place in a configuration with low forces needed.

Simulation times for this problem, with the discretization $N_t = 50$, $N_k = 26$, were around 10 minutes on an ordinary PC, even when a rather bad initial guess to the solution was introduced.

Muscular capacity restrictions are obviously important in weight-lifting. Simulations were performed with the above model for different combinations of capacity restrictions for the joint resultant moments, combined with different barbell masses. Here, only one example will be given. The solution in Figure 4 refers to a case with a barbell mass of $m_b = 75$ kg, and the restrictions on joint resultants: $-600 \leq c_1, c_2, c_3 \leq 600$ Nm, $-400 \leq c_4, c_5 \leq 400$ Nm; these restrictions were rather arbitrarily chosen, but based on values given in [54]. The time was here set as $T = 3.5$ s. The figure shows the calculated time variations of the joint angles, but also a stick-figure interpretation of the optimal movement for this case, when the moment capacities have started to affect the lifting movement, although not too severely. It is obvious that the

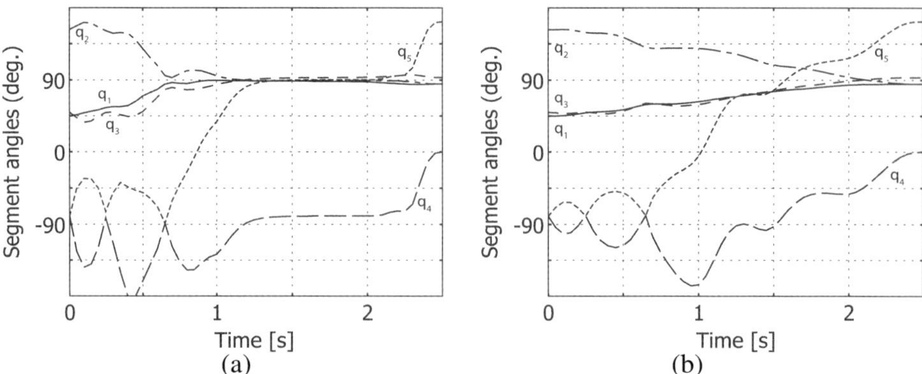


Figure 3: Simulations of 'clean-and-jerk' weightlifting, 50 kg, with $T = 2.5$ s prescribed. Displacement solutions obtained with (a) minimized joint moments ($\alpha_{cc} = 1$); (b) minimized joint moment time gradients ($\alpha_{ct} = 1$)

limits in forces give a very different movement pattern, with a much closer interaction of the different body parts. The initially very low body position, while arms are starting to move, and the final rising of the body fits reasonably well with real lifting movements.

# 5 Concluding remarks

The method proposed is attractive for the numerical simulation of optimal mechanism trajectories between two or several configurations, solving simultaneously the movement path and the forces required to produce it. In addition to a simplified problem specification, the consistence in time-integration allows general software components for solution of the optimization problem, with higher degrees of generality and robustness than more problem-specific forms. It also avoids the demanding and error-prone time integrations of the dynamical formulation, central in shooting type dynamic simulations. In particular, the efficiency and reliability aspects become troublesome when shooting forms are used for the numerical differentiation of several quantities with respect to a large number of unknown variables, *e.g.*, the discretized force variations for the involved controls. This robustness, however, is by necessity more demanding for simple cases, where the solution is to a high degree prescribed or predictable. This might for instance be the case when tracking optimization is considered, seeking a best-fit with a previous registration of movement, [27, 28, 54].

The experiments shown here, and simulations previously reported, show that the algorithm is useful for reasonably realistic problems of human movement. The simulations have in most cases converged from a rather crude initial guess to the solution, but occasionally one restart was necessary. No cases lead to non-global minima, when using the general optimization algorithm. The experiment also verified that the opti-

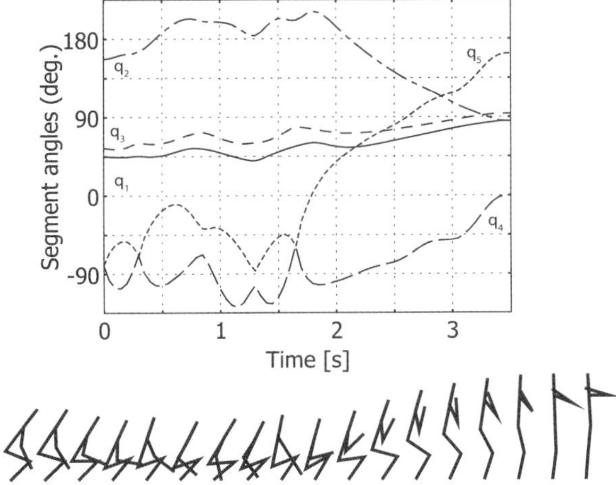


Figure 4: Evaluated optimal clean-and-jerk weightlifting, with minimized control force rates, and restricted joint moment resultants. Time-variation of joint angles, and interpretation of optimal movement obtained

mization algorithm used, with its good global convergence properties, is computationally rather expensive, even for good initial approximations to the minimal solution. Preliminary tests indicated that the computation time was essentially proportional to the square of the number of unknowns in $\boldsymbol{z}$. The generality of the formulation made the setup of the different cases very simple, once the dynamic equilibrium equations of the problem were derived. The simulation time was reasonable.

Numerical experiments verify that the optimality in the movement can be formulated through criteria based on either the displacements themselves, or the controls creating them, the latter being either individual or resultant forces. The choice of optimization criterion affects the solution obtained. It is for instance noted that solutions with minimized joint resultant moments are rather different from solutions with minimized individual muscle tensions, [53]. The differences between solutions are normally reduced with more severe restrictions. In experiments, the general conclusion has been that criteria minimizing controls are more numerically robust, but tend to give physiologically less relevant solutions, showing rapid variations of controls. Somewhat smoother solutions are obtained with a criterion based on the control gradients. Earlier experiments have also shown that there is a rather high similarity between solutions obtained for minimization of control gradients, jerks or accelerations, whereas the minimum control solution is more clearly distinguished. The five cost functions discussed can be used together, in arbitrary combinations. As the variables introduced are often of different magnitudes, the scaling of the $\alpha$ coefficients must always be considered, if more than one non-zero $\alpha$ is introduced. It must further be considered that a criterion based on control gradients is not well-defined for all mechanically redundant systems. Although the control gradients and the movement come out as identical, the values of antagonistic muscles could be varied, with the same cost. For redundant cases, it might therefore be necessary always to have at least a small coefficient $\alpha_{cc}$ present in the cost function. This fact must be noted when following recommendations in References [55, 56].

From experiments performed, it is verified that a more refined discretization will normally always leads to lower costs. Occasionally, a very coarse discretization lead to a slightly lower cost than the finer ones, but this was always verified to be an artifact, following from the fact that kinematic restrictions are not enforced within the intervals. If the discretization of controls is finer than the one used for displacements, the collocation procedure together with a minimization of the force cost $\Pi_{cc}$, Equation (33), will converge towards an un-physiological solution with short force impulses at the collocation points, and zero forces between these. The experiments thereby indicate that the displacements and controls should be discretized with either the same time intervals, or the control intervals double those of the displacements. This observation emphasizes the aspect mentioned above: the basic formulation, the discretization and interpolation, and the optimization criterion are interconnected, and must be chosen together.

The results show that the optimal cost functions always increase with the severity of the restrictions, in particular for restrictions in control capacity. The interpretation is that more severe kinetic restrictions will demand other, more expensive, movement strategies.

The results emphasize that with a long time defined for a movement, optimal solutions with respect to force magnitude will tend to use a 'pumping' or pendulum type of movement, when the forces can be infinitely quickly regulated. The experiment above also shows that the fixed-time definition is not a major problem, with respect to the solution obtained. With a time specified, which is longer than the optimal, the simulated time interval is partially used for resting at a configuration of small forces, causing unnecessarily extensive calculations, but not affecting results. This is, however, not necessarily true for all problem settings.

As the algorithm establishes very large systems of equations, computational efficiency is essential. At present, only rather limited efforts have been directed towards these aspects. More elaborate methods for the extremely sparse formulations are a main remaining challenge, [38]. The analytical differentials following from the consistent interpolation of variables, however, lead to significant reductions in computational demands.

The results are from a bio-mechanical viewpoint closely related to the numerical model stated for the problem. The present setting allows very general descriptions of the mechanism used to describe the human movement, but can only be seen as a motion planning tool, as it includes neither the neural activation of muscle actuators, nor the feed-back system regulating the movement in real-time. For an improved description of real movements, the descriptions have some obvious shortcomings, as they are not closely enough related to physiological behavior. A better representation of the history dependence in muscular force production is necessary for the simulations to be more predictive for real human movements. Activation variables have been used in several of the studies mentioned above, but only for limited settings. A general, analytical formulation — consistent with the idea of efficient global convergence of the optimization — demands further work. Similar efforts would also make possible the inclusion of other non-linear effects in the controls, *e.g.*, muscle forces or moment arms dependent on current configuration.

# References


[1] M. Stelzer, O. von Stryk, *Efficient forward dynamics simulation and optimization of human body dynamics*, ZAMM, 86, 828–840, 2006.

[2] B. A. Garner, M. G. Pandy, *Musculoskeletal model of the upper limb based on the visible human male dataset*, Comp. Meth. Biomech. Biomed. Eng., 4, 93–126, 2001.

[3] A. Eriksson, A. G. Tibert, *Redundant and force-differentiated systems in engineering and nature*, Comput. Methods Appl. Mech. Engrg., 195, 5437–5453, 2006.

[4] D. A. Hawkins, M. L. Hull, *A computer-simulation of muscle tendon mechanics*, Comp. Biol. Med., 21, 369–382, 1991.

[5] J. Duysens, H. W. A. A. Van de Crommert, B. C. M. Smits-Engelsman, F. C. T. Van der Helm, *A walking robot called human: lessons to be learnt from neural control of locomotion*, J. Biomech., 35, 447–453, 2002.

[6] P. Binding, A. Jinha, W. Herzog, *Analytic analysis of the force sharing among synergistic muscles in one- and two-degree-of-freedom models*, J. Biomech., 33, 1423–1432, 2000.

[7] F. C. Anderson, M. G. Pandy, *Static and dynamic optimization solutions for gait are practically equivalent*, J. Biomech., 34, 153–161, 2001.

[8] S. Heintz, E. Gutierrez-Farewik, A. Eriksson, *Evaluation of load-sharing and load capacity in force-limited muscle systems*, Comp. Meth. Biomech. Biomed. Eng. (submitted).

[9] S. Heintz, E. M. Gutierrez-Farewik, *Static optimization of muscle forces during gait in comparison to EMG-to-force processing approach*, Gait and Posture, 26, 279–288, 2007.

[10] G. K. Cole, A. J. van den Bogert, W. Herzog, K. G. M. Gerritsen, *Modelling of force production in skeletal muscle undergoing stretch*, J. Biomech., 29, 1091–1104, 1996.

[11] M. Forcinito, M. Epstein, W. Herzog, *Can a rheological muscle model predict force depression/enhancement?*, J. Biomech., 31, 1093–1099, 1998.

[12] D. L. Morgan, N. P. Whitehead, A. K. Wise, J. E. Gregory, U. Proske, *Tension changes in the cat soleus muscle following slow stretch or shortening of the contracting muscle*, J. Physiol – London, 522, 503–513, 2000.

[13] G. J. C. Ettema, K. Meijer, *Muscle contraction history: Modified Hill versus an exponential decay model*, Biol. Cyber., 83, 491–500, 2000.

[14] S. R. Bullimore, T. R. Leonard, D. E. Rassier, W. Herzog, *History-dependence*

*of isometric muscle force: Effect of prior stretch or shortening amplitude*, J. Biomech., 40, 1518–1524, 2007.

[15] N. Hogan, T. Flash, *Moving gracefully: Quantitative theories of motor coordination*, Trends in Neurosciences, 10, 170–174, 1987.

[16] T. Phanindra, S. Majumdar, *Selection of the cost function for determination of muscle forces*, in: B. H. V. Topping, C. A. Mota Soares (Eds.), Proc. 7th Int. Conf. Comp. Struct. Techn., Civil-Comp Press, Stirling, U.K., 2004.

[17] E. Todorov, *Optimality principles in sensorimotor control*, Nature Neurosci., 7, 907–915, 2004.

[18] E. Todorov, M. I. Jordan, *Optimal feedback control as a theory of motor coordination*, Nature Neurosci., 5, 1226–1235, 2002.

[19] A. Eriksson, *Structural instability analyses based on generalised path-following*, Comput. Methods Appl. Mech. Engrg., 156, 45–74, 1998.

[20] A. Eriksson, C. Pacoste, *Solution surfaces and generalised paths in non-linear structural mechanics*, Int. J. Struct. Stab. Dyn., 1, 1–30, 2001.

[21] J. Dul, M. A. Townsend, R. Shiavi, G. E. Johnson, *Muscular synergism — I. On criteria for load sharing between synergistic muscles*, J. Biomech., 17, 663–673, 1984.

[22] E. Mijangos, *An implementation of Newton-like methods on nonlinearly constrained networks*, Comp. Op. Res., 31, 181–199, 2004.

[23] A. Eriksson, *Analysis methodology based on temporal FEM for bio-mechanical simulations*, in: M. Ursino, C. A. Brebbia, G. Pontrelli, E. Magosso (Eds.), Modelling in Medicine and Biology VI, WIT Press, Southampton, 2005.

[24] M. G. Calkin, *Lagrangian and Hamiltonian Mechanics*, World Scientific, 1996.

[25] M. G. Pandy, F. C. Anderson, D. G. Hull, *A parameter optimization approach for the optimal control of large-scale musculoskeletal systems*, J. Biomech. Eng., 114, 450–460, 1992.

[26] R. D. Cook, D. S. Malkus, M. E. Plesha, R. J. Witt, *Concepts and Applications of Finite Element Analysis*, 4th edition, Wiley, 2002.

[27] M. L. Kaplan, J. H. Heegaard, *Predictive algorithms for neuromuscular control of human locomotion*, J. Biomech., 34, 1077–1083, 2001.

[28] M. L. Kaplan, J. H. Heegaard, *Second-order optimal control algorithm for complex systems*, Int. J. Numer. Meth. Eng., 53, 2043–2060, 2002.

[29] J. T. Betts, *Survey of numerical methods for trajectory optimization*, J. Guidance, Control and Dynamics, 21, 193–207, 1998.

[30] M. Kaphle, A. Eriksson, *Optimality in forward dynamics simulations*, J. Biomech., 41, 1213–1221, 2008.

[31] T. Flash, N. Hogan, *The coordination of arm movements: an experimentally confirmed mathematical model*, J. Neurosci., 5, 1688–1703, 1985.

[32] A. Eriksson, *Optimization for targeted movements*, in: C. A. Mota Soares (Ed.), Proc. III Eur. Conf. Comp. Mech.: Solids, structures and coupled problems in engineering, Lissabon, 2006.

[33] J. Nocedal, S. J. Wright, *Numerical Optimization*, Springer, 1999.

[34] A. Eriksson, *Temporal finite elements for target control dynamics of mechanisms*,

Comp. Struct., 85, 1399–1408, 2007.

[35] P. E. Gill, W. Murray, M. A. Saunders, User's guide for SNOPT version 7: software for large-scale nonlinear programming, Tech. rep., Stanford, CA., USA (2006).

[36] K. Svanberg, *Method of moving asymptotes — a new method for structural optimization*, Int. J. Numer. Meth. Eng., 24, 359–373, 1987.

[37] K. Svanberg, *A class of globally convergent optimization methods based on conservative convex separable approximations*, SIAM J. on Optimization, 12, 555–573, 2002.

[38] R. Beauwens, *Iterative solution methods*, Appl. Num. Math., 51, 437–450, 2004.

[39] A. Hill, *The heat of shortening and the dynamic constants of muscle*, Proc. Royal Soc. London B, 126, 136–195, 138.

[40] A. Hill, *First and Last Experiments in Muscle Mechanics*, Cambridge University Press, Cambridge, U.K., 1970.

[41] Y. C. Fung, *Biomechanics: Mechanical Properties of Living Tissues*, 2nd Edition, Springer, New York, 1993.

[42] U. Proske, J. E. Gregory, D. L. Morgan, P. Percival, N. S. Weerakkody, B. J. Canny, *Force matching errors following eccentric exercise*, Human Movement Sci., 23, 365–378, 2004.

[43] E. J. Perreault, C. J. Heckman, T. G. Sandercock, *Hill muscle model errors during movement are greatest within the physiologically relevant range of motor unit firing rates*, J. Biomech., 36, 211–218, 2003.

[44] W. Herzog, T. R. Leonard, *The role of passive structures in force enhancement of skeletal muscles following active stretch*, J. Biomech., 38, 409–415, 2005.

[45] W. Herzog, T. R. Leonard, J. Z. Wu, *The relationship between force depression following shortening and mechanical work in skeletal muscle*, J. Biomech., 33, 659–668, 2000.

[46] N. Kosterina, H. Westerblad, J. Lännergren, A. Eriksson, *Muscular force production after concentric contraction*, J. Biomech.(submitted).

[47] K. A. P. Edman, G. Elzinga, and M. I. M. Noble, *Residual force enhancement after stretch of contracting frog single muscle fibers*, J. General Physiol., 80, 769–784, 1982.

[48] W. Herzog and T. R. Leonard, *Force enhancement following stretching of skeletal muscle: a new mechanism*, J. Exp. Biol., 205, 1275–1283, 2002.

[49] N. P. Whitehead, N. S. Weerakkody, J. E. Gregory, D. L. Morgan, and U. Proske, *Changes in passive tension of muscles in humans and animals after eccentric exercise*, J. Physiol – London, 533, 593–604, 2001.

[50] G. Markovic and S. Jaric, *Scaling of muscle power to body size: The effect of stretch-shortening cycle*, Eur. J. Appl. Physiol., 95, 11–19, 2005.

[51] L. Malisoux, M. Francaux, H. Nielens, and D. Theisen, *Stretch-shortening cycle exercises: An effective training paradigm to enhance power output of human single muscle fibers*, J. Appl. Physiol., 100, 771–779, 2006.

[52] M. Günther, S. Schmitt, V. Wank, *High-frequency oscillations as a consequence of neglected serial damping in Hill-type muscle models*, Biol. Cyber., 97, 63–79,

2007.

[53] A. Eriksson, *Optimization in target movement simulations*, Comput. Methods Appl. Mech. Engrg.(submitted).

[54] K. G. M. Gerritsen, A. J. van den Bogert, M. Hulliger and R. F. Zernicke, *Intrinsic muscle properties facilitate locomotor control — a computer simulation study*, Motor control, 2, 206–220, 1998.

[55] Y. Uno, M. Kawato, R. Suzuki, *Formation and control of optimal trajectory in human multijoint arm movement. Minimum torque-change model*, Biol. Cyber., 61, 89–101, 1989.

[56] M. G. Pandy, B. A. Garner, F. C. Anderson, *Optimal control of non-ballistic muscular movements: A constraint-based performance criterion for rising from a chair*, J. Biomech. Eng., 117, 15–26, 1995.

©Saxe-Coburg Publications, 2008.
Trends in Engineering Computational Technology
B.H.V. Topping and M. Papadrakakis, (Editors)
Saxe-Coburg Publications, Stirlingshire, Scotland, 151-171.

# Chapter 8

# Computational Modelling of Reactive Porous Media in Hydrometallurgy

**C.R. Bennett[1], D. McBride[1], M. Cross[1], T.N. Croft[1] and J.E. Gebhardt[2]**
**[1] School of Engineering**
**Swansea University, United Kingdom**
**[2] Process Engineering Resources Inc**
**Salt Lake City UT, United States of America**

## Abstract

Heap leaching is becoming increasingly popular as an efficient hydrometallurgical method for the extraction of high value metals from complex ores. The process involves the interaction of a three phase transport system (a solid matrix, liquid solution and gas flow) involving suites of very complex reaction sequences, which for some systems are also closely coupled with the thermal conditions and the behaviour of key microbial populations. This paper provides an overview of the challenges associated with the development of a comprehensive computational model framework for the whole family of heap leaching processes. Validation is a key issue that is addressed, so that the models might be utilised with confidence both to characterise the process and also enable reliable predictions of future production.

**Keywords:** heap leaching, CFD modelling, variably saturated flow, reactive porous media, process modelling, model validation.

## 1 Introduction

In the race to recover the world's base metal resources as efficiently as possible, an increasingly popular method is based on the hydrometallurgical process of heap or stockpile leaching [1]. This process essentially involves the mining of ore through typical open pit mining methods. Mined ore, in the form of piles of run-of-mine ore or sometimes crushed and agglomerated ore, is placed on a "heap" or leach pad that is sprayed or exposed to a chemical solution. The solution dissolves the valuable metals as it percolates through the heap and is collected at the bottom by a geo-membrane lining placed under the heap prior to the construction of the heap pad. The metal-containing or pregnant solution is then processed to concentrate and remove the metal, and finally, to produce a metal product that might go to a refinery. For more details on the heap process, Bartlett [1] provides a good overview on the various aspects of solution mining. These processes mine millions of tonnes of ore to recover metals which are measured in less than 1% of the mined volumes and in the case of gold, the output is measured in parts per million!

The process essentially involves creating the conditions within the heap, so that the liquid solution (containing reactants) is able to contact the ore and enable the leach reactions to occur. The leaching process is extremely complex, especially for sulphide ores, and is a fine balance of a range of physical and chemical interactions involving microbial populations as catalysts. Hence, the leaching process might involve some or all of the following:

a) liquid solution flow down through the porous heap – this flow may well vary from unsaturated to fully saturated flow and this physics has to be captured carefully
b) air may be injected through the base and will essentially flow counter-current to the solution, providing an oxygen source to enable the reaction process
c) the ore may well contain pyrite, which is exothermic as it reacts and affects the heat balance of the system, and
d) microbial populations which reside on the surface of the solid complex and catalyse the leaching reactions.

The overall heap leach process typically works by the building of heaps in layers that are 8-15 m deep and the laying of solution application lines along the surface. The optimisation of these processes depends upon the chemical and physical characteristics of the ore, and although heap leaching is a fairly tolerant process, its optimisation over time is really not straightforward. Ironically, it is the recent requirements of accountants to enable a more discriminating and accurate assessment of the metal values both within the ore matrix and in solution within the heap that have provided a motivation to develop and use advanced 'physics' based models as the basis for tracking assets.

Given both the technical and the financial imperatives to capture the behaviour of the heap leaching process, it is not surprising that in the last few years there has been a substantial effort at the development of computational models that are sufficiently comprehensive to be exploited in optimising the industrial family of such operations. There have been a number of efforts at heap leach modelling and many of these were summarised by Dixon in 2003 [2]. Since then Dixon and co-workers [3-7] have published a number of papers on the development and exploitation of their HEAPSIM model on a variety of ores. Whilst this work attempts to capture the bio-chemistry as accurately as possible, the work of Leahy *et al.* [8-11] exploiting the CFD technology CFX, has focussed upon the thermo-fluid transport issues. The objective of this paper is to provide an overview of the work by the authors and their colleagues on the computational modelling of both sulphide and oxide ore complexes [12-16] arising from a sustained research and technology development programme over the last decade.

# 2 Computational Model formulation

The approach to the computational modelling described here takes a similar perspective to Leahy *et al.* [8-11] by tracking the gas and liquid flow through the exploitation of CFD transport software technology, but also focussing upon the details of the chemical reactions in a manner reminiscent of Dixon *et al.* [3-7]. The CFD technology employed here is the PHYSICA code [17], which employs a finite volume formulation of the transport equations on an unstructured mesh within three dimensional geometries, and provides an open environment for the inclusion of a range of process specific physical phenomena. From the perspective of the host CFD technology, the key challenge was to develop and embed a robust algorithm for capturing the variably saturated flow of the liquid solution.

## 2.1 Variably Saturated Liquid Flow

Heaps under leach are subject to an application of a solution of reactants and occasional rain events. Heaps internally may be made up of different ores leading to considerable heterogeneity, resulting in widely varying flow conditions and therefore levels of saturation. Other phenomena that can influence flow conditions include decrepitation of the substrate, compaction, precipitation of salts from the liquid phase and transport of fines (*e.g.* clay). Compaction can to a certain extent be modeled through appropriate use of heterogeneity. Decrepitation would only be an issue where large proportions of the solid phase are soluble, which is not usually typical for these kinds of problems. Precipitation is easily modelled and fines can be dealt with using appropriate rules.

Flow through variably saturated porous media is typically characterised by the Richards equation. Modelling this type of flow has presented something of a challenge to the computational community. Saturated and unsaturated flow have individually been well described but modelling systems containing both saturated and unsaturated regions offer considerable problems. Various versions of the classical Richards equation have been proposed in the past and used to provide the basis of specific numerical solution procedures. A transformed method, proposed by Pan & Wierenga [18] has been shown to have the potential to be a fast, numerically robust scheme. A computational procedure based on an extension of the method of Pan & Wierenga into a fully unstructured context has been implemented into the PHYSICA code [19]. This model allows for variably saturated flow conditions with variable material properties and arbitrarily complex three dimensional geometries. These kinds of flow are typical of those found in heap leaching processes and must be well described in any comprehensive model of this kind of operation.

The mixed form of the Richard's equation, is written in terms of two unknown variables, moisture content ($\theta$) and the pressure head ($h$), where $K$ is the hydraulic conductivity.

$$\frac{\partial \theta}{\partial t} = \nabla K(h)\nabla h + \frac{\partial K(h)}{\partial z} \quad (1)$$

The governing flow equations are non-linear and require an iterative solution procedure. The moisture content term is separated into two parts during the iterative solution process using the relationship $\frac{\partial \theta}{\partial t} = \frac{\partial \theta}{\partial h}\frac{\partial h}{\partial t}$, as proposed by Celia *et al.* [20], where the pressure head is solved as the primary variable. To reduce the non-linearity of the problem the equation is transformed using $h_t = \frac{h}{1+\beta h}$, as proposed by Pan & Wiengara [18]. Constitutive relationships for pressure head, moisture content, and hydraulic conductivity also need to be specified. The most commonly employed models by the community and used within in this work are those of van Genuchten [21], where the moisture, $\theta$, and hydraulic conductivity, $K$, are defined by:

$$\theta = \begin{cases} \theta_{res} + \dfrac{\theta_{sat} - \theta_{res}}{\left[1 + |\alpha h|^{n}\right]^{m}}, & h < 0 \\ \theta_{sat} & h \geq 0 \end{cases} \quad (2)$$

$$K = K_{sat}\left[\frac{\theta - \theta_{res}}{\theta_{sat} - \theta_{res}}\right]^{0.5}\left[1 - \left(1 - \left[\frac{\theta - \theta_{res}}{\theta_{sat} - \theta_{res}}\right]^{1/m}\right)^{m}\right]^{2} \quad (3)$$

where $\alpha$ and n are material parameters that affect the shape of the soil hydraulic functions and m = 1-1/n.

The equation for solute transport through porous materials has been described by Bear [22]. The transport algorithm is coupled with the flow module through the moisture content ($\theta$) and Darcy flux ($q_{x_i} = -k(h)\frac{\partial (h+z)}{\partial x_i}$), where z is the elevation head. The production and consumption of species is given by the reaction module and enters the transport equation as a source term ($S$). Thus, the continuity equation for convective-dispersive transport of multiple solutes in porous media can be described as,

$$\frac{\partial(\theta C^{i})}{\partial t} - \nabla(\theta D.\nabla C^{i}) + \nabla(qC^{i}) = S^{i} \quad (4)$$

where $C^i$ is the concentration of species i in the solution phase and $S^i$ the production or consumption of species i. $D_{ij}$ is the dispersion coefficient which only becomes significant in fully saturated flows.

Having implemented the above algorithm within PHYSICA, it was evaluated on a number of test problems [19]. Two test problems involving flow into initially very dry layered soil of sand and clay are shown. Figure 1 shows the flux velocity plot after 12.5 days for unsaturated flow. Figure 2 shows the flux velocity plot after 1 day for variably saturated flow with a developing water table. The contour plots of saturation and pressure head agree well with published results. A uniform mesh consisting of 3000 elements with dimensions 4 x 3 meters was used in the simulations. Simulating this perched water table problem often results in convergence problems and requires very small time step sizes. Using a transformation parameter of $\beta = -4\ m^{-1}$ gave fast converging solutions. The CPU time was 91 seconds with an average time step size of 1000 seconds for the unsaturated case and 29 seconds with an average time step size of 882 seconds on a AMD Athlon 1600+ 1.39Ghz processor.

Examples showing the three dimensional capability of the code are described by McBride *et al.* [19]. This algorithm is at the heart of the flow transport model for the full heap leach models described here.

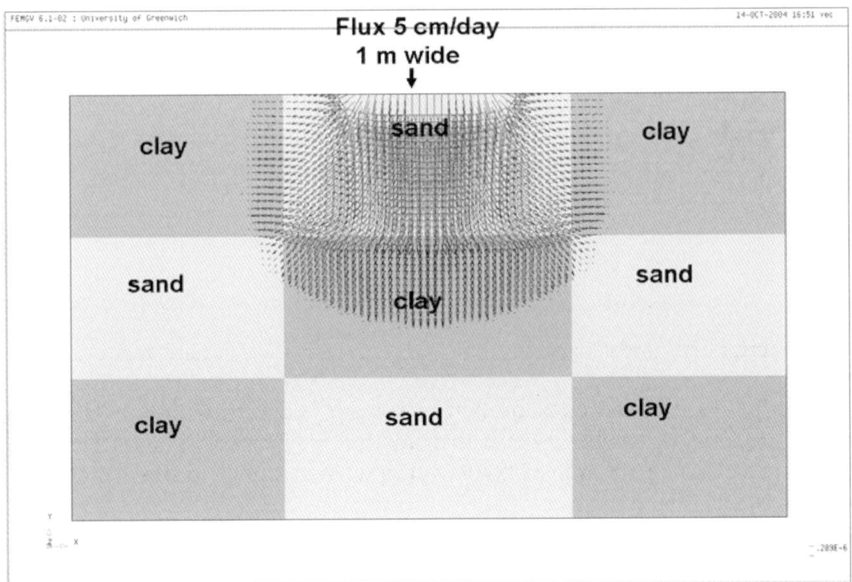


Figure 1. Velocity vector plot of unsaturated flow through heterogeneous media

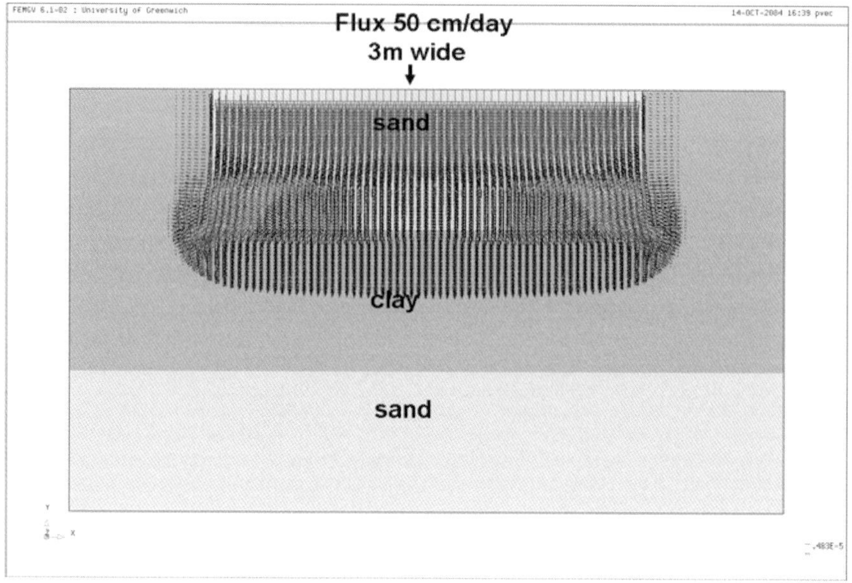


Figure 2. Velocity vector plot of perched water table problem

## 2.2 Gas phase transport

Oxygen is of huge importance in systems containing sulphide-type minerals, such as many copper-bearing ores containing pyrite. Oxygen mainly enters the heap through air injection or through air convection at exposed slopes. Air flow within the heap is subject to the same range of heterogenous flow conditions as the liquid phase but in addition must respond to liquid flow and saturation. There is a one way coupling between liquid and gas phase flow with liquid flow influencing gas flow but not the reverse.

The basic continuity equation for gas phase flow [23] is

$$\frac{\partial \rho_g}{\partial t} + div\left[\rho_g v_g\right] = S_g \tag{6}$$

where

$\rho_g$ is the gas density
$S_g$ is a source term for gas
$v_g$ is the gas velocity, given by [22]

$$v_g = -\frac{k_{in} k_g(S)}{\varepsilon_g \mu_g}\left(\nabla p^g + \rho_g g \nabla z\right) \tag{7}$$

where

$k_{in}$ is the intrinsic permeability of the porous media
$\varepsilon_g$ is the volume fraction of the gas phase

$\mu_g$ is the gas viscosity

$k_g(S)$ is the unsaturated permeability of the gas at liquid saturation (S), related to the liquid unsaturated permeability $k_l(S)$ by

$$k_g(S) = 1 - k_l(S) \tag{8}$$

Substitution of Equation (7) into Equation (6) gives the following continuity equation for the gas pressure,

$$\nabla^2 p^g = S_g \tag{9}$$

where

$$S_g = S_{thermal} + S_{vol} + S_{other} \tag{10}$$

The effects of temperature gradients on gas density can be modeled using the Boussinesq approximation [23], using

$$\rho^g = \rho_o^g (1 - \beta T) \tag{11}$$

which gives

$$S_{thermal} = \rho_o^g g \beta \frac{\partial T}{\partial z} \tag{12}$$

The volume source term $S_{vol}$ is equal to the mass of gas displaced by any change in volume in the element due to a change in liquid phase saturation.

Other source terms are due to boundary conditions, evaporation and condensation, and mass transfer between gas and liquid phases.

## 2.3 Heat Balance

Factors effecting heat balance in the heap are heat generated by chemical reactions, the heat of liquid and gas entering the heap, external temperature and solar radiation, and evaporation and condensation.

The heat balance in the heap can be solved using the temperature conservation equation:

$$\frac{\partial(\rho c_p T)}{\partial t} + div(\rho \underline{u} c_p T) = div\{K \nabla (T)\} + S_T \tag{13}$$

where $c_p$ is the specific heat,

ρ the density,
u̲ the velocity vector,
K the thermal conductivity,
T the temperature
$S_T$ captures any heat source, incorporating heats of reaction, evaporation and condensation, liquid solution and gas temperature, as well as, weather effects.

A single 'mixture' temperature is calculated for the whole system with properties dependent on average thermal properties based on the solid, liquid and gas fractions present in each element. Heat energy generated (or lost) from all of the implemented chemical reactions is then converted into a temperature rise.

The key thermal properties are density, specific heat and thermal conductivity. Density ρ is given by

$$\rho_{tot} = \sum_{i=g,l,s} VF_i \rho_i \tag{14}$$

where $VF_i$ is volume fraction of phase i.

Specific heat $c_p$ is given by

$$c_p = \sum_{i=g,l,s} \frac{m_i}{m_{tot}} c_{p_i} \tag{15}$$

where $m_i$ is the mass of phase i.

Thermal conductivity K is based on the inverse of the conductivities of the three phases.

$$K = \frac{m_{tot}}{\sum_{i=g,l,s} \frac{m_i}{K_i}} \tag{16}$$

## 2.4 Chemical Reactions

The main mineral reaction in sulphide ores is the oxidation of the sulphide by ferric ions ($Fe^{3+}$) to give a metal ion and ferrous ($Fe^{2+}$). For a metal *M* this typically takes the following form.

$$MS_{(solid)} + 2Fe^{3+} \rightarrow M^{2+}{}_{(aq)} + 2Fe^{2+} \tag{17}$$

The main reactant in the dissolution of many metal oxides is an acid

$$MO_{(solid)} + 2H^{+} \rightarrow M^{2+}{}_{(aq)} + H_2O \tag{18}$$

Gold dissolves in the presence of cyanide ($CN^-$)

$$4Au + 8CN^- + O_2 + 2H_2O = 4Au(CN)_2^- + 4OH^- \tag{19}$$

Mineral reactions are solved by dividing the ore into discrete size fractions with characteristic radii and mineral concentrations. Solid reaction kinetics are modeled using a shrinking core reaction to link pore diffusion and rate kinetics, both being heavily dependent upon the particle diameter, then solved for each mineral across a set of size fractions. The equation used to calculate the rate of dissolution of a particular mineral is given by the following equation:

$$\frac{dr_m}{dt} = -\frac{3r_m}{4\pi r_o^2}\frac{M_i}{\rho_{ore}x_i}\frac{D_{eff}c_o A_m}{\left[3D_{eff}r_o c_o + 2(r_o - r_m)r_m^2(1-\varepsilon_p)A_m\right]} \tag{20}$$

where

$r_o$ is the initial particle radius
$r_m$ is the current mineral radius
$A_m$ comes from the kinetic rate equation for the current mineral
R is the gas constant
T is the temperature in Kelvin
$D_{eff}$ is the effective rock diffusion coefficient
$\varepsilon_p$ is the rock voidage
$\rho_{ore}$ is the ore density
$M_i$ is the molecular weight of the mineral
$x_i$ is the mass fraction of the mineral

The value of $A_m$ comes from the general expression for the kinetic rate equations, such as those produced by Paul *et al.* [24], where the general form is

$$\frac{d\beta}{dt} = Ae^{\left(\frac{-B}{RT}\right)} \tag{21}$$

where

β is the fraction of mineral reacted
A and B are functions of the individual kinetic rate equation

Liquid phase reactions, including precipitation, are modelled using individual kinetics for each reaction.

## 2.5 Bacteria kinetics

In copper and pyrite systems bacteria drive the oxidation of ferrous ions to the primary reactant of ferric and also oxidise sulphur to form acid. There may be some resident population of bacteria, such as acidothiobacillus ferrooxidans, naturally occurring in the heap. In some heap leach operations, the applied solution is inoculated with certain types of bacteria.

There are numerous bacteria species which can catalyse these reactions so that it is difficult to model individual species. Instead bacteria are modelled as generic ferrous or sulphur oxidisers. Bacterial activity is also highly temperature dependent with families of bacteria broadly active at different temperatures. Mesophiles are active up to about 40°C, moderate thermophiles are most active at about 60°C and extreme thermophiles such as archea are active above 80°C. This gives 6 basic bacterial types for the model, though extreme thermophiles are unlikely to be encountered in standard heaps.

Each bacteria type can exist in the liquid phase or attached to mineral surfaces and can transfer between the two states, allowing the bacterial population to build up a healthy population and spread through the heap when the conditions are right for them. Growth and oxidation kinetics are separate for each bacterial type.

# 3 Computational model parameterisation and validation

The above general formulation has been applied in a number of contexts and so the validation discussion here will be performed with the example of copper sulphide ores. However, the validation and in particular the parameterisation of the ore specific model is non-trivial and cannot be separated from the overall task of validation.

## 3.1 Parameterisation

Particularly for sulphide ores, the parameterisation of the model is extremely challenging. There are a number of parameters that are not easy to specify. Laboratory based experiments are typically done at a number of scales – in columns with diameters and heights that vary from 15 cm to 2 m and 1.5 m to 10 m, respectively. These columns are set up to match the process conditions over the life of a lift, with regard to the particle size distribution, the packing (and so the flow resistance), the solution flow and any applied air flow rates. Hence, such experiments are frequently run from 120 to 400 days – experiments over such durations have to be carefully monitored over the entire timeframe and are neither quick nor cheap to acquire meaningful data from. Typically, the experimental procedure captures the ambient temperature together with the concentrations of the pregnant solution flowing out from the bottom of the heap. This is analysed at regular intervals to provide a temporal profile of the pregnant solution and its constituents.

Most parameterisation exercises are carried out against data from the smaller column experiments and then the validation exercise is carried out against the larger scale column experiments. The parameters to capture the flow, notably van Genutchen, are not trivial to evaluate. Fortunately, the flow model is not very sensitive to these parameters and they are normally set to ensure that the simulated break-through time (*i.e.* the time at which flow breaks the outflow surface at the heap base after solution application at the top of the lift) matches the observed value. Setting the reaction rate parameters is entirely another matter. Although the reaction rates at a site may be well characterised experimentally, the way they combine within a specific ore matrix means that key reaction rate parameters have to be identified for each separate ore matrix. Bennett *et al.* [25] have developed an optimisation based procedure to automate this parameter estimation process for copper sulphide ores.

## 3.2 Model validation

Model validation is normally pursued against data on the larger columns which are typically run for 200 - 400 days. Figure 3 below shows the results of one such

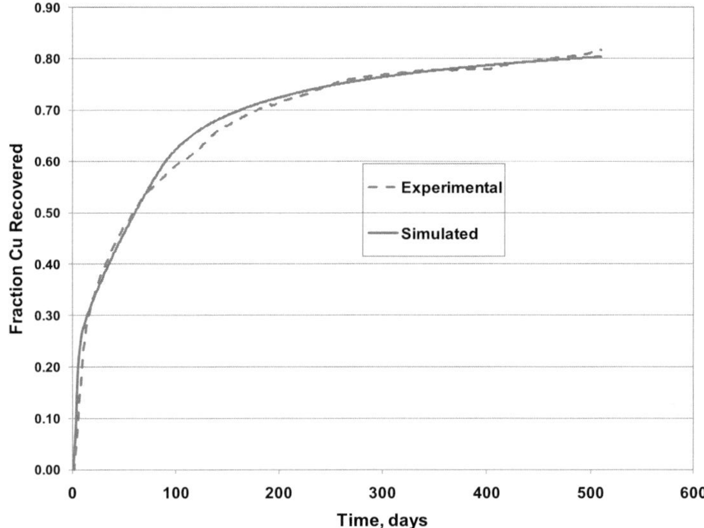


Figure 3. Fraction of copper recovered over time – experimental data vs model prediction

experiment whereby the data is simply provided for by particle size for the ore, together with flow rates etc. As is clear from the above figure, very close matches with the experimental data are generally made with the model once it has been properly parameterised.

## 3.3 The Nifty operation.

Some years ago the Nifty Copper Operation reported [26] on their results from a test heap, designated 6-4A, that contained 1.2% copper composed predominantly of chalcocite with 30% transitional ore, typically a mix of oxides, native copper, cuprite, and chalcocite. Particle crush size was reported as a P80 of 0.017 m with no other information on particle size distribution. Previous simulation work has shown that particle size distribution is a key factor in representing the recovery rate comprehensively. Hence, for the purposes of these simulations, an ore consisting of six particle size classes was used with a P100 of 0.019 m. Each particle size fraction is assumed to have the same copper and iron head grade, though experience suggests that the smaller size fractions generally tend to have slightly higher copper amounts. Nifty reports the total iron in the raffinate as 1 g/l, and this is assumed to be almost entirely ferric.

Other assumptions for data input to the model include: lift height of 4 m and fully aerated conditions for the initial calibration. Average properties for a moderate mesophilic ferric-oxidizing bacteria were used for the model. Average hydrological properties were also assumed using the particle size distribution as an indicator of typical Van Genuchten properties of 1.8 and 0.04 for n and alpha, respectively. A saturated conductivity of 1E-04 m/s was set for the material.

The Nifty Copper Operation reported a measure of bacteria in the outflow or pregnant leach solution (PLS) stream – with numbers ranging from 1 to 10 x $10^7$. An initial population was assumed for the model, and the raffinate was assumed to carry 1.0E6 bacteria/ml. The leach cycle is 180 days with solution applied at a constant rate. Results for the heap operation are shown in Figure 4 tracking the amount of copper recovery, iron concentrations, and pH of the PLS solution. Approximately 70% of the contained copper was recovered. Although the full details of the operation are not available, enough information is in the public domain to enable some meaningful analysis of the process using the suitable model [25].

There are some features in the Nifty results that deserve comment. In the initial days of the leach cycle, there is rapid recovery of copper oxide. Temperatures in the range of 35-60°C were recorded. Temperatures can be driven up by pyrite dissolution, but in this case there were other contributing factors. Significant iron is recorded in the PLS, at concentrations of up to 4 g/l in the PLS.

This suggests that there are either significant quantities of pyrite available for reaction or that there is another non-specified source or iron. Moreover, we see that initially there is little ferric in the PLS, but as the acidity increases then ferric concentrations increase, yet the PLS still has a significant ferrous iron concentration.

### 3.3.1 General heap model behaviour

The initial leach phase is dominated by fast reactions in smaller particles where reactions are dominated by chemical kinetics rather than diffusion of reactants to the minerals dispersed in the larger ore particles. There are low bacteria counts and most of the acid is consumed by the presence and leaching of copper oxide. Copper minerals are preferentially leached to pyrite. Low bacteria counts and low acid levels lead to high ferrous concentrations in the PLS.

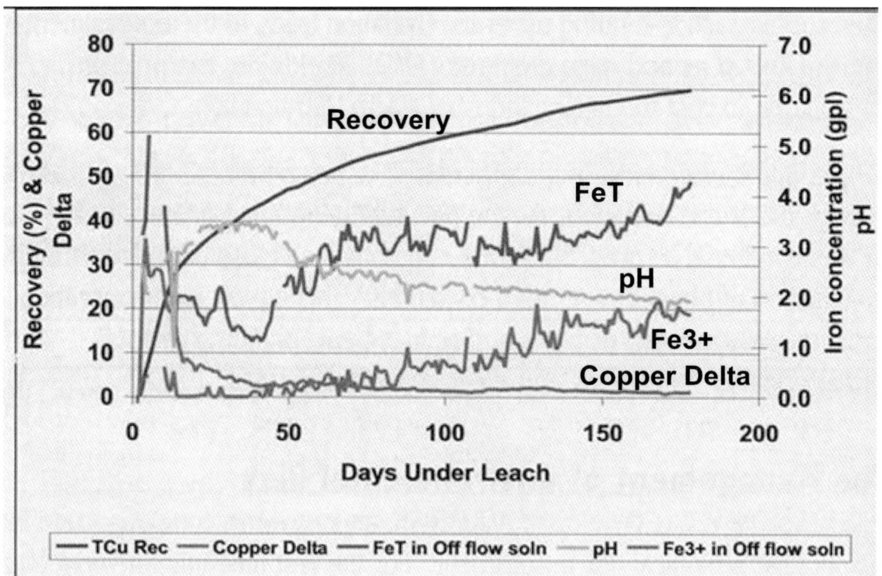


Figure 4. Heap data, as reported for the Nifty Copper Operation [26]

During the mid-phase of the process, copper dissolution (chalcocite and copper oxide) slows down with an increase in the rate of pyrite dissolution, leading to increased acidity. Maximum growth in bacteria population leads to increasing ferric levels in the PLS. As the aeration point is near the exit point for the PLS, there are high levels of oxygen available. When the rate of pyrite oxidization is highest, temperatures can reach quite high levels, 40°C in the 1D model simulations and even higher in the 2D results.

Toward the end of the leach cycle, the rate of copper and pyrite extraction slows to a near steady state dominated by diffusion in the larger particle size fractions. Temperatures are reduced, as excess heat is removed primarily through transport by the raffinate through the heap.

### 3.3.2 Model Validation - 1D Simulations

Simulation results for the model run in 1D are shown in Figures 5 - 7 indicating the predictions for species concentrations in the PLS and the overall recovery of the copper. From these results, it can be observed that although the model captures the overall recovery and the pH levels quite well, the simulation does not capture ferrous levels that agree with the Nifty values over time. That is, all the iron in the PLS reports as ferric in the simulation results. This is because the solution across the base is well aerated, and so all the ferrous converts to ferric in the PLS recorded by the model. This is reinforced by the levels of precipitation that are occurring as the process progresses. From Figures 5 and 6(b), it is clear that the iron re-dissolves back into the solution after about 130 days, but this all appears as ferric in the PLS because of the degree of aeration that the 1D simulation imposes across the base element.

It is also worth noting that:

- As the temperature of the PLS rises to just above 35 °C, the bacteria become more active, but when the major dissolution reactions have progressed then the bacteria population returns to its stable level.
- The bacterial population in the PLS compares reasonably well with that reported by Nifty
- The temperature of the PLS is at the low end of that indicated in the Nifty results
- No ferrous is reported in the PLS, for the reasons explained above.

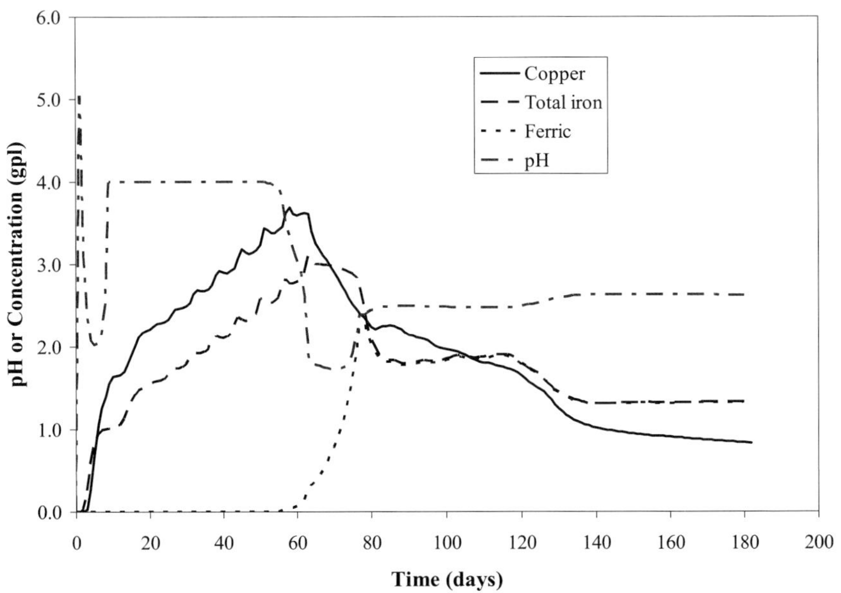


Figure 5. Simulated heap leach model results after calibration

Assuming a healthy bacterial population and sufficient acid, most of the available ferrous is likely to be converted to ferric near the base of the heap where oxygen is introduced. This must call into question the veracity of using a 1D 'column' style simulation which necessarily assumes uniform air injection across the base. A more likely scenario for the full heap is that oxygen from the air inlet is unlikely to reach all parts of the heap, resulting in areas where the solution will not be exposed to significant bacterial populations. A consequence would be the incomplete re-generation of ferrous ions and the appearance of the ferrous species in the PLS.

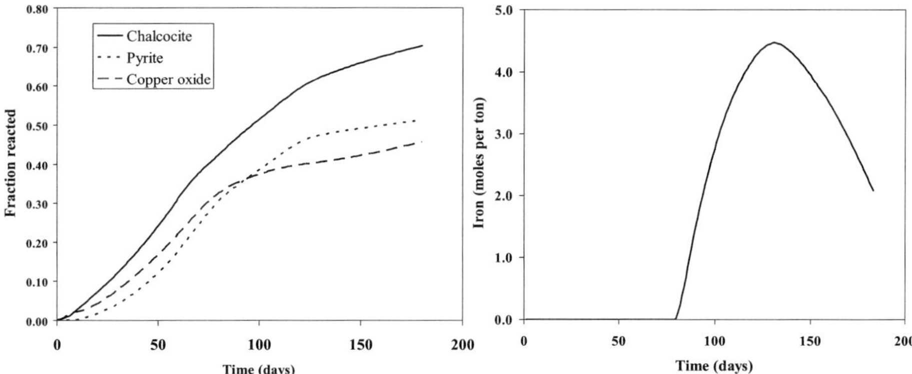


Figure 6. Model results showing (a) progress of the leach reactions and (b) the levels of iron precipitation over time

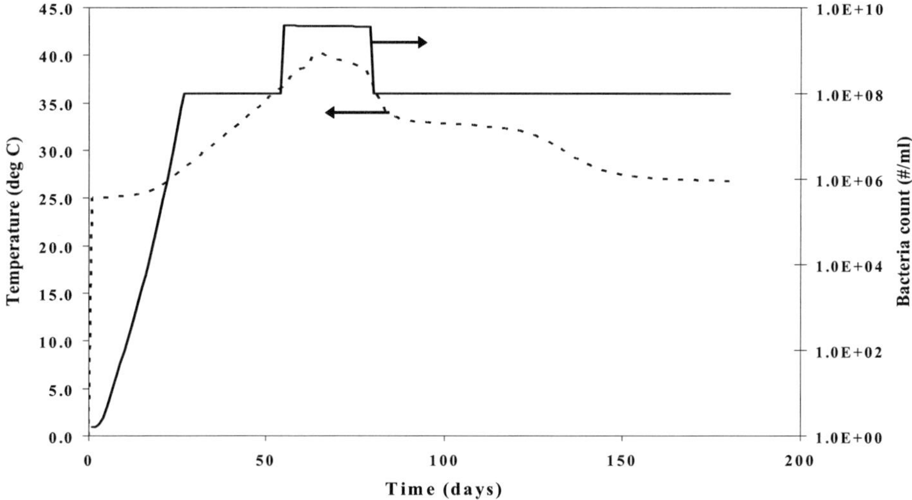


Figure 7. Model results for the temperature and bacterial population in the PLS

### 3.3.3 Model Validation - 2D Simulations

Simulation results in two dimensions for the Nifty copper heap conditions are reported in Figures 8 and 9. The lack of fit of the 1D results for the ferrous concentration in the PLS was the motivation to examine to what extent the flow variability across the heap, primarily from the air, might affect the levels of aeration and, as a result, the ferrous levels in the heap. The 2D simulation geometry was a 7-m wide by 4-m tall heap section with an air injection point in the middle at the base. All of the air was injected through this location. It is clear from the 2D model results in Figures 8 and 9 that the copper recovery and pH levels are still similar to the Nifty data, and the model now predicts:

- Ferrous species in the PLS initially, and virtually no ferric because of high pH conditions
- Ferric concentrations emerging in the PLS after about 70 days, similar to that measured at the Nifty operation
- Although not shown, a rise in temperature is observed near the center of the domain where 42°C is reached in the PLS

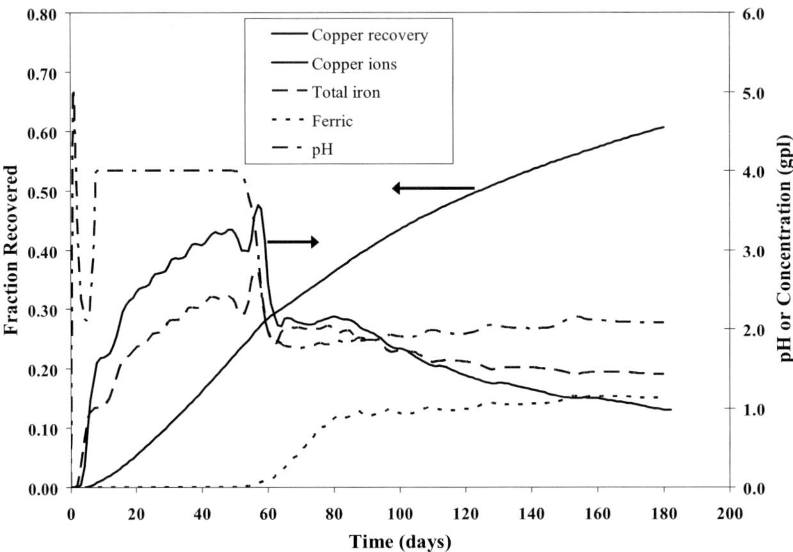


Figure 8. Total copper recovery in the PLS of the Nifty heap with the 2D simulation

In Figures 9 (a) – (d), the distributions of oxygen concentration, copper remaining in the solid, temperature, and ferric levels are shown in the heap at 80 days into the leach cycle. In these results, the significant variation in each of the process variables across the width of heap can be observed. Even when the lixiviant is supplied uniformly across the top of the heap, variations across the heap are induced by the spot air injection location. The temperature (c) at this stage appears to be highest at the base of the heap, but this is not always the case and will depend mainly on where the pyrite is reacting most aggressively. The ferric distribution shows a reaction front moving throughout the heap driving the leach reaction process.

It would require more specific process input data in order to make a more comprehensive comparison to the Nifty copper operational data. The model simulation results shown thus far capture a range of quite subtle behaviours with respect to the interactions between the macro flow, transport, and thermal interactions with the micro-scale effects at the reaction sites on and within the particles. These results provide a fairly high degree of confidence in the use of the model in further process assessment.

# 4 Computational model application

The above computational modelling technology has been adapted and used to model gold-silver-copper oxide ore in Peru. The nature of the topography of the mine *etc.* means that to capture the full effects of all the physical phenomena active in the process then a full 3D version of the model is required. In Figure 10 we show an outline of how a growing heap is captured in the model. Layers are added to the heap over time across what can be complex geographical topographies; each full layer is called a lift. Each part of the lift which is loaded and then activated at the same time is called a cell. To capture all the relevant information for simulating the growth and management of the heap requires a good deal of housekeeping that is consistent with what is delivered in practice. This tends to play a significant role in any project with mining companies.

A key application of the 3D simulation is in the planning of operations into the near future [27]. Mine planners have a number of options when trying to work out the most effective way to develop a mine over a long period. They conventionally try to develop the mining and loading plans for one to two years in advance. In so doing they are trying to balance a number of factors, including:

- the challenge of recovering the rubble ore from the mine
- the most efficient way to lay out the lift design and to break it up into the monthly cell cycles
- the efficient use of process solution and its recycle

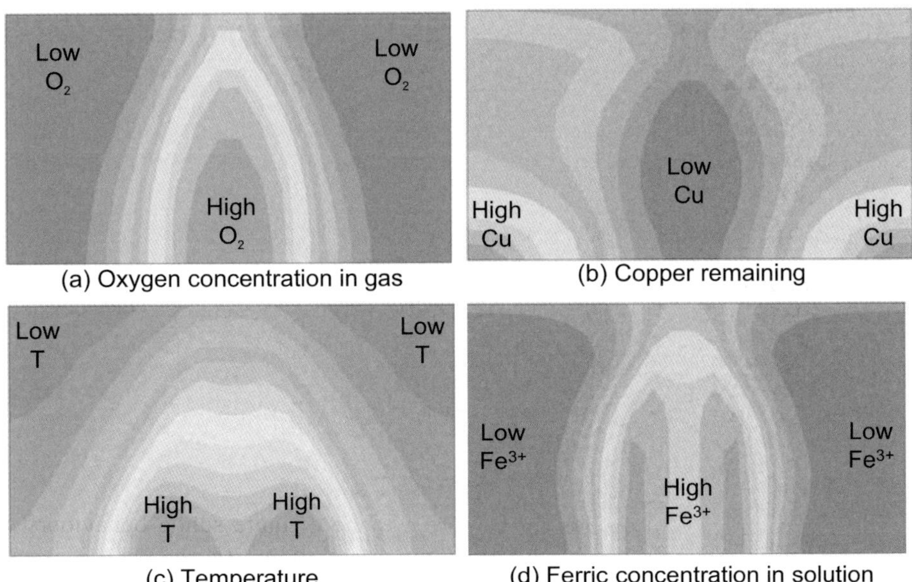


Figure 9. Distribution within the heap after 80 days from 2D model simulation

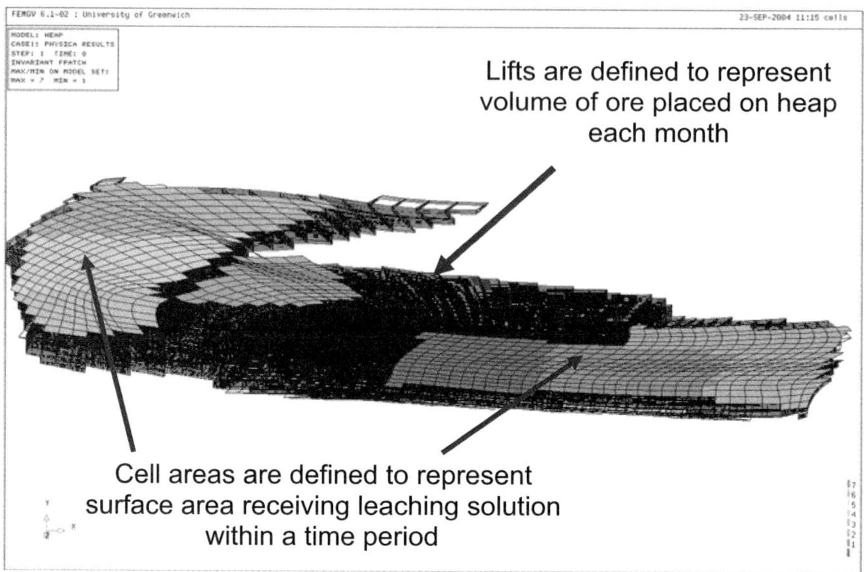


Figure 10. An illustration of how the growing three dimensional geometry is captured within the computational model

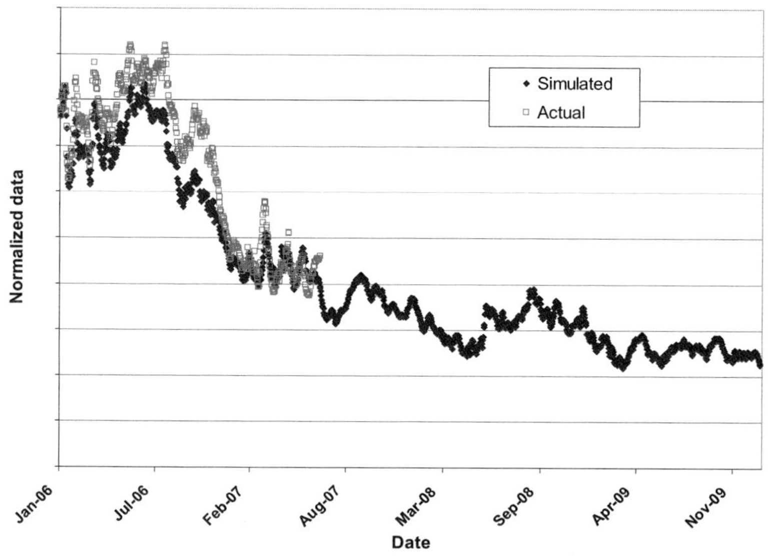


Figure 11. Example of simulated forecast of heap process.

The plant metallurgists are interested in developing the heap topography in such a way that the secondary solution passing through the lower lifts might exploit the surplus reactants and so recover the remaining ore that still resides in that part of the heap. Figure 11 shows the use of the model in planning for a couple of years from mid 2007 onwards. Note the good comparison with the recoveries from the mine for the previous 18 months. At this point the planners have decided they must recover ore from a part of the mine that contains material that has a lower ore grade and is not so easy to recover (*i.e.* less reactive). It is clear that the implication of this proposed strategy will reduce the total recovery from the mine over a couple of years. Needless to say, this work and its predictions of reduced annual production allows the company to re-visit the mining and heap building strategies to determine the most efficient way to exploit the mineable ore base. These strategies can be periodically updated in the face of changing commodity market prices.

# 5 Conclusions

Although mining of high value metals from ore using hydrometallurgical processes is extremely well established, and reasonably well understood by the relevant engineering community, it has become clear that in these days of inexorably high demand for their products, advanced computational models have a role to play. In heap leaching, the suite of processes that are involved are all rather complex and present challenges to the computational engineering community.

In this paper we have provided an overview of the core computational modelling approach, its implementation within the host numerical modelling software technology, the challenges of parameterisation and also those of validation. This means a substantial investment is required before the company is in a position to be able to exploit the arising simulation technology in the analysis and optimisation of their operation. However, since even small improvements in recovery efficiencies lead to major increases in income and reductions in environmental impact then the impact of using computational models of these processes can be quite significant – this is without the role of accountants who are now demanding independent assessments of production rates for the future and of inventories within the heaps – both within the ore complex and in the solution.

# References

[1] R.W. Bartlett, *Solution Mining*, 2nd Edition, Gordon & Breach Science Publishers, Amsterdam, The Netherlands, 1998.

[2] D.G. Dixon, "Heap Leach Modeling – The Current State of the Art", *Hydrometallurgy 2003*, C.A. Young *et al.*, Eds., The Minerals, Metals & Materials Society, Warrendale, PA, Vol. 1, 289-314, 2003.

[3] D.G Dixon, "Analysis of Heat Conservation During Copper Sulphide Heap Leaching", *Hydrometallurgy*, Vol. 58, 27-41, 2000.

[4] D.G Dixon, J. Petersen, "Comprehensive Modeling Study of Chalcocite Column and Heap Bioleaching", *Copper – Cobre 2003*, Vol. VI,

Hydrometallurgy of Copper (Book 2), P.A. Riveros *et al.*, Eds., 493-515, 2003.

[5] J. Petersen, D.G. Dixon, "Competitive Bioleaching of Pyrite and Chalcopyrite", *Hydrometallurgy*, Vol 83, 40-49, 2006

[6] S.C. Bouffard, B.F. Rivera-Vasquez, D.G. Dixon, "Leaching Kinetics and Stoichiometry of Pyrite Oxidation from a Pyrite-Marcasite Concentrate in Acid Ferric Sulfate Media", *Hydrometallurgy*, Vol 84, 225-238, 2006.

[7] J.G. Petersen, D.G. Dixon, "Modelling zinc heap bioleaching", *Hydrometallurgy*, Vol. 85, 127-143, 2007

[8] M. Leahy, M. Schwarz, M. Davidson, "An Air Sparging CFD Model for Heap Bioleaching of Copper-Sulphide", *Proceedings of the 3rd CFD Minerals and Process Industries Conf.*, M. P. Schwarz, Ed., CSIRO, Melbourne, Australia, 581-586, 2003.

[9] M. Leahy, M. Davidson, M. Schwarz, "A Model for Heap Bioleaching of Chalcocite with Heat Balance: Bacterial Temperature Dependence", *Minerals Engineering,* Vol. 18, 1239-1252, 2005.

[10] M J Leahy, P Schwarz, M Davidson, "An air sparging CFD model for heap bioleaching of chalcocite", *Appl Math Modelling*, Vol 30, 1428-1444, 2006

[11] M J Leahy, M R Davidson, M R Schwarz, "A model of heap bioleaching for chalcocite with heat balance: mesophiles and moderate thermophiles", *Hydrometallurgy*, Vol 85, 24-41, 2007

[12] C.R. Bennett, M. Cross, T.N. Croft, J.L. Uhrie, C.R. Green, J.E. Gebhardt, "A Comprehensive Copper Stockpile Leach Model: Background and Model Formulation," *Hydrometallurgy 2003,* C.A. Young *et al.*, Eds., The Metallurgical Society, Warrendale, PA, 315- 319, 2003.

[13] C.R. Bennett, M. Cross, T.N. Croft, J.L. Uhrie, C.R. Green, J.E. Gebhardt, "A Comprehensive Copper Stockpile Leach Model: Background and Model Sensitivity", *Copper 2003 – Cobre 2003*, P.A. Riveros, D. Dixon, D. Dreisinger, and J. Menacho, Eds., MetSoc, Quebec, 563-579, 2003.

[14] M. Cross, C.R. Bennett, D. McBride, J. Gebhardt, D. Taylor, "Computational Modelling of Primary Metal Ore Heap Leaching", *Minerals Engineering,* Vol. 19, 1098-1108, 2006.

[15] C.R. Bennett, D. McBride, M. Cross, J. Gebhardt, D. Taylor, "Simulation Technology to Support Primary Ore Heap Leaching", *Minerals Processing and Extractive Metallurgy*, Vol. 115, 41-48, 2006.

[16] J.E. Gebhardt, A. Hernandez, M. Cross, C.R. Bennett, D. McBride, "Analysis of Heap Leach Processes with Modeling and Simulation Tools", in *Copper 2007-Cobre 2007 The John Dutrizac International Symposium on Copper Hydrometallurgy*, ed. P.A. Riveros, D.G. Dixon, D.B. Dreisinger, M.J. Collins, CIM, Montreal, Quebec, Vol. IV (Book 1), 231-244, 2007.

[17] PHYSICA, see htpp://www.physica.co.uk

[18] L. Pan, P.J. Wierenga, "A transformed pressure head-based approach to solve Richards' equation for variably saturated soils" *Water Resources Research.*, Vol 31, 925–931, 1995

[19] D. McBride, M. Cross, T. Croft, C. Bennett, J. Gebhardt, "Computational Modelling of Variably Saturated Flow in Porous Media with Complex Three

Dimensional Geometries", *Int. J. Numer. Meth. Fluids*, Vol. 50, 1085-1117, 2006.

[20] M.A. Celia, E.T. Bouloutas, R.L. Zarba, "A general mass-conserved numerical solution for the unsaturated flow equation", *Water Resources Research*, Vol 26, 1483-1496, 1990.

[21] M.T. van Genuchten, "A closed form equation for predicting the hydraulic conductivity of unsaturated soils", *Soil Sci. Am. J.,* Vol 44, 892-898, 1980.

[22] J. Bear, *Dynamics of Fluids in Porous Media*, Elsevier, New York, 1972

[23] H.K. Versteeg, W. Malalasekera, *An Introduction To Computational Fluid Dynamics: The Finite Volume Method*, Longman Scientific & Technical, Essex, England, John Wiley & Sons, New York, NY, 257 pp., 1995

[24] B.C. Paul, H.Y. Sohn, M.K. McCarter, "Model for Ferric Sulfate Leaching of Copper Ores Containing a Variety of Sulfide Minerals: Part II. Process Modeling of In Situ Operations," *Met. Trans.,* **23B**, 549-555, 1992.

[25] C.R. Bennett, M. Cross, J.E. Gebhardt, "Process Analysis with a Multi-Physics Copper Heap Leach Model", *Hydrometallurgy 2008 Conference*, TMS/SME, August 2008.

[26] D. Readett, L. Sylwestrzak, "Bioleaching at Nifty Copper Corporation", *AusIMM Bulletin*, pp. 12-18, Dec 2002.

[27] J E Gebhardt, M Cross, "Heap Leach Simulation and Software Tools", Chapter in *Metallurgical Accounting*, in press.

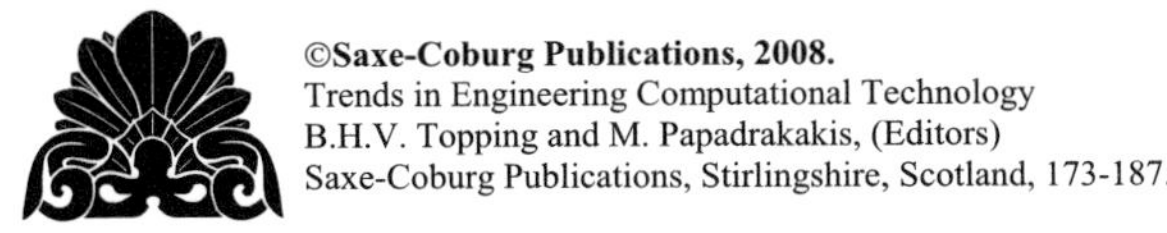
©Saxe-Coburg Publications, 2008.
Trends in Engineering Computational Technology
B.H.V. Topping and M. Papadrakakis, (Editors)
Saxe-Coburg Publications, Stirlingshire, Scotland, 173-187.

# Chapter 9

# An Enrichment-Based Multiscale Partition of Unity Method

**M. Macri[1] and S. De[2]**
**[1] Army Research Lab**
**APG, MD, United States of America**
**[2] Department of Mechanical, Aerospace & Nuclear Engineering**
**Rensselaer Polytechnic Institute, Troy NY, United States of America**

## Abstract

In this paper we present a multiscale partition of unity method developed using an enrichment technique for the modelling of heterogeneous media in the presence of singularities such as cracks. In order to compute the microscopic fields near the crack edge within the macroscale computations, a structural enrichment-based homogenization method is introduced in which the approximation space at the macroscopic scale is enriched by functions generated at the microscopic scale using the asymptotic homogenization technique.

**Keywords:** enrichment, partition of unity, multiscale, composites, cracks, homogenization.

## 1 Introduction

In order to create materials for the basis of new products, engineers must learn to model and design the macroscale by taking into account the relevant micro structural features. A full finite element solution modelling the microstructure and spanning the entire macroscale is computationally not feasible. Hence, the goal of multiscale modelling is to capture the microscopic phenomena, while still preserving the macroscopic conservation principles.

The mathematical theory of asymptotic homogenization [1] which uses asymptotic expansions of field variables about macroscopic values is a well-known hierarchical multiscale technique [2-4]. The asymptotic homogenization method provides overall effective properties as well as microscopic stress and strain values. However, the homogenization method suffers from a major limitation stemming from its basic assumptions, *viz.* (a) uniformity of the macroscopic fields within each RVE and (b) local spatial periodicity of the RVE. Hence, this method breaks down in critical regions of high gradients such as cracks. To extend the homogenization method to non-periodic problems techniques have been developed, such as the s-version of the finite element method [5] which requires cumbersome quadrature techniques and various multigrid-like bridging scale methods [6,7] that require remeshing to capture evolving fine scale features.

To overcome the drawbacks of the existing methods we propose an enrichment method based on the principles of partition of unity [8]. The major advantage of the partition of unity-based enrichment strategies, in contrast to others such as those based on the moving least squares techniques [9], is that the enriched local function space may be easily varied from one node to the other. In techniques such as the extended finite element methods the enrichment is applied to a set of nodes in the vicinity of the crack tip and certain "transition elements" are necessary to connect the enriched finite elements to the nonenriched ones which result in a degradation of performance [10]. To overcome this problem we have developed "geometry independent" enrichment strategies for the meshfree method of finite spheres [11] in which localized "bubble functions" are used to apply the enrichment independent of the underlying geometric discretization without the need for transition regions.

To overcome the problem of predicting the correct displacement response at the microscopic scale, we have developed a new structural enrichment based method in which we enrich the approximation in the vicinity of the cracks with specialized functions derived using asymptotic homogenization theory which allow macro-scale computations to be performed with the micro-structural features explicitly considered.

We begin by reviewing the partition of unity method in Section 2. In Section 3 we present a structural enrichment based method. In Section 4 several examples are provided to demonstrate the effectiveness of our approach for both one and two-dimensional problems.

## 2 Partition of Unity method

We begin by defining the domain $\Omega$. For simplicity, we assume that the domain is a square in 2D or a cube in 3D. The domain is subdivided into a grid with nodes placed at the intersections, as shown in Figure 1a. The "support" of node $I, \overline{B}_I$, is chosen to be a square, centered at the node and the shape functions are compactly supported on $\overline{B}_I$. Some of the supports are shown in Figure 1b.

The partition of unity shape functions are generated using the partition of unity paradigm [8] which is satisfied with Shepard functions [12]. We define, at each node '$I$ ', a weighting function $W_I$ which is compactly supported on $\overline{B}_I$ and has the following properties:

$$\begin{array}{lll} 1. & W_I(\mathbf{x}) \in C_0^s(\overline{B}_I) & s \geq 0 \\ 2. & W_I(\mathbf{x}) \geq 0 & \forall \mathbf{x} \in \overline{B}_I \\ 3. & W_I(\mathbf{x}) = 0 & elsewhere \end{array} \tag{1}$$

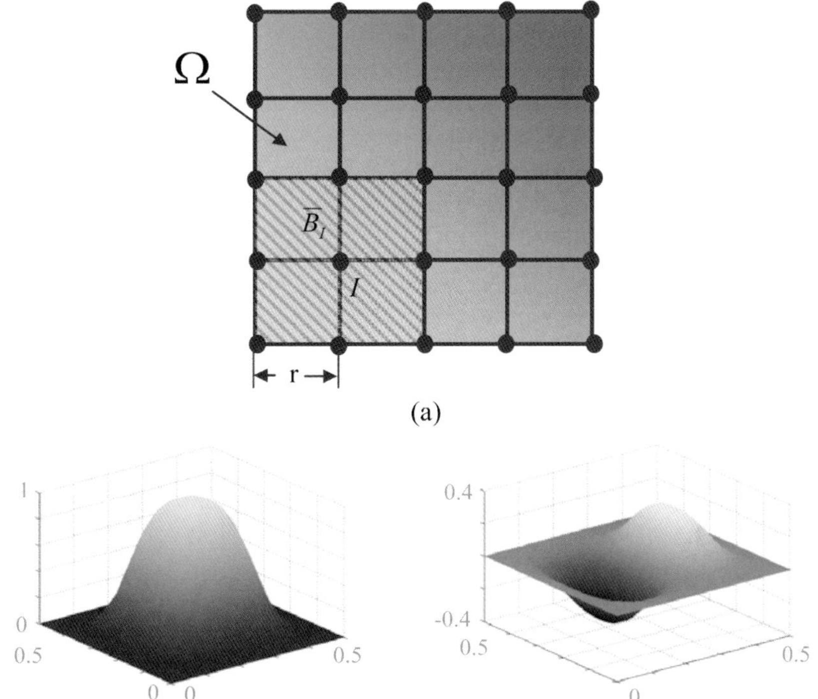


Figure 1: (a) The discretized domain. (b) Some shape functions generated using the partition of unity method for node $I$

The symbol $C_0^s\left(\overline{B}_I\right)$ stands for the space of functions that are compactly supported on $\overline{B}_I$ which have continuous derivates of order $s$. Typically, the weighting functions take on the form of a spline or a finite element shape function. In this work the weighting function is based on the quartic spline in each direction, such that:

$$W_I^A(\mathbf{x}) = \prod_{i=1}^{d} \Phi_I^Q(x_i) \tag{2}$$

where,

$$\Phi_I^Q(x_i) = \begin{cases} 1 - 6s^2(x_i) + 8s^3(x_i) - 3s^4(x_i) & s \le 1 \\ 0 & s > 1 \end{cases} \tag{3}$$

$$s(x_i) = \left|x_i - (x_i)_0\right| / r_I \tag{4}$$

and $d$ is the dimensionality of the mode.

We define the Shepard partition of unity function at each node '$I$', using a simple normalization procedure

$$\varphi_I^0(\mathbf{x}) = \frac{W_I(\mathbf{x})}{\sum_{J=1}^{N} W_J(\mathbf{x})} \tag{5}$$

which ensures $\sum_{J=1}^{N} \varphi_I^0(\mathbf{x}) = 1 \, \forall \mathbf{x} \in \Omega$ *i.e.*, rigid body modes are exactly satisfied.

We develop global approximation spaces with higher order consistency by defining, at each node '*I* ', a local approximation space

$$V_I^{h,p} = span_{m \in \xi}\{p_m(\mathbf{x})\} \subset H^1(\overline{B}_I \cap \Omega) \tag{6}$$

where *h* is a measure of the size of the element, *p* is the polynomial order, $\xi$ is an index set, $H^1$ is the first order Hilbert space, and $p_m(\mathbf{x})$ is a polynomial or other function. For instance, a quadratic accurate local approximation space in 2D has $V_I^{h,p} = span\{1, x, y, x^2, y^2, xy\}$. The global approximation space is then generated as follows

$$V^{h,p} = \sum_I^N \varphi_I^0 V_I^{h,p} \subset H^1(\Omega) \tag{7}$$

Hence, any function $v^{h,p} \in V^{h,p}$ can be written as

$$v^{h,p}(\mathbf{x}) = \sum_{I=1}^{N} \sum_{m \in \xi} h_{\mathrm{Im}}(\mathbf{x}) \alpha_{\mathrm{Im}} \tag{8}$$

where

$$h_{\mathrm{Im}}(\mathbf{x}) = \varphi_I^0(\mathbf{x}) p_m(\mathbf{x}) \tag{9}$$

is the shape function at node *I* corresponding to the $m^{th}$ degree of freedom, and $\alpha_{\mathrm{Im}}$ is an unknown constant.

## 3 Structurally Based Enrichment

The advantage of using partition of unity to model the response of heterogeneous materials such as particulate composites is that it provides a route to local enrichment using specialized functions in a straightforward manner. To ensure that correct displacement fields are predicted at the microscale, an enrichment function is implemented using the asymptotic homogenization method to generate functions for enriching the macroscale.

## 3.1 Introduction to Homogenization

In the theory of asymptotic homogenization, a macroscopic scale and a microscopic scale are designated and represented by coordinates **x** and **y**, respectively (Figure 2) and a function $f^{\gamma}$ implies

$$f^{\gamma}(\mathbf{x}) = f(\mathbf{x}, \mathbf{y}(\mathbf{x})) \tag{10}$$

where the coordinates $x_i$ and $y_j$ are related by $\mathbf{y} = \mathbf{x}/\gamma$, $\gamma$ being a scalar scale factor that denotes the ratio of length scales at the two scales.

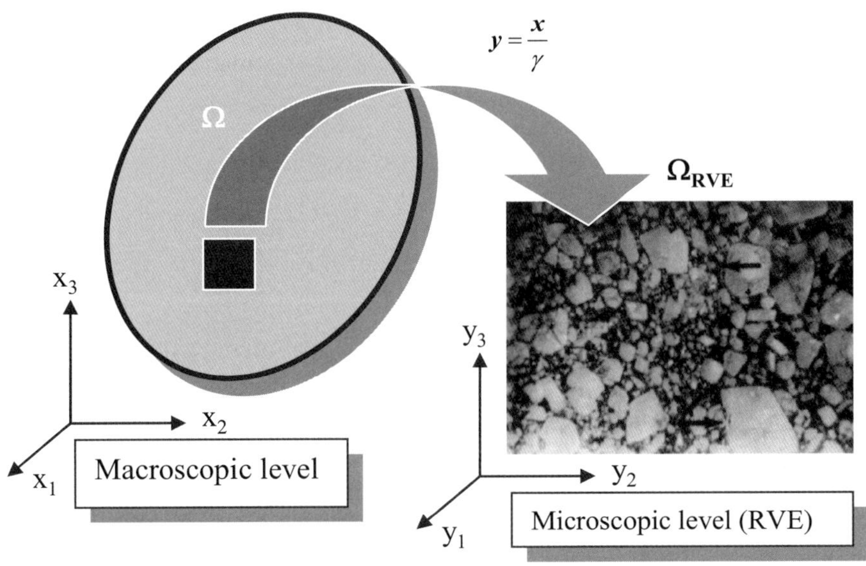


Figure 2: Relationship between macro (Ω) and micro ($\Omega_{RVE}$) coordinate systems

The assumption of *Y-periodicity* of the representative volume element (RVE) ensures

$$f(\mathbf{x}, \mathbf{y}) = f(\mathbf{x}, \mathbf{y} + k\mathbf{Y}) \tag{11}$$

where ***Y*** is the size of the RVE.

For homogenization, one assumes an asymptotic expansion of the displacement field in powers of the scale factor γ. We can therefore expand the displacements $u_i^{\gamma}$, strains $\varepsilon_{ij}^{\gamma}$ and stresses $\sigma_{ij}^{\gamma}$ to

$$u_i^\gamma(\mathbf{x},\mathbf{y}) = u_i^0(\mathbf{x},\mathbf{y}) + \gamma u_i^1(\mathbf{x},\mathbf{y}) + O(\gamma^2)$$
$$\varepsilon_{ij}^\gamma(\mathbf{x},\mathbf{y}) = \gamma^{-1}\varepsilon_{ij}^{-1}(\mathbf{x},\mathbf{y}) + \varepsilon_{ij}^0(\mathbf{x},\mathbf{y}) + \gamma\varepsilon_{ij}^1(\mathbf{x},\mathbf{y}) + O(\gamma^2) \quad (13)$$
$$\sigma_{ij}^\gamma(\mathbf{x},\mathbf{y}) = \gamma^{-1}\sigma_{ij}^{-1}(\mathbf{x},\mathbf{y}) + \sigma_{ij}^0(\mathbf{x},\mathbf{y}) + \gamma\sigma_{ij}^1(\mathbf{x},\mathbf{y}) + O(\gamma^2)$$

We assume that the constitutive behaviour for the RVE is linear elastic. Furthermore, we define the following:

$$\sigma_{ij}^a = C_{ijkl}\varepsilon_{kl}^a$$
$$\varepsilon_{ij}^a = \varepsilon_{ijx}^a + \varepsilon_{ijy}^{a+1}$$
$$\varepsilon_{ijx}^a = \frac{1}{2}\left(u_{i,x_j}^a + u_{j,x_i}^a\right) \quad (14)$$
$$\varepsilon_{ijy}^a = \frac{1}{2}\left(u_{i,y_j}^a + u_{j,y_i}^a\right)$$

Substituting the asymptotic expansions into the linear elastic equilibrium equations yields:

$$\gamma^{-2}\left(\sigma_{ij,y_j}^{-1}\right) + \gamma^{-1}\left(\sigma_{ij,x_j}^{-1} + \sigma_{ij,y_j}^0\right) + \gamma^0\left(\sigma_{ij,x_j}^0 + \sigma_{ij,y_j}^1 + b_i\right) + O(\gamma) = 0 \quad (15)$$

where $b_i$ is the body force term.

Examining theleading order equilibrium equations individually results in the following *effective* equation on the macroscale:

$$\left(\tilde{C}_{ijmn}\varepsilon_{mnx}^0\right)_{,x_j} + \tilde{b}_i = 0 \quad (16)$$

where:

$$\tilde{C}_{ijmn} = \frac{1}{Y}\int_Y C_{ijmn}\left(I_{mnkl} + \Psi_{mnkl}\right)dY \quad (17)$$

$$\tilde{b}_i = \frac{1}{Y}\int_Y b_i dy \quad (18)$$

The term $\Psi_{mnkl}$ is defined by deriving from (15) the RVE problem. We begin by assuming the relation:

$$u_k^1 = H_{mnk}(\mathbf{y})\varepsilon_{mnx}^0(\mathbf{x}) \quad (19)$$

where $H_{mnk}(\mathbf{y})$ is a Y-periodic third rank structural tensor that relates the strains at the macroscopic level to the first order displacement perturbation term $u_i^1(\mathbf{x},\mathbf{y})$.

Some examples of the $H_{mnk}(\mathbf{y})$ functions for some one and two-dimensional RVEs are shown in Figures 4 and 8, respectively.

The strains within the microstructure are

$$\varepsilon^1_{kly} = \Psi_{mnkl}\varepsilon^0_{mnx} \tag{20}$$

Where $\Psi_{mnkl}$ is now defined as the symmetric part of the spatial gradient of the $H_{mnk}(\mathbf{y})$ tensor:

$$\Psi_{mnkl} = \frac{1}{2}\left(H_{mnl,y_k} + H_{mnk,y_l}\right) \tag{21}$$

Hence, the RVE problem is defined as

$$\left(C_{ijkl}\left(I_{klmn} + \Psi_{mnkl}\right)\right)_{,y_j} = 0 \tag{22}$$

where $\Psi_{mnkl}$ is solved for using a weak formulation and $I_{klmn}$ is a fourth order identity tenser. We have used the finite element method to solve the RVE problem with periodic boundary conditions.

## 3.2 Enrichment Function

From equation (15) we can derive that $u^0_i$ is solely a function of $\boldsymbol{x}$, such that the displacement field on the RVE, spanning $\boldsymbol{y}$, is a constant value: $u^\gamma_i(\mathbf{x},\mathbf{y}) = u^0_i(\mathbf{x})$. To resolve this problem, we present a structural enrichment based homogenization technique where functions, derived using asymptotic homogenization, are used to enrich the macroscale approximation space.

Using partition of unity shape functions, the displacement field over the entire domain is approximated as:

$$u^0_i(\mathbf{x}) = \sum_{I=1}^{N}\sum_{m\in\zeta} h_m(\mathbf{x})\alpha^i_{\mathrm{Im}} \tag{23}$$

In [11], we have shown that enrichment of a carefully selected sub-domain near the crack tip improves the solution. Let us assume that a sub-domain $\Omega^*\subset\Omega$ is to be enriched (Figure 3a) so that the displacement approximation in $\Omega^*$ is

$$u_i^{enriched}(\mathbf{x}) = u^0_i(\mathbf{x}) + \sum_{J\in\Theta}^{N^{enriched}} \varphi^0_J(\mathbf{x})F(\mathbf{x})\beta^i_J \tag{24}$$

where $N^{enriched} < N$ are enriched nodes, $\Theta$ is an index set with $card(\Theta) = N^{enriched}$, $F$ is the enrichment function being used and $\beta_J^i$ is an array of constant unknown associated with the enrichment.

To eliminate the need for a transition region, a positive weight function (W) is defined which is compactly supported on $\Omega^*$, *i.e.*, $\text{supp(W)} = \Omega^*$. The enrichment function $F$ is multiplied by $W$ to generate a compactly supported function such that $\text{supp}(WF) = \Omega^*$ (Figure 3b).

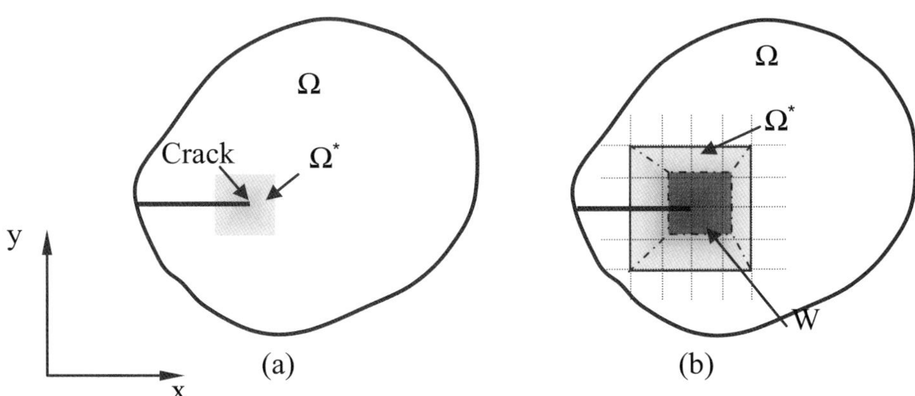


Figure 3: For applying enrichment to a sub-domain Ω* of Ω (a) a weight function is defined which is compactly supported on Ω* (b)

To capture the phenomena on the micro-scale we use $H_{mnk}(\mathbf{y})$ as the enrichment function to capture $u_i^1$ such that the displacement field within $\Omega^*$ can be written as

$$u_i^{enriched}(\boldsymbol{x}) = u_i^0(\boldsymbol{x}) + \sum_{J\in\Theta}^{N^{enriched}} \sum_{n\in\zeta} \sum_{k,q,r\in d} h_{Jn}(\boldsymbol{x}) W H_{kqr} \beta_{Jnpqr}^i \tag{25}$$

where $d$ is the dimensionality of the problem. Some enriched shape functions are shown in Figures 5 and 6. The multiscale enrichment based partition of unity method developed in [13] is similar in approach in the context of the extended finite element method.

# 4 Numerical Examples

In the following section we discuss specific numerical examples demonstrating the effectiveness of the enrichment method . In Section 4.1 a one-dimensional example of a composite bar is presented and in Section 4.2 a center-cracked tension (CCT) specimen is discussed.

## 4.1 A heterogeneous bar in $R^1$

We consider a heterogeneous bar of unit length composed of a periodic array of two linear elastic, homogeneous and isotropic constituents (Young's module $E_1$=1 and $E_2$=10) with perfect interfaces, as illustrated in Figure 4. The bar is fixed at one end and is subjected to a body force of the following form

$$b(x) = -\left(6x + \left(5000 - \left(\frac{1-2x}{0.0004}\right)^2\right) e^{-\left(\frac{x-0.5}{0.02}\right)^2}\right) \tag{26}$$

which results in a sharply peaked response at $x = 0.5$. The microscale problem may be solved analytically with periodic boundary conditions. It is straightforward to see that the solution for $H_{111}$ is of the following form

$$H = \begin{cases} \left(\dfrac{10}{9\alpha+1} - 1\right) y & y < \alpha \\ \dfrac{9\alpha - y}{9\alpha - 1} - y & y \geq \alpha \end{cases} \tag{27}$$

where $\alpha$ is the fraction of the RVE composed of the material with Young's modulus E1. The function $H_{111}$ is plotted in Figure 4b with $\alpha$=0.5.

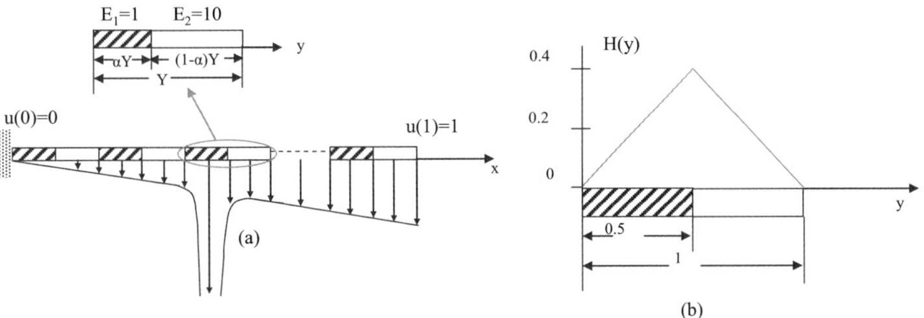


Figure 4: (a) Heterogeneous bar of unit length;
(b) the $H$ function on an RVE for $\alpha = 0.5$

We compare all results to an analytical solution such that the modelled domain consists of 24 unit cells. The effective stiffness of the bar may be computed as $C = 1.818$. To compute the actual displacement field within the microstructure as part of the global solution the following approximation is used to solve the problem:

$$u(x) = \sum_{I=1}^{N} \sum_{m=0}^{5} h_{\mathrm{Im}}(x) \alpha_{\mathrm{Im}}$$
$$\begin{aligned} h_{I0} &= \varphi_I^0 & h_{I3} &= \varphi_I^0 H \\ h_{I1} &= \varphi_I^0 \left[ (x - x_I) / r \right] & h_{I4} &= \varphi_I^0 H \left[ (x - x_I) / r \right] \\ h_{I2} &= \varphi_I^0 \left[ (x - x_I) / r \right]^2 & h_{I5} &= \varphi_I^0 H \left[ (x - x_I) / r \right]^2 \end{aligned} \tag{28}$$

For the 1D problem we assume that the entire domain is enriched and $W$ has a value of 1 throughout the model. The shape functions are shown in Figures 5 where the support of node $I$ contains 2 (Figure 5a) and 8 (Figure 5b) RVEs.

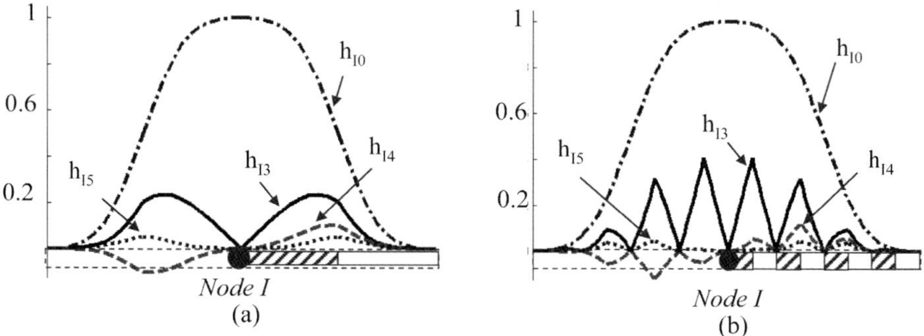


Figure 5: The shape functions for node $I$ whose support contains (a) two RVEs and (b) eight RVEs

We compare in Figure 6, using structural enrichment throughout the domain with eight RVEs per support, versus no enrichment with asymptotic homogenization. In both cases we discretize the domain using 7 nodes. Figures 6a and b show the macroscopic displacement and strain fields, respectively. In Figures 6c and d, we zoom into the RVE located at x=0.354 and plot the displacement (u) and strain (ε) fields.

## 4.2 Centered crack tension (CCT) specimen in $R^2$

In two-dimensional analysis, we consider a center cracked tension (CCT) specimen in plane stress. The specimen is of unit thickness and is loaded as shown in Figure 7a. Since the problem is symmetric, we only examine the upper right quarter of the model. We consider a simple RVE consisting of a circular inclusion embedded in a matrix, shown in Figure 7b.

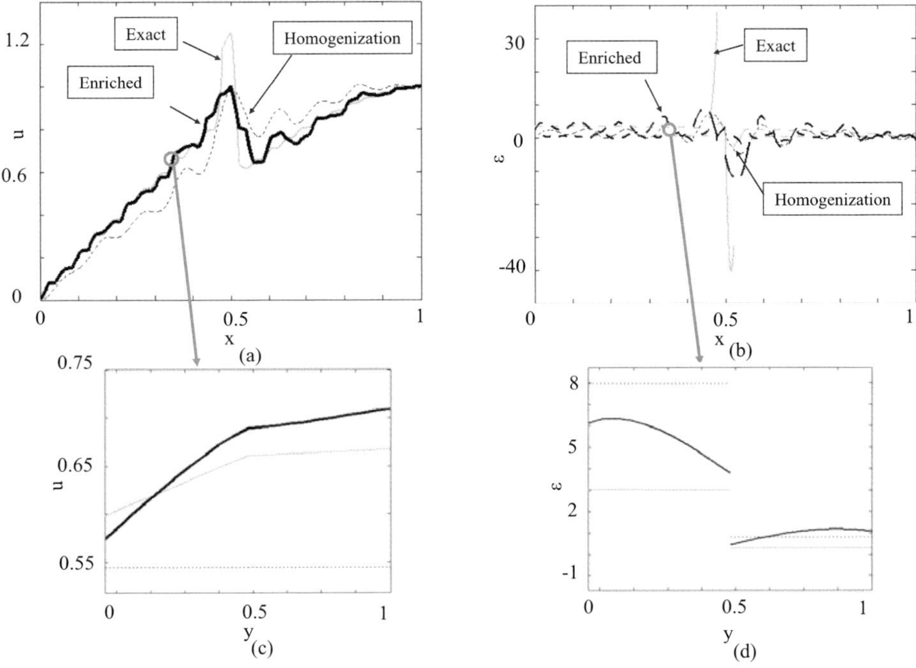


Figure 6: Displacement (u) and strain (ε) response on the macroscale of the heterogeneous bar

In all the cases we use a linear basis for nodes that are not enriched. All results are compared to a 'fine mesh solution' generated using the commercial finite element program FEMLAB® using approximately 57,000 six-noded triangular elements. We assume a finite size of the RVE such that the modelled domain consists of 16×16 unit cells. Asymptotic homogenization is used to compute the effective properties as $C_{11} = 1.630429 psi$, $C_{12} = 0.393643 psi$ and $C_{33} = 0.560592 psi$, where $C_{11}$, $C_{12}$, and $C_{33}$ are the components of the elasticity matrix.

The structural-based enrichment has been applied only to a single layer of elements adjacent to the crack surface. Figure 8 shows the third rank tensor $H_{ijk}$ for the microstructure in Figure 7b computed using a finite element mesh of 1000 six-noded triangular elements.

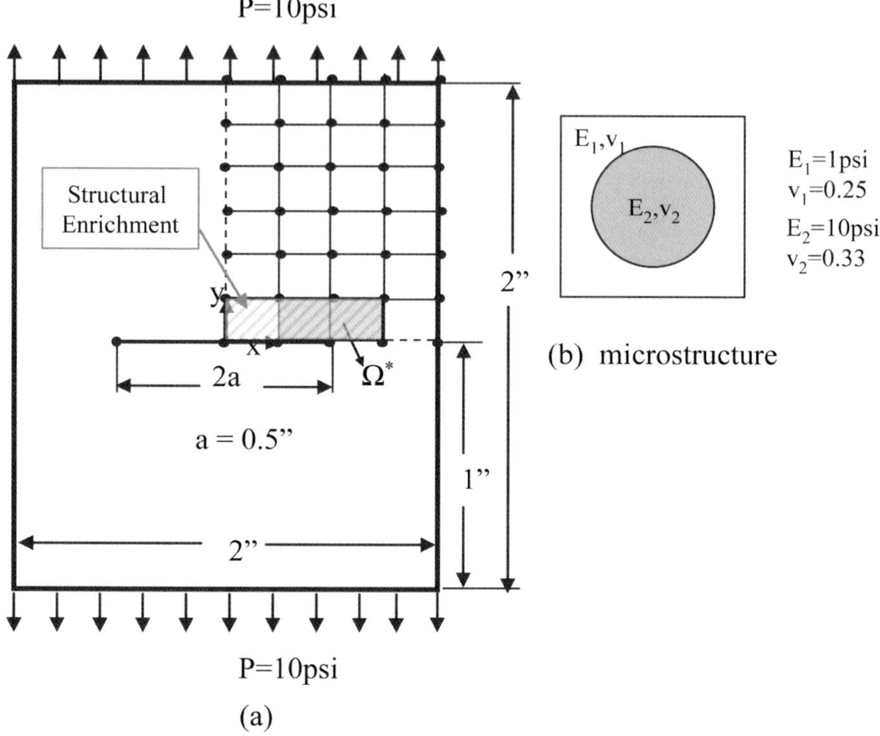


Figure 7: (a) A center cracked tension (CCT) specimen with the microstructure shown in (b)

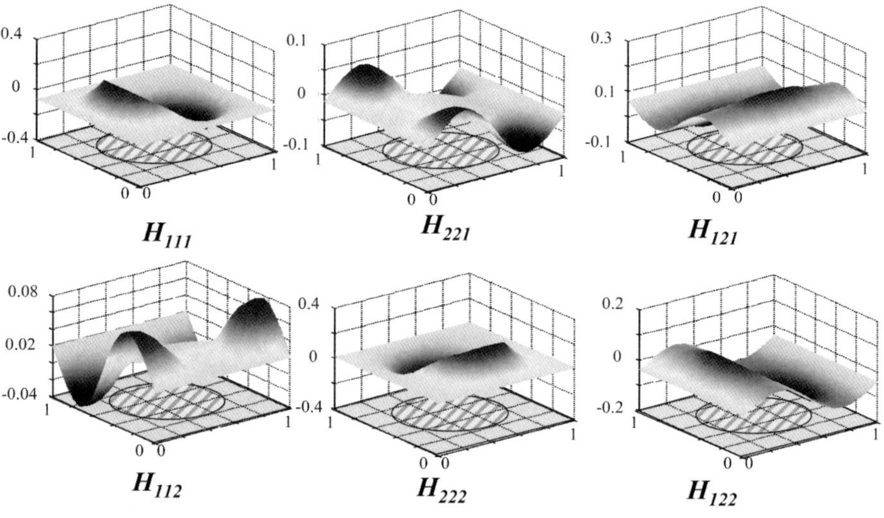


Figure 8: The six components of Hijk

The following enriched shape functions are used

$$u_j(x)=\sum_{I=1}^{N}\sum_{m=0}^{12}h_{\mathrm{Im}\,j}(x)\alpha_{\mathrm{Im}\,j}\qquad j=1,2$$

$$\begin{array}{lll}
h_{I01}=h_{I02}=\varphi_I^0 & h_{I11}=h_{I12}=\varphi_I^0(x-x_I)/r_I & h_{I21}=h_{I22}=\varphi_I^0(y-y_I)/r_I \\
h_{I31}=\varphi_I^0 H_{111} & h_{I41}=\varphi_I^0 H_{111}(x-x_I)/r_I & h_{I51}=\varphi_I^0 H_{111}(y-y_I)/r_I \\
h_{I32}=\varphi_I^0 H_{112} & h_{I42}=\varphi_I^0 H_{112}(x-x_I)/r_I & h_{I52}=\varphi_I^0 H_{112}(y-y_I)/r_I \\
h_{I61}=\varphi_I^0 H_{221} & h_{I71}=\varphi_I^0 H_{221}(x-x_I)/r_I & h_{I81}=\varphi_I^0 H_{221}(y-y_I)/r_I \\
h_{I62}=\varphi_I^0 H_{222} & h_{I72}=\varphi_I^0 H_{222}(x-x_I)/r_I & h_{I82}=\varphi_I^0 H_{222}(y-y_I)/r_I \\
h_{I91}=\varphi_I^0 H_{121} & h_{I101}=\varphi_I^0 H_{121}(x-x_I)/r_I & h_{I111}=\varphi_I^0 H_{121}(y-y_I)/r_I \\
h_{I92}=\varphi_I^0 H_{122} & h_{I102}=\varphi_I^0 H_{122}(x-x_I)/r_I & h_{I112}=\varphi_I^0 H_{122}(y-y_I)/r_I
\end{array}\tag{29}$$

where the structural enrichment shape functions are direction dependent. Several examples of the shape functions are shown in Figure 9 where the support of node $I$ contains 4 RVEs.

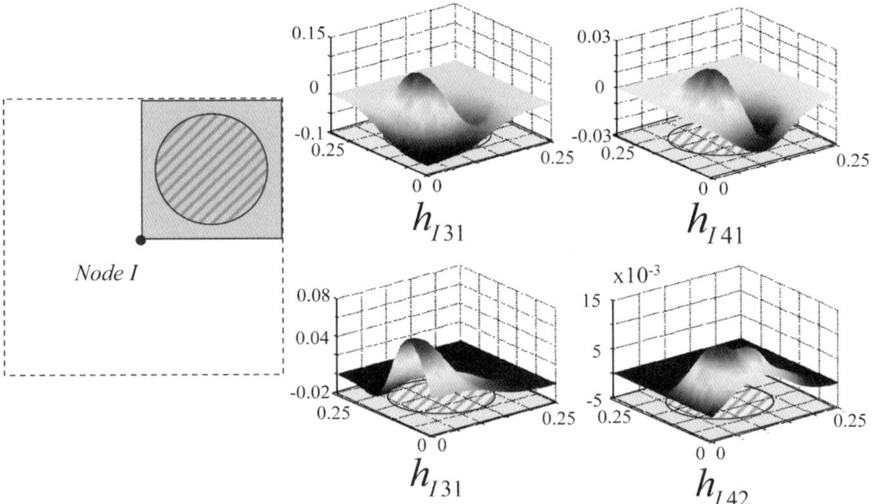


Figure 9: Enriched shape functions for a 2D analysis are shown when a quarter of the support for node $I$ coincides with a single RVE

The displacement in the y-direction (v) is plotted within the RVE located at the crack tip in Figure 10 and the displacement in the x-direction (u) is plotted within the RVE located at the crack edge in Figures 11. The OctPUM with structural enrichment-based homogenization is seen to capture the correct variation of the displacement field within the RVE. In Figure 12, we show that the normal strain ($\varepsilon_{yy}$) in the RVE located at the crack tip is also similar to what the fine mesh solution provides.

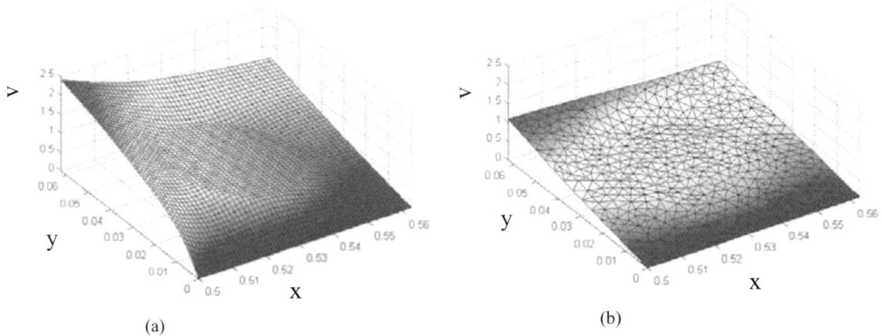


Figure 10: For the RVE in Figure 7b located at the crack tip, the displacement in the y-direction (v) is plotted for (a) the fine mesh and (b) with structural enrichment-based homogenization

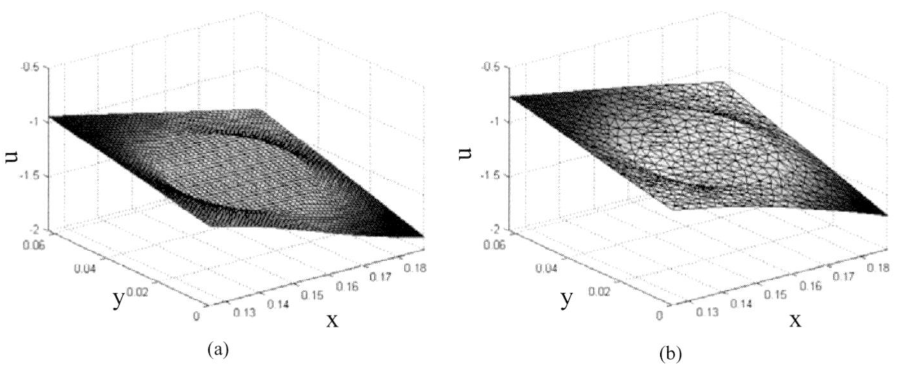


Figure 11: For the RVE in Figure 7b located along the crack edge, the displacement in the x-direction (u) is plotted for (a) the fine mesh and (b) with structural enrichment-based homogenization

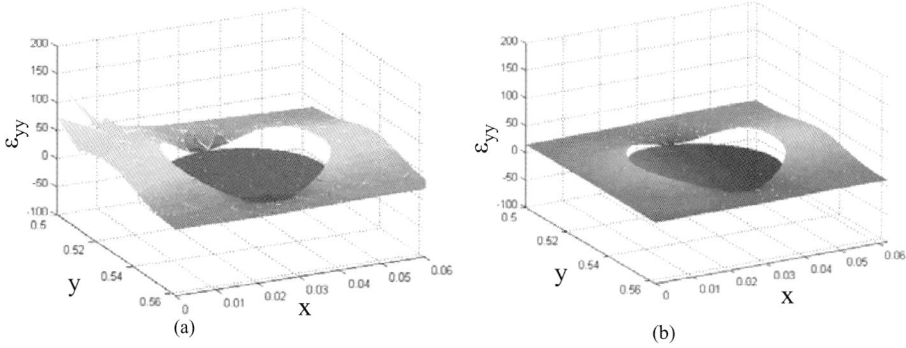


Figure 12: For the RVE in Figure 7b located at the crack tip, strain ($\varepsilon_{yy}$) is plotted for (a) the fine mesh and (b) with structural enrichment-based homogenization

## 5 Summary

A structural enrichment-based homogenization method has been developed to compute the response at the microscopic scale in localized regions of the computational domains such as along the crack edge while performing macroscale simulations. In this technique the enrichment functions are derived from the asymptotic homogenization method.

## References

[1] A. Bensoussan, J.L. Lions, G. Papanicolaou, "Asymptotic analysis for periodic structures", Amsterdam: Noth-Holland, 1978.

[2] A. Tolenado, H. Murakami, "A high-order mixture model for periodic particulate composites", International Journal of Solids and Structures, 23, 989-1002, 1987.

[3] S.J. Hollister, N. Kikuchi, "A comparison of homogenization and standard mechanics analyses for periodic porous composites", Computational Mechanics, 10, 73-95, 1992.

[4] J. Fish, Q. Yu, K. Shek, "Computational damage mechanics for composite materials based on mathematical homogenization" International Journal for Numerical Methods in Engineering, 45, 1657-1679, 1999.

[5] J. Fish, "The s-version of the finite element method. Computers and Structures", 43, 539-547, 1992.

[6] J. Fish, V. Belsky, "Multigrid method for periodic heterogeneous medium. Part I: Convergence studies for one-dimensional case", Computer Methods in Applied Mechanics and Engineering, 126, 1-16, 1995.

[7] J. Fish, V. Belsky "Multigrid method for periodic heterogeneous medium. Part II: Multiscale modeling and quality control in multidimensional case." Computer Methods in Applied Mechanics and Engineering, 126, 17-38, 1995.

[8] K. Yosida, "Functional analysis 5th Edn.", Springer-Verlag: Berlin Heidelberg, 1978.

[9] P. Lancaster, K. Salkauskas, "Surfaces generated by moving least squares methods", Mathematics of Computation, 37, 141-58, 1981.

[10] M. Fleming, Y. Chu, B. Moran, T. Belytschko, "Enriched element-free Galerkin methods for crack tip fields", International Journal for Numerical Methods in Engineering, 40, 1483-1504, 1997.

[11] M. Macri, S. De, "Enrichment of the method of finite spheres using geometry independent localized scalable bubbles", Int. J. Num. Meth. Eng., 69, 1-32, 2006.

[12] D. Shepard, "A two-dimensional interpolation function for irregularly spaced data", Proc. 23rd Nat. Conf. ACM, 517-524, 1968.

[13] J. Fish, Z. Yuan, "Multiscale enrichment based partition of unity" International Journal for numerical methods in engineering, 62, 1341-1359, 2005.

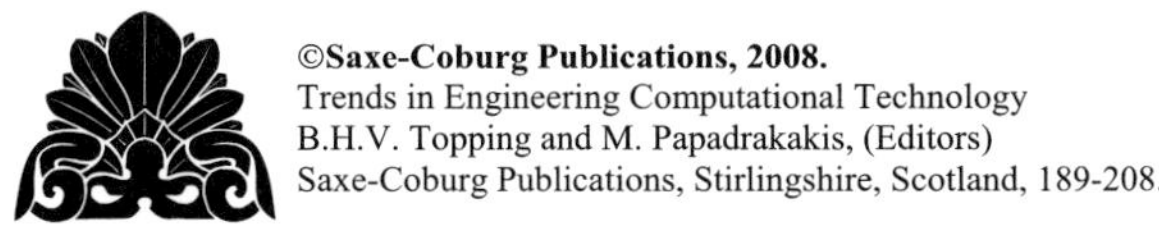
©Saxe-Coburg Publications, 2008.
Trends in Engineering Computational Technology
B.H.V. Topping and M. Papadrakakis, (Editors)
Saxe-Coburg Publications, Stirlingshire, Scotland, 189-208.

# Chapter 10

## A Multi-Scale Formulation of Gradient Elasticity and Its Finite Element Implementation

**H. Askes[1], T. Bennett[1], I.M. Gitman[1] and E.C. Aifantis[2,3]**
**[1] Department of Civil and Structural Engineering**
**University of Sheffield, United Kingdom**
**[2] School of Engineering**
**Aristotle University of Thessaloniki, Greece**
**[3] Center for Mechanics of Materials and Instabilities**
**Michigan Technological University, Houghton, United States of America**

## Abstract

In this chapter, we describe the formulation and implementation of a novel gradient elasticity theory that can be used to describe wave dispersion. To provide microstructural background for the model, second-order homogenisation of a representative volume element (RVE) is considered. The microstructural equations of motion are considered with appropriate series expansions of the state variables and the material properties. In the resulting macrostructural equations of motion, additional inertia gradients and strain gradients appear. The accompanying additional constitutive parameters can be expressed in terms of the size of the RVE. In order to enable the use of standard $\mathscr{C}^0$-continuous finite element interpolations, the fourth-order partial differential equations are rewritten in terms of a symmetric set of fully coupled second-order partial differential equations, whereby the primary unknowns can be identified as the microstructural displacements and the macro-structural displacements. Thus, the developed model is a true multi-scale model in which the two scales of observation interact.

**Keywords:** gradient elasticity, representative volume element, homogenisation, internal length scale, multi-scale formulation.

# 1 Introduction

Dispersive wave propagation occurs when the different harmonic components of a wave propagate with different velocities. This phenomenon has been observed on a wide range of scales. For instance, dispersive propagation of phonons through a Bismuth lattice has been reported experimentally in the 1960s [1], whereas elastic waves in rails were found to be dispersive due the interaction with the sleepers [2]. Heterogeneity of the structure or the material causes dispersion, and for a physically relevant modelling of wave dispersion this heterogeneity must be accounted for. Rather than modelling every heterogeneity individually, it is more efficient to formulate enriched continuum models, such as for instance *gradient elasticity* theories.

Various formats of gradient elasticity exist. In the gradient elasticity theory formulated by Aifantis and coworkers [3, 4] the stresses depend not only on the strains but also on the *strain gradients*. This format of gradient elasticity was motivated for use in statics and it has successfully been applied in the removal of strain singularities at sharp crack tips [4] and dislocation cores [5]. In another format of gradient elasticity, the equations of motion are expanded with additional *inertia gradients*. The latter theory was motivated for use in dynamics and it has successfully been used in the description of dispersive wave propagation [6]. A unified theory of gradient elasticity includes both types of gradients, *i.e.* strain gradients as well as inertia gradients [7–12]. The equations of motion of such a model can be written in terms of the displacements $u$ as

$$\rho \left(\ddot{u}_i^{\mathrm{M}} - \ell_m^2 \ddot{u}_{i,mm}^{\mathrm{M}}\right) = C_{ijkl}\left(u_{k,jl}^{\mathrm{M}} - \ell_s^2 u_{k,jlmm}^{\mathrm{M}}\right) \tag{1}$$

where $\rho$ is the mass density, $C_{ijkl}$ contains the elastic moduli and a superscript M denotes a macrostructural quantity. Furthermore, two internal length scale parameters appear, namely a length scale $\ell_m$ associated with inertia gradients and a length scale $\ell_s$ associated with strain gradients.

Several procedures have been derived to link the two length scale parameters $\ell_m$ and $\ell_s$ to the microstructural properties. The continualisation of a discrete model of masses and springs was carried out in [9] (in one dimension) and in [12] (in two dimensions). The two length scales of Equation (1) were obtained in terms of the interparticle distance and a model parameter that sets the kinematic coupling between discrete model and equivalent continuum model. In the present contribution, an alternative method of deriving Equation (1) will be pursued, namely the homogenisation of a Representative Volume Element (RVE). Two different homogenisation schemes exist, namely first-order and second-order homogenisation [13, 14]. In a first-order homogenisation scheme only the average quantities of the relevant variables within a RVE are considered, whereas the average quantities as well as the derivatives are taken into account in a second-order homogenisation scheme. It has been shown that the second-order homogenisation leads to the occurrence of length scale effects on the macrolevel [13, 14]. More recently, it was established that the apparent length scale is proportional to the size of the RVE — similar results were obtained in [15] and in [16] for the static case, although a homogeneous material was assumed in [16] for which the RVE size is theoretically equal to zero. These results have also been extended to the dynamic case, and indeed a model of the format of Equation (1) was obtained [17]. That is to say, if a second-order homogenisation scheme is applied to the microstructural equations of motion within an RVE, the additional gradients in Equation (1) are obtained automatically and the two length scale parameters $\ell_m$ and $\ell_s$ are found to be proportional to the size of the RVE.

The numerical implementation of Equation (1) is hampered by the $\mathscr{C}^1$-continuity requirements imposed on the interpolation functions, as dictated by the fourth-order spatial derivatives. To overcome this drawback, an operator split in the spirit of the

Ru-Aifantis theorem [4] has been developed in [18, 19] by which the original equations are rewritten as a series of second-order equations in terms of the true displacements as well as a set of auxiliary displacements. Since this operator split involves one further assumption, it is necessary to verify the energetic background of the new $\mathscr{C}^0$-formulation. However, the real advantage of rewriting the model as a set of second-order equations is that the usual $\mathscr{C}^0$-continuous finite elements can be used for discretisation and implementation.

In this contribution, we will discuss the various physical and mathematical backgrounds of the model given in Equation (1). In Section 2, it will be shown how the two length scale parameters $\ell_m$ and $\ell_s$ can be linked to the size of the RVE in dynamics and statics, respectively. Next, in Section 3 the operator split is discussed by which the original fourth-order equations are rewritten as a set of second-order equations. The energetic backgrounds and the format of the boundary conditions are discussed in Section 4. We have also compared the presented model (both in its fourth-order format and in its second-order format) with Mindlin's theory of elasticity with microstructure — this is presented in Section 5 and one of the conclusions is that the two displacement fields of the second-order formulation can be interpreted as the macrostructural and the microstructural displacements. In Section 6 we demonstrate how the two length scales affect the propagation of compressive waves and shear waves. Section 7 treats the finite element equations and in Section 8 we show a number of examples. Some closing remarks can be found in Section 9.

## 2 Homogenisation of the response of a Representative Volume Element

The gradient elasticity theory of Equation (1) can be motivated from the homogenised response of a sample of a heterogeneous material. For the elastic response of a material, normally a so-called *Representative Volume Element* (RVE) can be defined. Several definitions of RVEs exist, but usually the RVE is understood as a unit cell of minimum size such that the statistical deviation of the mechanical response is smaller than a user-defined threshold. The response of the RVE must be homogenised, and below we will illustrate how a gradient elasticity theory can be obtained by using a so-called *second-order homogenisation* scheme, whereby we adopt the terminology introduced in [13, 14, 16].

The equations of motion on the micro-level are written as

$$\rho^{\mathrm{m}} \ddot{u}_i^{\mathrm{m}} = \sigma_{ij,j}^{\mathrm{m}} \tag{2}$$

where $\sigma$ is the stress tensor, an index following a comma denotes a spatial derivative and a superscript 'm' indicates that the associated quantity is a microstructural quantity. Equation (2) is assumed to hold throughout the RVE and it is integrated over the

RVE. Changing the order of differentiation and integration then gives

$$\frac{1}{V_{\text{RVE}}}\int_{V_{\text{RVE}}} \rho^{\text{m}}\ddot{u}_i^{\text{m}}\,\mathrm{d}V = \frac{\partial}{\partial x_j}\left(\frac{1}{V_{\text{RVE}}}\int_{V_{\text{RVE}}} C_{ijkl}^{\text{m}}\varepsilon_{kl}^{\text{m}}\,\mathrm{d}V\right) \tag{3}$$

where the microstructural elastic constitutive relation $\sigma_{ij}^{\text{m}} = C_{ijkl}^{\text{m}}\varepsilon_{kl}^{\text{m}}$ is substituted.

## 2.1 Second-order homogenisation

The micro-structural quantities are developed in series around their values at the centre of the RVE, *i.e.*

$$C_{ijkl}^{\text{m}} = C_{ijkl}^{\text{M}} + C_{ijkl,o}^{\text{M}}x_o \tag{4}$$

$$\varepsilon_{kl}^{\text{m}} = \varepsilon_{kl}^{\text{M}} + \varepsilon_{kl,p}^{\text{M}}x_p \tag{5}$$

$$\rho^{\text{m}} = \rho^{\text{M}} + \rho_{,o}^{\text{M}}x_o \tag{6}$$

$$\ddot{u}_i^{\text{m}} = \ddot{u}_i^{\text{M}} + \ddot{u}_{i,p}^{\text{M}}x_p \tag{7}$$

The values of the microscopic fields at the centre of the RVE can also be identified as the average values or macroscopic values and are denoted with a superscript 'M'. A first-order homogenisation scheme is obtained if all derivatives in Equations (4–7) are ignored, whereas the inclusion of these derivatives yields a second-order homogenisation scheme [17]. Following the latter assumption, Equation (3) can be expanded as

$$\frac{1}{V_{\text{RVE}}}\int_{V_{\text{RVE}}} \left(\rho^{\text{M}}\ddot{u}_i^{\text{M}} + \rho_{,o}^{\text{M}}\ddot{u}_i^{\text{M}}x_o + \rho^{\text{M}}\ddot{u}_{i,p}^{\text{M}}x_p + \rho_{,o}^{\text{M}}\ddot{u}_{i,p}^{\text{M}}x_ox_p\right)\mathrm{d}V =$$
$$\frac{\partial}{\partial x_j}\frac{1}{V_{\text{RVE}}}\int_{V_{\text{RVE}}} \left(C_{ijkl}^{\text{M}}\varepsilon_{kl}^{\text{M}} + C_{ijkl,o}^{\text{M}}\varepsilon_{kl}^{\text{M}}x_o + C_{ijkl}^{\text{M}}\varepsilon_{kl,p}^{\text{M}}x_p + C_{ijkl,o}^{\text{M}}\varepsilon_{kl,p}^{\text{M}}x_ox_p\right)\mathrm{d}V \tag{8}$$

The second and third terms of both left-hand-side and right-hand-side of Equation (8) consist of odd functions integrated over a symmetric domain, hence they cancel. The last term on either side can be rewritten using integration by parts, that is

$$\int_{V_{\text{RVE}}} \rho_{,o}^{\text{M}}\ddot{u}_{i,p}^{\text{M}}x_ox_p\,\mathrm{d}V = \oint_S \rho^{\text{M}}\ddot{u}_{i,p}^{\text{M}}n_ox_ox_p\,\mathrm{d}S$$
$$-\int_{V_{\text{RVE}}} \left(\rho^{\text{M}}\ddot{u}_{i,op}^{\text{M}}x_ox_p + \rho^{\text{M}}\ddot{u}_{i,p}^{\text{M}}x_{o,o}x_p + \rho^{\text{M}}\ddot{u}_{i,p}^{\text{M}}x_ox_{p,o}\right)\mathrm{d}V \tag{9}$$

and

$$\int_{V_{\text{RVE}}} C_{ijkl,o}^{\text{M}}\varepsilon_{kl,p}^{\text{M}}x_ox_p\,\mathrm{d}V = \oint_S C_{ijkl}^{\text{M}}\varepsilon_{kl,p}^{\text{M}}n_ox_ox_p\,\mathrm{d}S$$
$$-\int_{V_{\text{RVE}}} \left(C_{ijkl}^{\text{M}}\varepsilon_{kl,op}^{\text{M}}x_ox_p + C_{ijkl}^{\text{M}}\varepsilon_{kl,p}^{\text{M}}x_{o,o}x_p + C_{ijkl}^{\text{M}}\varepsilon_{kl,p}^{\text{M}}x_ox_{p,o}\right)\mathrm{d}V \tag{10}$$

In these expressions, the boundary integrals cancel due to the fact that the origin of the coordinate system is positioned in the centre of the RVE. The first and second terms on the right-hand-sides of Equations (9) and (10) again consist of odd functions integrated over a symmetric domain, by which they vanish. Furthermore, in a three-dimensional context $V_{\text{RVE}} = L^3$ and

$$\int_{V_{\text{RVE}}} x_o x_p \, \mathrm{d}V = \int_{-\frac{1}{2}L}^{\frac{1}{2}L} \int_{-\frac{1}{2}L}^{\frac{1}{2}L} \int_{-\frac{1}{2}L}^{\frac{1}{2}L} x_o x_p \, \mathrm{d}x_1 \, \mathrm{d}x_2 \, \mathrm{d}x_3 = \frac{1}{12} L^5 \delta_{op} \tag{11}$$

where $\delta_{op}$ is the Kronecker delta. As a result, Equation (8) can be rewritten as

$$\rho^{\mathrm{M}} \left( \ddot{u}_i^{\mathrm{M}} - \frac{1}{12} L^2 \ddot{u}_{i,oo}^{\mathrm{M}} \right) = C_{ijkl}^{\mathrm{M}} \left( \varepsilon_{kl,j}^{\mathrm{M}} - \frac{1}{12} L^2 \varepsilon_{kl,joo}^{\mathrm{M}} \right) \tag{12}$$

## 2.2 Static and dynamic RVE sizes

To keep the above derivations transparent, the same RVE has been used for the inertia terms as well as for the stiffness terms. However, the implied volume averaging does not have to be taken over identical volumes. For instance, let $V_2 > V_1 = V_{\text{RVE}}$, then for any microscopic quantity $a$ it holds that

$$\frac{1}{V_1} \int_{V_1} a \, \mathrm{d}V = \frac{1}{V_2} \int_{V_2} a \, \mathrm{d}V \tag{13}$$

within a given error tolerance. A certain volume can be a Representative Volume $V_{\text{RVE}}$ for one quantity which may be larger than the Representative Volume for another quantity. In the context of Equation (12), the $L$ appearing in the stiffness term on the left-hand-side does not have to be equal to the $L$ appearing in the inertia term on the right-hand-side. With this observation in mind, Equation (12) can be written so as to comply with the format of Equation (1), that is

$$\rho^{\mathrm{M}} \left( \ddot{u}_i^{\mathrm{M}} - \frac{1}{12} \ell_{\text{dyn}}^2 \ddot{u}_{i,oo}^{\mathrm{M}} \right) = C_{ijkl}^{\mathrm{M}} \left( \varepsilon_{kl,j}^{\mathrm{M}} - \frac{1}{12} \ell_{\text{stat}}^2 \varepsilon_{kl,joo}^{\mathrm{M}} \right) \tag{14}$$

where $\ell_{\text{dyn}}$ and $\ell_{\text{stat}}$ are the RVE size in dynamics and statics, respectively. In the context of Equation (1), we thus have $\ell_{\text{dyn}} = \ell_m \sqrt{12}$ and $\ell_{\text{stat}} = \ell_s \sqrt{12}$.

# 3 Operator split

While the model of Equation (1) has been provided with a microstructural background through the links of $\ell_s$ and $\ell_m$ to the RVE sizes of stiffness and inertia, respectively, the drawback of Equation (1) is that fourth-order spatial derivatives of the displacements appear. As a consequence, numerical implementations require $\mathscr{C}^1$-continuity of the interpolation functions. To avoid this, the recently developed operator split of [18, 19] is employed. An auxiliary displacement field $u^a$ is introduced according to

$$u_i^a = u_i^{\mathrm{M}} - \ell_s^2 u_{i,mm}^{\mathrm{M}} \tag{15}$$

Equation (1) can then be rewritten as

$$\rho\left(\ddot{u}_i^{\mathrm{M}} - \ell_m^2 \ddot{u}_{i,mm}^{\mathrm{M}}\right) = C_{ijkl} u_{k,jl}^a \tag{16}$$

The system of equations is fully coupled as both variables $u^a$ and $u_{\mathrm{M}}$ appear in both expressions, Equations (15) and (16). However, the coefficient of $u^a$ in Equation (15) is not equal to the coefficient of $u^{\mathrm{M}}$ in Equation (16), which makes it difficult to identify the energy functionals underlying these expressions. In order to arrive at a symmetric formulation, the following manipulations are made.

1. The second time derivative of Equation (15) is taken:

$$\ddot{u}_i^a = \ddot{u}_i^{\mathrm{M}} - \ell_s^2 \ddot{u}_{i,mm}^{\mathrm{M}} \tag{17}$$

2. Equation (17) is multiplied with $\ell_m^2/\ell_s^2$, by which the $\ddot{u}_{i,mm}^{\mathrm{M}}$ term in Equation (16) can be replaced.
3. Equation (17) is multiplied with $\rho(1-\ell_m^2/\ell_s^2)$ and used instead of Equation (15).

With these substitutions, the symmetric set of coupled equations reads

$$\rho\left(\frac{\ell_m^2}{\ell_s^2}\,\ddot{u}_i^a - \frac{\ell_m^2-\ell_s^2}{\ell_s^2}\,\ddot{u}_i^{\mathrm{M}}\right) = C_{ijkl} u_{k,jl}^a \tag{18}$$

together with

$$\rho\left(-\frac{\ell_m^2-\ell_s^2}{\ell_s^2}\,\ddot{u}_i^a + \frac{\ell_m^2-\ell_s^2}{\ell_s^2}\,\ddot{u}_i^{\mathrm{M}} - \left(\ell_m^2-\ell_s^2\right)\ddot{u}_{i,mm}^{\mathrm{M}}\right) = 0 \tag{19}$$

# 4 Energy potentials and boundary conditions

The symmetric form of Equations (18) and (19) facilitates the identification of the kinetic energy density $\mathscr{U}^{\mathrm{kin}}$ and the potential energy density $\mathscr{U}^{\mathrm{pot}}$, that is

$$\mathscr{U}^{\mathrm{kin}} = \frac{1}{2}\,\rho\left(\left(\dot{u}_i^a\right)^2 + \frac{\ell_m^2-\ell_s^2}{\ell_s^2}\left(\dot{u}_i^{\mathrm{M}}-\dot{u}_i^a\right)^2 + \left(\ell_m^2-\ell_s^2\right)\left(\dot{u}_{i,m}^{\mathrm{M}}\right)^2\right) \tag{20}$$

$$\mathscr{U}^{\mathrm{pot}} = \frac{1}{2}\,\varepsilon_{ij}^a\, C_{ijkl}\, \varepsilon_{kl}^a \tag{21}$$

where $\varepsilon^a$ is the auxiliary strain related to $u^a$. Note that both energy densities are positive definite provided that $\ell_m > \ell_s$. The kinetic and potential energy densities contain at most first-order spatial derivatives of the primary unknowns $u^a$ and $u^{\mathrm{M}}$. Therefore, $\mathscr{C}^0$-continuous interpolations suffice for a numerical implementation.

It was shown in [19] that Equations (18) and (19) can be obtained from Equations (20) and (21) using the standard Hamilton-Ostrogradsky principle. Furthermore, the associated boundary conditions were found as

$$\begin{array}{llll} \text{either prescribe} & u_i^a & \text{or prescribe} & n_j C_{ijkl} \varepsilon_{kl}^a \\ \text{either prescribe} & u_i^{\mathrm{M}} & \text{or prescribe} & n_j \rho \left(\ell_m^2 - \ell_s^2\right) \ddot{u}_{i,j}^{\mathrm{M}} \end{array} \tag{22}$$

where $n$ is the vector normal to the boundary $\Gamma$ of the domain $\Omega$. These boundary conditions should be compared to those that accompany Equation (1). The latter can be written as

$$\begin{array}{llll} \text{either prescribe} & u_i^{\mathrm{M}} & \text{or prescribe} & n_j C_{ijkl} \left( \varepsilon_{kl}^{\mathrm{M}} - \ell_s^2 \varepsilon_{kl,mm}^{\mathrm{M}} \right) + n_j \rho \ell_m^2 \ddot{u}_{i,j}^{\mathrm{M}} \\ \text{either prescribe} & n_j u_{i,j}^{\mathrm{M}} & \text{or prescribe} & n_j n_m C_{ijkl} \varepsilon_{kl,m}^{\mathrm{M}} \end{array} \tag{23}$$

where for transparency we have assumed that the boundaries of the domain are aligned with the Cartesian coordinate axes — more general expressions can be found in [20], for instance. The higher-order boundary conditions given in the second row of expression (23) are normally taken as homogeneous natural boundary conditions, *i.e.*

$$n_j n_m C_{ijkl} \varepsilon_{kl,m}^{\mathrm{M}} = 0 \tag{24}$$

This has been motivated from the point of view of energy conservation, in the sense that with such boundary conditions the gradient elasticity model merely redistributes the energy of the associated classical elasticity model without loss or gain of energy [20].

The question is now to which extent Equation (24) can be emulated by appropriate formats of expression (22). Since the two natural boundary conditions in expression (22) contain derivatives of the displacement, not derivatives of the strain, we suggest using a tying between the two essential boundary conditions, in particular $u_i^a = u_i^{\mathrm{M}}$. In combination with Equation (15) this implies that $u_{i,jj}^{\mathrm{M}} = 0$. Whilst this is not an identical replication of Equation (24), we have found its performance to be satisfactory. Furthermore, the condition $u_i^a = u_i^{\mathrm{M}}$ can also be interpreted as preserving compatibility between microscale and macroscale on the boundary, as explained in Section 5. As regards prescribed tractions, these can be substituted into either natural boundary condition of expression (22) — however, we will not explore this issue further here and equilibrate externally applied tractions with the natural boundary conditions associated with $u^a$.

# 5 Comparison with Mindlin's theory

To provide context for the presented model, we will compare it with the theory of elasticity with microstructure ascribed to Mindlin [7]. This will be done for the fourth-order formulation of Equation (1) as well as for the second-order formulation of Equations (18) and (19). The latter comparison will also provide a physical identification of the auxiliary displacements $u^a$. Mindlin's theory distinguishes between macrostructural deformation and microstructural deformation; the difference between them was denoted as relative deformation. Mindlin also considered particularisations of the general theory in which the relative deformation is zero. The case with zero relative deformation will be used to interpret the fourth-order formulation of Equation (1), while the case of finite relative deformation is used for the interpretation of the second-order formulation of Equations (18) and (19).

## 5.1 Fourth-order formulation

For the particular case of isotropic elasticity, $C_{ijkl} = \lambda\,\delta_{ij}\delta_{kl} + \mu\left(\delta_{ik}\delta_{jl} + \delta_{il}\delta_{jk}\right)$ where $\lambda$ and $\mu$ are the Lamé constants. In this case, Equation (1) can be rewritten as

$$(\lambda+\mu)\left(u^{\mathrm{M}}_{j,ij} - \ell_s^2 u^{\mathrm{M}}_{j,ijkk}\right) + \mu\left(u^{\mathrm{M}}_{i,jj} - \ell_s^2 u^{\mathrm{M}}_{i,jjkk}\right) = \rho\left(\ddot{u}^{\mathrm{M}}_i - \ell_m^2 \ddot{u}^{\mathrm{M}}_{i,kk}\right) \tag{25}$$

In its general format, whereby macrostructural and microstructural deformations are included separately, Mindlin's theory has 18 independent additional coefficients. However, further simplification was made by setting the relative deformation equal to zero. In particular for the so-called "Format I" the equations of motion can be written as [7]

$$(\lambda+\mu)\,u^{\mathrm{M}}_{j,ij} + \mu u^{\mathrm{M}}_{i,jj} - 2\,(a_1+a_2+a_5)\,u^{\mathrm{M}}_{j,ijkk} - 2\,(a_3+a_4)\,u^{\mathrm{M}}_{i,jjkk} = \rho\ddot{u}^{\mathrm{M}}_i - b_1\ddot{u}^{\mathrm{M}}_{i,jj} - b_2\ddot{u}^{\mathrm{M}}_{j,ij} \tag{26}$$

where the various $a$ and $b$ constants are material parameters. In order to relate the two theories, it should hold that $2(a_1+a_2+a_5) = (\lambda+\mu)\ell_s^2$ and $2(a_3+a_4) = \mu\ell_s^2$ as concerns the stiffness terms. For the inertia terms, it should hold that $b_1 = \rho\ell_m^2$ and, in particular, $b_2 = 0$. The gradient elasticity theory of Equation (1) can thus be perceived as a particular version of Mindlin's theory.

## 5.2 Second-order formulation

Equation (20) contains contributions in terms of the velocities as well as a contribution in terms of the velocity gradients. This is similar to Mindlin's theory of elasticity with microstructure, in which

$$\mathscr{U}^{\mathrm{kin}} = \frac{1}{2}\rho\left((\dot{u}_i)^2 + \frac{1}{12}\ell^2\left(\dot{u}'_{i,j'}\right)^2\right) \tag{27}$$

where $u$ and $u'$ are defined as the macrostructural and microstructural displacements, respectively, and $\ell$ is the size of the microstructural unit cell. The spatial derivative in the higher-order term is taken with respect to the microstructural spatial coordinates $x'_j$. We will assume that $x'_j = x_j + \text{constant}$, therefore derivatives with respect to macroscopic and microscopic coordinates are identical.

The meaning of $u^{\mathrm{M}}$ in Equation (20) as well as the meaning of $u$ in Equation (27) is that of the macroscopic displacement. Extending this analogy between the two theories, we seek a relation between the auxiliary displacement $u^a$ of Equation (20) and the microscopic displacement $u'$ of Equation (27). Equation (20) is rewritten using Equation (15), *i.e.* [21]

$$\begin{aligned}
\mathscr{U}^{\text{kin}} &= \frac{1}{2}\rho\left((\dot{u}_i^a)^2 + \frac{\ell_m^2-\ell_s^2}{\ell_s^2}\left(\dot{u}_i^{\mathrm{M}}-\dot{u}_i^a\right)^2 + \left(\ell_m^2-\ell_s^2\right)\left(\dot{u}_{i,m}^{\mathrm{M}}\right)^2\right) \\
&= \frac{1}{2}\rho\left(\left(\dot{u}_i^{\mathrm{M}}-\ell_s^2\dot{u}_{i,jj}^{\mathrm{M}}\right)^2 + \frac{\ell_m^2-\ell_s^2}{\ell_s^2}\left(\ell_s^2\dot{u}_{i,jj}^{\mathrm{M}}\right)^2 + \left(\ell_m^2-\ell_s^2\right)\left(\dot{u}_{i,m}^a+\ell_s^2\dot{u}_{i,jjm}^{\mathrm{M}}\right)^2\right) \\
&= \frac{1}{2}\rho\left(\left(\dot{u}_i^{\mathrm{M}}\right)^2 - 2\ell_s^2\dot{u}_i^{\mathrm{M}}\dot{u}_{i,jj}^{\mathrm{M}} + \left(\ell_m^2-\ell_s^2\right)\left(\dot{u}_{i,m}^a\right)^2 + \mathscr{O}\left(\ell^4\right)\right)
\end{aligned} \tag{28}$$

In an infinite volume it holds that

$$-\int_\Omega \dot{u}_i^{\mathrm{M}}\dot{u}_{i,jj}^{\mathrm{M}}\,\mathrm{d}V = \int_\Omega \left(\dot{u}_{i,j}^{\mathrm{M}}\right)^2\,\mathrm{d}V \tag{29}$$

For arbitrary displacement fields, we can then write pointwise that

$$-\dot{u}_i^{\mathrm{M}}\dot{u}_{i,mm}^{\mathrm{M}} = \left(\dot{u}_{i,m}^{\mathrm{M}}\right)^2 = \left(\dot{u}_{i,m}^a+\ell_s^2\dot{u}_{i,jjm}^{\mathrm{M}}\right)^2 \tag{30}$$

where once more Equation (15) was used, so that

$$\mathscr{U}^{\text{kin}} = \frac{1}{2}\rho\left(\left(\dot{u}_i^{\mathrm{M}}\right)^2 + \left(\ell_m^2+\ell_s^2\right)\left(\dot{u}_{i,m}^a\right)^2 + \mathscr{O}\left(\ell^4\right)\right) \tag{31}$$

Comparing Equations (27) and (31) indicates that $u^a$ is similar to $u'$ and can thus be interpreted as the *microstructural displacement*. Hence, also the gradient elasticity theory of Equations (18) and (19) can be seen as a particular case of Mindlin's theory.

# 6 Dispersion properties

Next, the dispersive properties of Equation (1) are investigated. In view of the results of the previous Section, we will make the substitution $u^a = u^{\mathrm{m}}$, where an uppercase m denotes a microstructural quantity. We will also assume $C_{ijkl} = \lambda\delta_{ij}\delta_{kl} + \mu\delta_{ik}\delta_{jl} + \mu\delta_{il}\delta_{jk}$ as well as a two-dimensional plane strain configuration. Equations (18) and (19) are taken as the starting point. Both displacement fields $u^{\mathrm{m}}$ and $u^{\mathrm{M}}$ will be written as the derivatives of a dilatant potential $\varphi$ and a distortional potential $\psi$, that is

$$u_x^{\mathrm{m}} = \varphi_{,x}^{\mathrm{m}} + \psi_{,y}^{\mathrm{m}} \tag{32}$$

$$u_y^{\mathrm{m}} = \varphi_{,y}^{\mathrm{m}} - \psi_{,x}^{\mathrm{m}} \tag{33}$$

$$u_x^{\mathrm{M}} = \varphi_{,x}^{\mathrm{M}} + \psi_{,y}^{\mathrm{M}} \tag{34}$$

$$u_y^{\mathrm{M}} = \varphi_{,y}^{\mathrm{M}} - \psi_{,x}^{\mathrm{M}} \tag{35}$$

With these substitutions, Equation (18) is rewritten as

$$\begin{bmatrix} 0 \\ 0 \end{bmatrix} = \begin{bmatrix} \dfrac{\partial}{\partial x} \\ \dfrac{\partial}{\partial y} \end{bmatrix} \left\{ \frac{\ell_m^2}{\ell_s^2} \rho \ddot{\varphi}^{\mathrm{m}} - \frac{\ell_m^2 - \ell_s^2}{\ell_s^2} \rho \ddot{\varphi}^{\mathrm{M}} - (\lambda + 2\mu) \left( \varphi_{,xx}^{\mathrm{m}} + \varphi_{,yy}^{\mathrm{m}} \right) \right\}$$
$$+ \begin{bmatrix} \dfrac{\partial}{\partial y} \\ -\dfrac{\partial}{\partial x} \end{bmatrix} \left\{ \frac{\ell_m^2}{\ell_s^2} \rho \ddot{\psi}^{\mathrm{m}} - \frac{\ell_m^2 - \ell_s^2}{\ell_s^2} \rho \ddot{\psi}^{\mathrm{M}} - \mu \left( \psi_{,xx}^{\mathrm{m}} + \psi_{,yy}^{\mathrm{m}} \right) \right\} \tag{36}$$

whereas Equation (19) renders

$$\begin{bmatrix} 0 \\ 0 \end{bmatrix} = \begin{bmatrix} \dfrac{\partial}{\partial x} \\ \dfrac{\partial}{\partial y} \end{bmatrix} \left\{ -\frac{\ell_m^2 - \ell_s^2}{\ell_s^2} \rho \ddot{\varphi}^{\mathrm{m}} + \frac{\ell_m^2 - \ell_s^2}{\ell_s^2} \rho \ddot{\varphi}^{\mathrm{M}} - \left( \ell_m^2 - \ell_s^2 \right) \rho \left( \ddot{\varphi}_{,xx}^{\mathrm{M}} + \ddot{\varphi}_{,yy}^{\mathrm{M}} \right) \right\}$$
$$+ \begin{bmatrix} \dfrac{\partial}{\partial y} \\ -\dfrac{\partial}{\partial x} \end{bmatrix} \left\{ -\frac{\ell_m^2 - \ell_s^2}{\ell_s^2} \rho \ddot{\psi}^{\mathrm{m}} + \frac{\ell_m^2 - \ell_s^2}{\ell_s^2} \rho \ddot{\psi}^{\mathrm{M}} - \left( \ell_m^2 - \ell_s^2 \right) \rho \left( \ddot{\psi}_{,xx}^{\mathrm{M}} + \ddot{\psi}_{,yy}^{\mathrm{M}} \right) \right\} \tag{37}$$

Equations (36) and (37) hold if the bracketed expressions vanish individually. Compressive waves are found from the two coupled expressions in terms of $\varphi^{\mathrm{m}}$ and $\varphi^{\mathrm{M}}$ as

$$\frac{\ell_m^2}{\ell_s^2} \rho \ddot{\varphi}^{\mathrm{m}} - \frac{\ell_m^2 - \ell_s^2}{\ell_s^2} \rho \ddot{\varphi}^{\mathrm{M}} - (\lambda + 2\mu) \left( \varphi_{,xx}^{\mathrm{m}} + \varphi_{,yy}^{\mathrm{m}} \right) = 0 \tag{38}$$

$$-\frac{\ell_m^2 - \ell_s^2}{\ell_s^2} \rho \ddot{\varphi}^{\mathrm{m}} + \frac{\ell_m^2 - \ell_s^2}{\ell_s^2} \rho \ddot{\varphi}^{\mathrm{M}} - \left( \ell_m^2 - \ell_s^2 \right) \rho \left( \ddot{\varphi}_{,xx}^{\mathrm{M}} + \ddot{\varphi}_{,yy}^{\mathrm{M}} \right) = 0 \tag{39}$$

Simultaneous trial solutions for $\varphi^{\mathrm{m}}$ and $\varphi^{\mathrm{M}}$ are used as

$$\varphi^{\mathrm{m}} = A \exp\left(\mathrm{i}(k_x x + k_y y - \omega t)\right) \tag{40}$$
$$\varphi^{\mathrm{M}} = B \exp\left(\mathrm{i}(k_x x + k_y y - \omega t)\right) \tag{41}$$

The coupling between the two displacement fields is manifested by the same wave numbers $k_x$ and $k_y$ and the same angular frequency $\omega$ appearing in the two trial solutions, but the two amplitudes $A$ and $B$ are assumed to be different. Substitution of these trial solutions into Equation (39) yields a relation between the two amplitudes $A$ and $B$ as

$$A\omega^2 - B\omega^2 \left( 1 + \ell_s^2 k^2 \right) = 0 \quad \text{by which} \quad A = B \left( 1 + \ell_s^2 k^2 \right) \tag{42}$$

where $k^2 = k_x^2 + k_y^2$. Next, the two trial solutions are substituted into Equation (38), which gives

$$-A\omega^2 \frac{\ell_m^2}{\ell_s^2} \rho + B\omega^2 \frac{\ell_m^2 - \ell_s^2}{\ell_s^2} \rho + Ak^2(\lambda + 2\mu) = 0 \tag{43}$$

Division by $k^2(\lambda+2\mu)$ and substitution of Equation (42) results in

$$\frac{c^2}{c_p^2}\left(-\frac{\ell_m^2}{\ell_s^2}(1+\ell_s^2k^2)+\frac{\ell_m^2-\ell_s^2}{\ell_s^2}\right)+1+\ell_s^2k^2=0 \tag{44}$$

which can be rearranged as

$$\frac{c^2}{c_p^2}=\frac{1+\ell_s^2k^2}{1+\ell_m^2k^2} \tag{45}$$

where $c=\omega/k$ is the phase velocity and $c_p=\sqrt{(\lambda+2\mu)/\rho}$ is the long wave limit of the compressive wave velocity. Following a similar procedure for the shear wave expressions, it can be verified that

$$\frac{c^2}{c_s^2}=\frac{1+\ell_s^2k^2}{1+\ell_m^2k^2} \tag{46}$$

where $c_s=\sqrt{\mu/\rho}$ is the long wave limit of the shear wave velocity. As is thus demonstrated, the compression waves and shear waves are governed by the same dispersion laws. Moreover, the same dispersion expressions were obtained earlier for a model with higher-order inertia and higher-order stiffness but without applying the operator split of Section 3, see [12]. Thus, it can be concluded that the dispersion characteristics of the body waves are not affected by the mathematical manipulations in going from Equation (1) to Equations (18) and (19).

## 7 Finite element equations

Next, a finite element discretisation of the field equations will be developed. Adopting Voigt's notation and again $u^a=u^{\mathrm{m}}$, the weak form of Equation (18) reads

$$\int_\Omega \delta\boldsymbol{u}^T\rho\left(\frac{\ell_m^2}{\ell_s^2}\,\ddot{\boldsymbol{u}}^{\mathrm{m}}-\frac{\ell_m^2-\ell_s^2}{\ell_s^2}\,\ddot{\boldsymbol{u}}^{\mathrm{M}}\right)\mathrm{d}V-\int_\Omega \delta\boldsymbol{u}^T\mathbf{C}\mathbf{L}\boldsymbol{\varepsilon}^{\mathrm{m}}\,\mathrm{d}V=0 \tag{47}$$

where $\delta\boldsymbol{u}$ is a vector of test functions and $\mathbf{L}$ is a differential operator. Integrating the last term of Equation (47) by parts leads to

$$\int_\Omega \delta\boldsymbol{u}^T\rho\left(\frac{\ell_m^2}{\ell_s^2}\,\ddot{\boldsymbol{u}}^{\mathrm{m}}-\frac{\ell_m^2-\ell_s^2}{\ell_s^2}\,\ddot{\boldsymbol{u}}^{\mathrm{M}}\right)\mathrm{d}V+\int_\Omega \delta\boldsymbol{\varepsilon}^T\mathbf{C}\boldsymbol{\varepsilon}^{\mathrm{m}}\,\mathrm{d}V=\int_{\Gamma_n}\delta\boldsymbol{u}^T\boldsymbol{t}\,\mathrm{d}S \tag{48}$$

where the prescribed tractions $\boldsymbol{t}$ have been substituted on the relevant part of the boundary $\Gamma_n$. Similarly, the weak form of Equation (19) reads

$$\int_\Omega \delta\boldsymbol{v}^T\rho\left(-\frac{\ell_m^2-\ell_s^2}{\ell_s^2}\,\ddot{\boldsymbol{u}}^{\mathrm{m}}+\frac{\ell_m^2-\ell_s^2}{\ell_s^2}\,\ddot{\boldsymbol{u}}^{\mathrm{M}}-\left(\ell_m^2-\ell_s^2\right)\nabla^2\ddot{\boldsymbol{u}}^{\mathrm{M}}\right)\mathrm{d}V=0 \tag{49}$$

with $\delta\boldsymbol{v}$ a vector of test functions. The last term is integrated by parts, by which

$$\int_\Omega \delta\boldsymbol{v}^T \rho \left( -\frac{\ell_m^2 - \ell_s^2}{\ell_s^2}\, \ddot{\boldsymbol{u}}^{\mathrm{m}} + \frac{\ell_m^2 - \ell_s^2}{\ell_s^2}\, \ddot{\boldsymbol{u}}^{\mathrm{M}} \right) \mathrm{d}V$$

$$+ \sum_{\xi=x,y,z} \int_\Omega \frac{\partial \boldsymbol{v}^T}{\partial \xi} \rho \left(\ell_m^2 - \ell_s^2\right) \frac{\partial \ddot{\boldsymbol{u}}^{\mathrm{M}}}{\partial \xi}\, \mathrm{d}V \;=\; 0 \tag{50}$$

As discussed in Section 4, homogeneous natural boundary conditions have been assumed, by which the boundary integral cancels.

Shape functions $\mathbf{N}_{\mathrm{m}}$ are used to discretise $\delta\boldsymbol{u}$ and $\boldsymbol{u}^{\mathrm{m}}$, whereas $\delta\boldsymbol{v}$ and $\boldsymbol{u}^{\mathrm{M}}$ are discretised with shape functions $\mathbf{N}_{\mathrm{M}}$. For the time discretisation the constant acceleration variant of the Newmark scheme is used, whereby the accelerations are taken as the unknowns. The fully discretised system of equations can then be written as

$$\begin{bmatrix} \mathbf{M}_{\mathrm{mm}} + \frac{1}{4}\Delta t^2 \mathbf{K}_{\mathrm{mm}} & -\mathbf{M}_{\mathrm{mM}} \\ -\mathbf{M}_{\mathrm{mM}}^T & \mathbf{M}_{\mathrm{MM}} \end{bmatrix} \begin{bmatrix} \ddot{\mathbf{u}}^{\mathrm{m}}_{(t+\Delta t)} \\ \ddot{\mathbf{u}}^{\mathrm{M}}_{(t+\Delta t)} \end{bmatrix} = \begin{bmatrix} \mathbf{f}^{\mathrm{ext}}_{(t+\Delta t)} - \mathbf{f}^{\mathrm{int}}_{(t)} \\ 0 \end{bmatrix} \tag{51}$$

where $\Delta t$ is the applied time step and

$$\mathbf{f}^{\mathrm{int}}_{(t)} = \mathbf{K}_{\mathrm{mm}} \left( \mathbf{u}^{\mathrm{m}}_{(t)} + \Delta t \dot{\mathbf{u}}^{\mathrm{m}}_{(t)} + \frac{1}{4}\Delta t^2 \ddot{\mathbf{u}}^{\mathrm{m}}_{(t)} \right) \tag{52}$$

Furthermore, the following matrices are introduced:

$$\mathbf{M}_{\mathrm{mm}} = \int_\Omega \mathbf{N}_{\mathrm{m}}^T \rho \frac{\ell_m^2}{\ell_s^2} \mathbf{N}_{\mathrm{m}}\, \mathrm{d}V \tag{53}$$

$$\mathbf{M}_{\mathrm{mM}} = \int_\Omega \mathbf{N}_{\mathrm{m}}^T \rho \frac{\ell_m^2 - \ell_s^2}{\ell_s^2} \mathbf{N}_{\mathrm{M}}\, \mathrm{d}V \tag{54}$$

$$\mathbf{M}_{\mathrm{MM}} = \int_\Omega \mathbf{N}_{\mathrm{M}}^T \rho \frac{\ell_m^2 - \ell_s^2}{\ell_s^2} \mathbf{N}_{\mathrm{M}}\, \mathrm{d}V + \sum_{\xi=x,y,z} \int_\Omega \frac{\partial \mathbf{N}_{\mathrm{M}}^T}{\partial \xi} \rho \left(\ell_m^2 - \ell_s^2\right) \frac{\partial \mathbf{N}_{\mathrm{M}}}{\partial \xi}\, \mathrm{d}V \tag{55}$$

$$\mathbf{K}_{\mathrm{mm}} = \int_\Omega \mathbf{B}_{\mathrm{m}}^T \mathbf{C} \mathbf{B}_{\mathrm{m}}\, \mathrm{d}V \tag{56}$$

$$\mathbf{f}^{\mathrm{ext}} = \int_{\Gamma_n} \mathbf{N}_{\mathrm{m}}^T \boldsymbol{t}\, \mathrm{d}S \tag{57}$$

and $\mathbf{B}_{\mathrm{m}} = \mathbf{L}\mathbf{N}_{\mathrm{m}}$.

# 8 Examples

In this Section we present a set of numerical results. Dispersive wave propagation in a one-dimensional bar is simulated. The test set-up is depicted in Figure 1, whereby

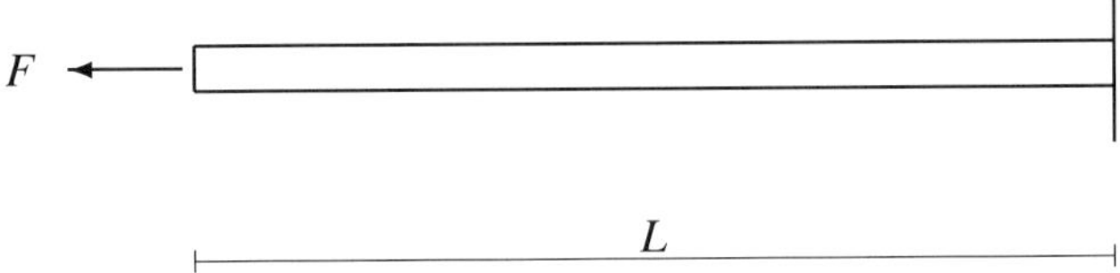


Figure 1: One-dimensional dynamic bar problem — geometry and loading conditions

the length of the bar $L = 100$ m and its cross-sectional area is 1 m$^2$. The material parameters are taken as $E = 1$ N/m$^2$ and $\rho = 1$ kg/m$^3$. The right end of the bar is fixed. Linear finite element shape functions have been used for $u^{\mathrm{m}}$ as well as for $u^{\mathrm{M}}$.

## 8.1 Influence of boundary conditions

Firstly, the influence of the higher-order natural boundary conditions is shown. Section 4 suggested imposing a tying between the two displacements $u^{\mathrm{m}}$ and $u^{\mathrm{M}}$ on the boundary. We will compare the case where this tying is not used (denoted as "incorrect boundary conditions") with the case where this tying is applied (indicated with "correct boundary conditions"). The length scales $\ell_m = 2$ m and $\ell_s = 1$ m, and a force $F = 1$ N is applied from time $t = 0$ s onwards. The bar is discretised with 200 linear finite elements, and for the time integration a time step size of 0.5 s is used.

With the incorrect boundary conditions, the microscopic and macroscopic strain profiles as depicted in Figure 2 are obtained, whereby the two strain fields are plotted along the bar for successive time instants. It can be seen that the microscopic strain (left of Figure 2) adopts realistic values at the left boundary, but the macroscopic strain (depicted on the right) unrealistically tends to zero at the left end. With the correct boundary conditions, the results of Figure 3 are obtained. For this case, the value of the macroscopic strain at the left end is realistic, which demonstrates the relevance of the tying and, thus, of emulating homogeneous higher-order natural boundary conditions. Note that the microscopic strains are affected as well: compared to Figure 2, the microscopic strains in Figure 3 are much smoother.

## 8.2 Dispersive properties

Next, we demonstrate the capability of the model to simulate dispersive wave propagation. Wave propagation is triggered by applying a force of magnitude 1 N at the left end of the bar during the first 1 s of the simulation. The bar has been discretised with 100 elements. The Newmark algorithm for time integration has been used with $\Delta t = 0.2$ s. Figure 4 shows the evolution in time of the microstructural displacement $u^{\mathrm{m}}$ and the macrostructural displacement $u^{\mathrm{M}}$ at times $t = 20$ s, $t = 40$ s and $t = 60$ s. Three values for $\ell_m$ were taken, namely $\ell_m = 1.1$ m, $\ell_m = 2$ m and $\ell_m = 5$ m. In all

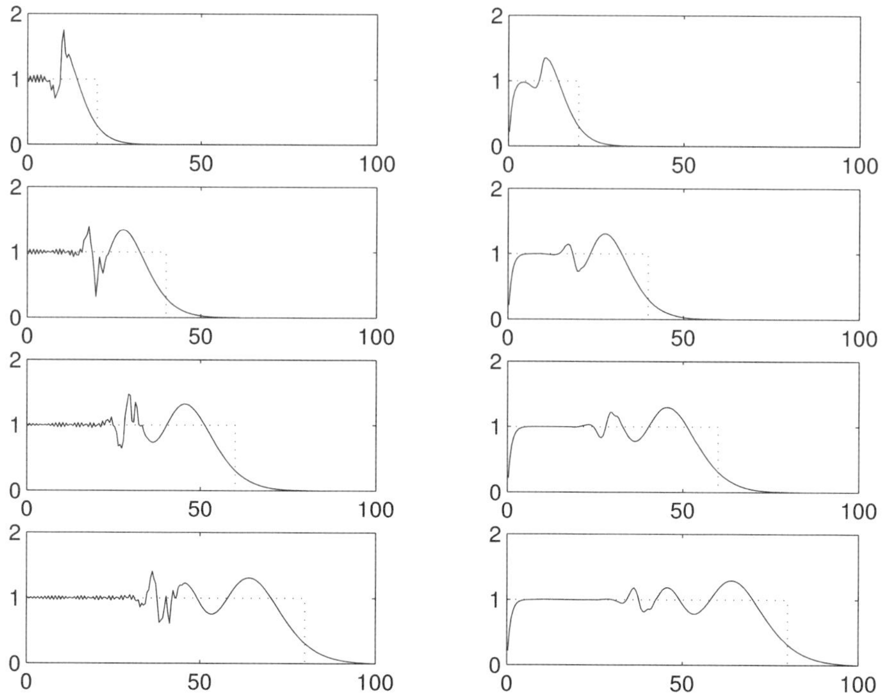


Figure 2: Influence of boundary conditions — microstructural strain (left) and macrostructural strain (right) across bar for times 20 s, 40 s, 60 s and 80 s with incorrect boundary conditions. Dotted lines indicate the analytical solution for classical elasticity

cases, $\ell_s = 1$ m. It can be seen that the three sets of curves diverge more and more with an ongoing propagation of the wave. As explained in [19], the dispersive effects are stronger for larger ratios of $\ell_m/\ell_s$. This holds for both displacement fields, *i.e.* $u^{\mathrm{m}}$ as well as $u^{\mathrm{M}}$, although in general the macrostructural displacement $u^{\mathrm{M}}$ is smoother than the microstructural displacement $u^{\mathrm{m}}$.

## 8.3 Wave propagation in a periodic laminate

Finally, we will present some results for dispersive wave propagation in periodic laminates. The studied laminate is depicted in Figure 5. The laminate consists of two materials with volume fractions $\alpha$ and $(1-\alpha)$, and the length of its periodic unit cell is denoted with $\ell$. The effective mass density $\rho$ and the effective Young's modulus $E$

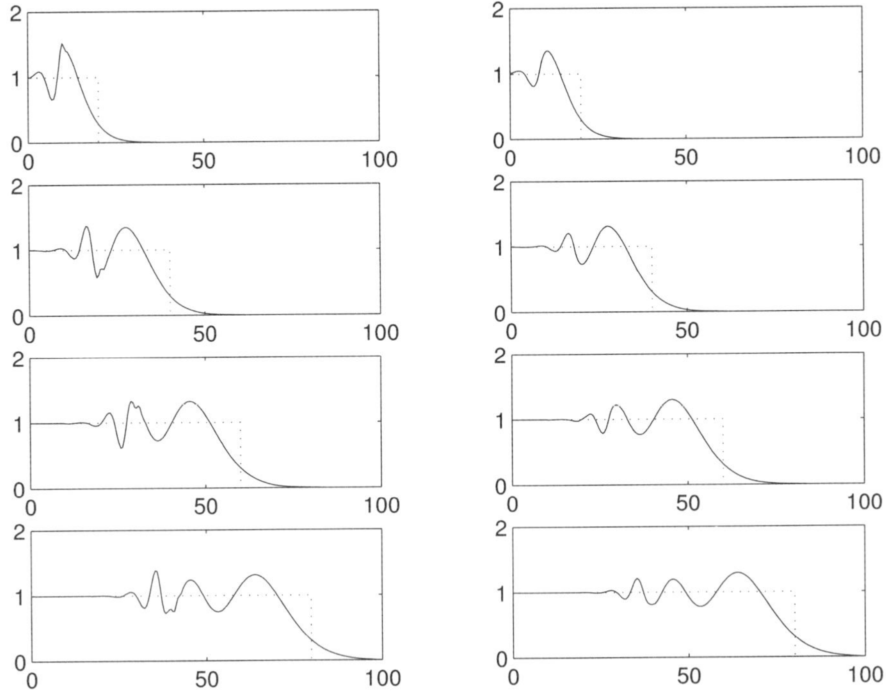


Figure 3: Influence of boundary conditions — microstructural strain (left) and macrostructural strain (right) across bar for times 20 s, 40 s, 60 s and 80 s with correct boundary conditions. Dotted lines indicate the analytical solution for classical elasticity

are given in terms of the component properties as [6]

$$\rho = \alpha\rho_1 + (1-\alpha)\rho_2 \tag{58}$$

$$E = \frac{E_1 E_2}{(1-\alpha)E_1 + \alpha E_2} \tag{59}$$

In [21] a gradient elasticity model for such a laminate was derived as

$$\rho\ddot{u} - \frac{1}{12}(\eta+1)\ell^2\rho\ddot{u}_{,xx} = Eu_{,xx} - \frac{1}{12}\ell^2 Eu_{,xxxx} \tag{60}$$

where

$$\eta = \left(\frac{\alpha(1-\alpha)(E_1\rho_1 - E_2\rho_2)}{(1-\alpha)\rho E_1 + \alpha\rho E_2}\right)^2 \tag{61}$$

For the simulations we have taken a 300 m long bar that is subjected to a unit-pulse force. The unit cell size $\ell = 1$ m and the volume fractions are set by $\alpha = \frac{1}{2}$. The solutions for the macrostructural displacement are compared with the case where the

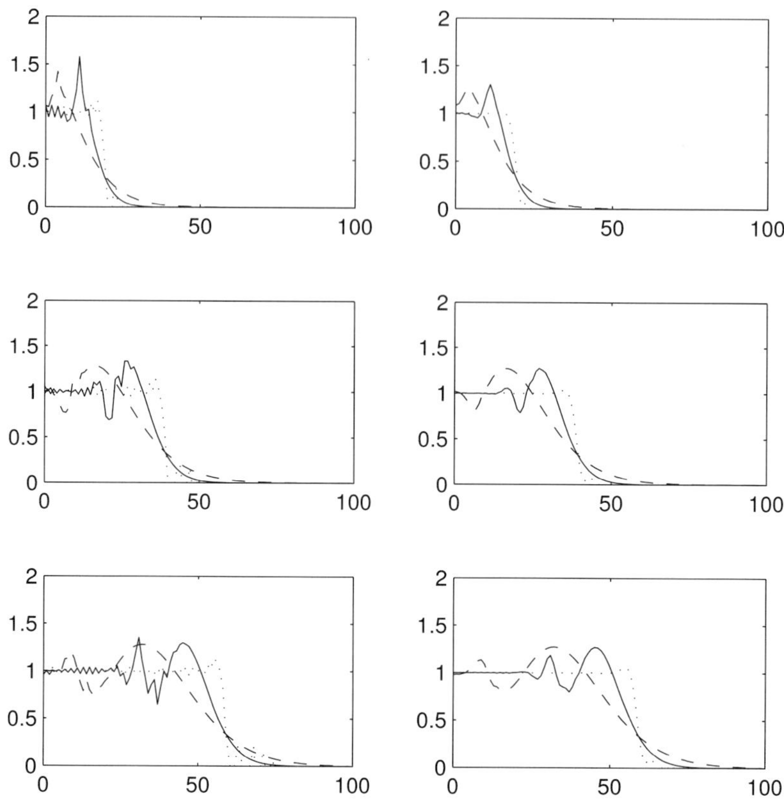

Figure 4: Dispersive wave propagation — microstructural displacement (left) and macrostructural displacement (right) across bar at times $t = 20$ s (top), $t = 40$ s (middle) and $t = 60$ s (bottom). The three line types indicate three values for $\ell_m$, namely $\ell_m = 1.1$ m (dotted), $\ell_m = 2$ m (solid) and $\ell_m = 5$ m (dashed)

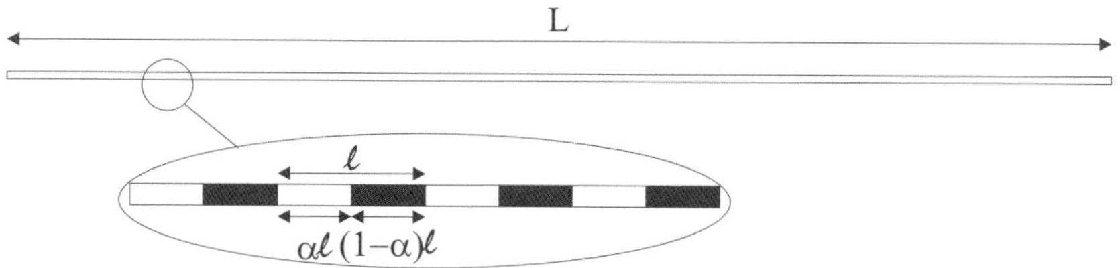


Figure 5: One-dimensional representation of a two-component laminate

laminates are modelled explicitly. The latter solution has been obtained using 6000 linear elements of length 0.05 m and a time step size of 0.05 s. Alternating groups of 10 elements have been assigned material properties $E_1$, $\rho_1$ and $E_2$, $\rho_2$. The second set of material properties ($E_2$ and $\rho_2$) are computed for given values of $E_1$ and $\rho_1$ from Equations (58) and (59) in order that the macroscopic material properties of the bar are $E = 1$ N/m$^2$ and $\rho = 1$ kg/m$^3$. Two sets of material parameters are taken. A strong variation of material properties is obtained through $E_1 = 10^6$ N/m$^2$ and $\rho_1 =$ 1.9999 kg/m$^3$, and a moderate variation of material properties is taken via $E_1 = 10$ N/m$^2$ together with $\rho_1 = 1.2$ kg/m$^3$. For the gradient elastic numerical simulations, 1500 linear elements are employed with a time step size of 0.2 s. Figure 6 shows the profile of the macrostructural displacement $u^{\mathrm{M}}$ of the wave front at $t = 280$ s. The corresponding wave profile (obtained numerically) for a homogeneous elastic bar without gradient effects is also shown. The larger the ratios $E_1/E_2$ and $\rho_1/\rho_2$, the more the wave front deviates from the case of homogeneous elasticity, thus indicating a more pronounced effect of dispersion. When examining the front of the wave, it can be seen that the gradient elasticity solution and the heterogeneous elastic solution show a good agreement.

# 9 Concluding remarks

In this paper a gradient elasticity theory is presented that is equipped with strain gradients as well as inertia gradients. The model has been derived through the second-order homogenisation of the response of a Representative Volume Element (RVE). Two additional terms appear in the equations of motion: the Laplacian of the usual inertia term, accompanied by a coefficient in terms of the dynamic RVE size, and the Laplacian of the usual stiffness term, accompanied by a coefficient in terms of the static RVE size. The resulting model is a fourth-order partial differential equation and would, thus, require $\mathscr{C}^1$-continuous interpolations for numerical simulations. To facilitate finite element implementations, we have rewritten the fourth-order equation as two fully coupled and symmetric second-order equations, in terms of two distinct displacement fields. We have studied the following aspects of the resulting formulation:

- **Energy functionals:** The kinetic and potential energy densities have been identified. Positive-definiteness of these functionals gives a relation between the two length scales (namely $\ell_m > \ell_s$ or $\ell_{\mathrm{dyn}} > \ell_{\mathrm{stat}}$). Furthermore, to emulate the effects of the commonly applied zero higher-order natural boundary condition of the original fourth-order formulation, we suggest using a tying between the two displacement fields on the boundary.

- **Comparison with Mindlin theory:** Both the original fourth-order formulation and the new second-order formulation can be perceived as special cases of Mindlin's 1964 theory of elasticity with microstructure. An important observation concerning the second-order formulation is that the two displacement fields can in fact be interpreted as the *microstructural displacement* and the *macrostructural displacement.* The second-order formulation is therefore a true multi-scale formulation with two interacting scales of observations.

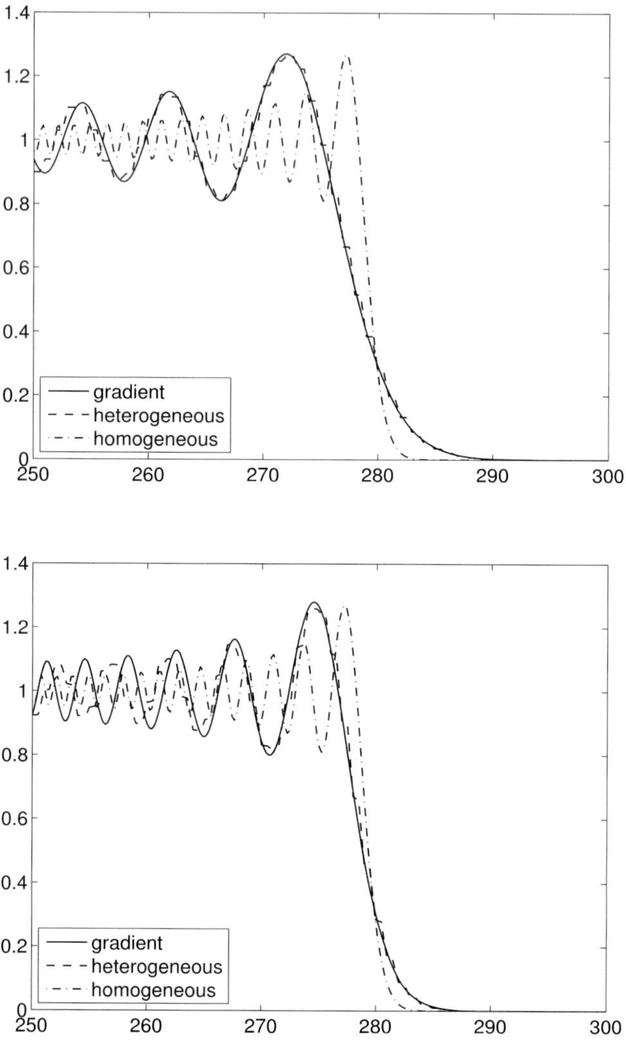


Figure 6: Laminated bar — profiles of the macrostructural displacement across bar for strong (top) and moderate (bottom) variation of material properties

- **Dispersion of body waves:** The dispersion of compression waves and shear waves has been investigated analytically. It was found that the two wave types obey the same dispersion relation; moreover, this holds for the fourth-order formulation as well as the second-order formulation. Thus, the dispersion characteristics of the model have not been affected by the mathematical manipulations in going from the fourth-order to the second-order formulation.

- **Numerical simulations:** We presented the finite element equations of the second-order formulation, for which the commonly used $\mathscr{C}^0$-continuous shape functions suffice. Numerical simulations of dispersive wave propagation have been carried out for a number of simple tests. We demonstrated the relevance of using boundary conditions with a tying between microstructural and macrostructural displacements. The effects of the two length scales have been shown as well. Finally, a further quantification of the length scales has been made for the particular case of a periodic laminate, and we found an excellent agreement between the response of our gradient elasticity model and the response of a model in which all microstructural components are modelled explicitly.

# References

[1] Y.L. Yarnell, J.L. Warren, R.G. Wenzel, S.H. Koenig, "Phonon dispersion curves in Bismuth", *IBM J. Res. Developm.*, 8:234–240, 1964.

[2] F. Lanza di Scalea, J. McNamara, "Measuring high-frequency wave propagation in railroad tracks by joint time-frequency analysis", *J. Sound Vibr.*, 273:637–651, 2004.

[3] E.C. Aifantis, "On the role of gradients in the localization of deformation and fracture", *Int. J. Engng. Sci.*, 30:1279–1299, 1992.

[4] C.Q. Ru, E.C. Aifantis, "A simple approach to solve boundary-value problems in gradient elasticity", *Acta Mech.*, 101:59–68, 1993.

[5] M.Y. Gutkin, E.C. Aifantis, "Dislocations in the theory of gradient elasticity", *Scripta Mater.*, 40:559–566, 1999.

[6] W. Chen, J. Fish, "A dispersive model for wave propagation in periodic heterogeneous media based on homogenization with multiple spatial and temporal scales", *ASME J. Appl. Mech.*, 68:153–161, 2001.

[7] R.D. Mindlin, "Micro-structure in linear elasticity", *Arch. Rat. Mech. Analysis*, 16:52–78, 1964.

[8] H.G. Georgiadis, I. Vardoulakis, G. Lykotrafitis, "Torsional surface waves in a gradient-elastic half-space", *Wave Mot.*, 31:333–348, 2000.

[9] A.V. Metrikine, H. Askes, "One-dimensional dynamically consistent gradient elasticity models derived from a discrete microstructure. Part 1: Generic formulation", *Eur. J. Mech. A/Solids*, 21:555–572, 2002.

[10] H. Askes, A.V. Metrikine, "One-dimensional dynamically consistent gradient elasticity models derived from a discrete microstructure. Part 2: Static and dynamic response", *Eur. J. Mech. A/Solids*, 21:573–588, 2002.

[11] H.G. Georgiadis, "The mode III crack problem in microstructured solids governed by dipolar gradient elasticity: static and dynamic analysis", *ASME J. Appl. Mech.*, 70:517–530, 2003.

[12] A.V. Metrikine, H. Askes, "An isotropic dynamically consistent gradient elasticity model derived from a 2D lattice", *Phil. Mag.*, 86:3259–3286, 2006.

[13] V. Kouznetsova, M.G.D. Geers, W.A.M. Brekelmans, "Multi-scale constitutive

modelling of heterogeneous materials with a gradient-enhanced computational homogenization scheme", *Int. J. Numer. Meth. Engng.*, 54:1235–1260, 2002.

[14] V.G. Kouznetsova, M.G.D. Geers, W.A.M. Brekelmans, "Multi-scale second-order computational homogenization of multi-phase materials: a nested finite element solution strategy", *Comp. Meth. Appl. Mech. Engng.*, 193:5525–5550, 2004.

[15] I.M. Gitman, H. Askes, L.J. Sluys, "Representative Volume size as a macroscopic length scale parameter", In V.C. Li et al., editor, *Fract. Mech. Concr. Struct.*, pages 483–489. Ia-FraMCoS, 2004.

[16] V.G. Kouznetsova, M.G.D. Geers, W.A.M. Brekelmans, "Size of a Representative Volume Element in a second-order computational homogenization framework", *Int. J. Multiscale Comp. Engng.*, 2:575–598, 2004.

[17] I.M. Gitman, H. Askes, E.C. Aifantis, "The representative volume size in static and dynamic micro-macro transitions", *Int. J. Fract.*, 135:L3–L9, 2005.

[18] H. Askes, E.C. Aifantis, "Gradient elasticity theories in statics and dynamics — a unification of approaches", *Int. J. Fract.*, 139:297–304, 2006.

[19] H. Askes, T. Bennett, E.C. Aifantis, "A new formulation and $C^0$-implementation of dynamically consistent gradient elasticity", *Int. J. Numer. Meth. Engng.*, 72:111–126, 2007.

[20] C. Polizzotto, "Gradient elasticity and nonstandard boundary conditions", *Int. J. Solids Struct.*, 40:7399–7423, 2003.

[21] T. Bennett, I.M. Gitman, H. Askes, "Elasticity theories with higher-order gradients of inertia and stiffness for the modelling of wave dispersion in laminates", *Int. J. Fract.*, 148:185–193, 2007.

©Saxe-Coburg Publications, 2008.
Trends in Engineering Computational Technology
B.H.V. Topping and M. Papadrakakis, (Editors)
Saxe-Coburg Publications, Stirlingshire, Scotland, 209-228.

# Chapter 11

# Multiscale Assessment of Low-Temperature Performance of Flexible Pavements

**E. Aigner[1], R. Lackner[2], M. Wistuba[3], J. Eberhardsteiner[1] and H.A. Mang[1]**
**[1] Institute for Mechanics of Materials and Structures**
**Vienna University of Technology, Austria**
**[2] FG Computational Mechanics**
**Technical University of Munich, Germany**
**[3] Braunschweig Pavement Engineering Centre**
**Technical University of Braunschweig, Germany**

## Abstract

The increase of heavy-load traffic within Europe requires the development of appropriate tools for the assessment of existing and new road infrastructure. In this paper, such a tool is presented, combining multiscale material modeling of asphalt with structural analysis of flexible pavements at low temperatures. At this temperature regime, rapid cooling of the road surface in consequence of temperature drops may result in so-called top-down cracking. These cracks, when propagating further into the base layer, significantly reduce the service life of road infrastructure. Within the proposed multiscale model for asphalt, the temperature-dependent viscoelastic properties of asphalt are related to the binder material (bitumen), the only material phase exhibiting thermorheological behavior, accounting for

- the large variability of asphalt mixtures, resulting from different mix design, different constituents (*e.g.*, bitumen, filler, aggregate,), and the allowance of additives, and
- changing material behavior in consequence of thermal, chemical, and mechanical loading.

In the case of multiscale modeling of the viscoelastic nature of bituminous mixtures, the parameters of the underlying viscoelastic model are obtained from upscaling from the respective parameters identified at the bitumen-scale towards the macroscale. Hereby, the viscoelastic behavior of bitumen serves as input and the effect of the addition of aggregates, *i.e.*, filler, sand, and stone is considered. Upscaling of viscoelastic properties is performed in the framework of continuum micromechanics, employing the Mori-Tanaka scheme [1]. Within the elastic-viscoelastic correspondence principle, the elastic shear compliance in the equations employed for upscaling of elastic properties is replaced by the respective Laplace-Carson transform [2] of the viscoelastic compliance [3]. The so-obtained macroscopic model parameters are employed in the numerical analysis of flexible pavements, giving access to stresses resulting from (i) a sudden drop of the surface temperature in consequence of changing weather conditions and (ii) traffic loading. Comparison of the so-obtained surface stresses with the tensile strength of asphalt at the respective surface temperature allows the assessment of the risk of top-down cracking in flexible pavements.

**Keywords:** bitumen, asphalt, pavement, viscoelasticity, cracking, multiscale model, upscaling, identification, validation, FEM, thermal loading.

# 1 Material model for asphalt

The thermorheological behavior of asphalt mainly defines the risk of so-called top-down cracking (see Figure 1) under combined thermal-traffic loading at low temperatures [4]. If these cracks propagate further into the base layer, the service life of the

Figure 1: Low-temperature cracking

road infrastructure is significantly reduced. In order to assess and finally optimize the underlying bituminous materials (surface layer and base layer) the multiscale framework of material description is proposed, allowing us to relate the asphalt behavior at the so-called structural scale (macroscale) to finer-scale characteristics, such as material composition and behavior of the constituents (see Figure 2) [5].

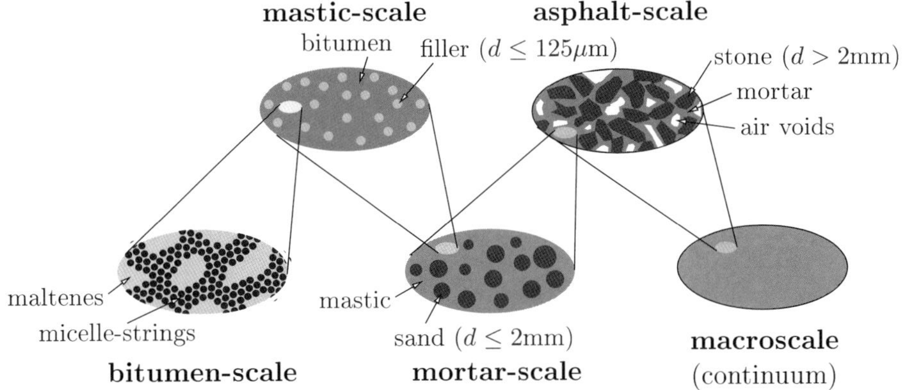


Figure 2: Multiscale model for asphalt

## 1.1 Rheological model and finite element implementation

Over the last decades, a large number of material models for the description of the viscoelastic behavior of asphalt have been developed. Starting with the Maxwell and the Burger model, the experimentally-observed nonlinearity in the creep deformation with time was taken into account by considering nonlinear dashpots (parabolic creep), which led to the so-called Huet model. Whereas the Huet model is well-suited for the description of monotonic loading of asphalt, it fails to predict the behavior under cyclic loading. As a remedy, the Huet-Sayegh model [7], obtained from adding a linear spring in parallel to the Huet model, was introduced. Finally, discrepancies between experimental data and the model answer of the Huet-Sayegh model in the low-frequency domain led to the extension of the latter towards the 2S2P1D model [8]. In addition to these models, generalized models (generalized Maxwell, generalized Kelvin-Voigt) are commonly employed to describe the viscoelastic behavior of asphalt, representing the nonlinear material behavior in a discretized manner (see, *e.g.*, [9]). In this paper, modes of implementation of the power-law (PL) model into the finite element program FEAP is presented. This model is well-suited for the simulation of the response of asphalt in the low-temperature regime, as it is the case during the assessment of the risk of low-temperature cracking.

The PL model (see Figure 3) consists of a linear spring element and a nonlinear dashpot connected in series, with the strain rate being composed of an elastic and a viscous part, reading

$$\dot{\varepsilon} = \dot{\varepsilon}^e + \dot{\varepsilon}^v = \frac{\dot{\sigma}}{E} + H \left(\frac{t}{\bar{\tau}}\right)^p \sigma \tag{1}$$

In Equation (1), $E$ [MPa] is Young's modulus, and $H$ [MPa$^{-1}$ s$^{-1}$] and $p$ [-] are parameters describing the nonlinear dashpot. $\bar{\tau}$ represents the characteristic time of the underlying creep process. In case of the parameters not changing with time, Equation

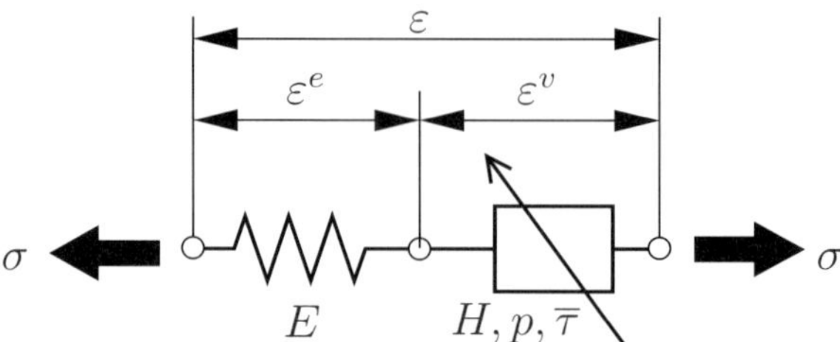


Figure 3: PL model

(1) may be integrated, giving the creep compliance $J = \varepsilon/\sigma_0$, where $\sigma_0$ is a constant stress, as

$$J_{PL}(t-\tau) = \frac{1}{E} + \frac{H}{\overline{\tau}^p\,(p+1)}\,(t-\tau)^{p+1} = \frac{1}{E} + J_a\left(\frac{t-\tau}{\overline{\tau}}\right)^k \tag{2}$$

where $\tau$ is the time instant of loading. In Equation (2), $J_a = H\,\overline{\tau}/(p+1)$ [MPa$^{-1}$] and $k = p+1$.

In case of a variable load history, the strain $\varepsilon(t)$ is given by the convolution integral

$$\varepsilon(t) = \int_0^t J(t-\tau)\frac{\partial\sigma}{\partial\tau}d\tau \tag{3}$$

In numerical simulations, the load (time) history is divided into load (time) increments. Accordingly, the strain at the end of the time increment $(n+1)$ is obtained from Equation (3) as

$$\varepsilon(t_{n+1}) = \varepsilon_{n+1} = \sum_{j=1}^{n+1}\Delta\varepsilon_j \qquad \text{with} \qquad \Delta\varepsilon_j = \int_{t_{j-1}}^{t_j} J(t_{n+1}-\tau)\frac{\partial\sigma}{\partial\tau}d\tau \tag{4}$$

where $\Delta\varepsilon_j$ represents the contributions to the viscoelastic strain $\varepsilon_{n+1}$ in consequence of stress changes within the time increment $j$, with $t_{j-1} \le t \le t_j$.

For the computation of the integral appearing in Equation (4), the stress history $\sigma(\tau)$ may be approximated by (see Figure 4):

1. A linear function (Figure 4(b)), with

$$\sigma(t) = \frac{t-t_{j-1}}{t_j-t_{j-1}}\,(\sigma_j-\sigma_{j-1}) = \frac{t-t_{j-1}}{\Delta t_j}\,\Delta\sigma_j \tag{5}$$

where $\Delta t_j = t_j - t_{j-1}$ and $\Delta\sigma_j = \sigma_j - \sigma_{j-1}$. The contribution to the viscoelastic strain $\varepsilon_{n+1}$ from changes in the stress state within the $j$-th increment is obtained as

$$\Delta\varepsilon_j = \int_{t_{j-1}}^{t_j} J(t_{n+1}-\tau)\frac{\partial\sigma}{\partial\tau}d\tau = \frac{\Delta\sigma_j}{\Delta t_j}\int_{t_{j-1}}^{t_j} J(t_{n+1}-\tau)d\tau \tag{6}$$

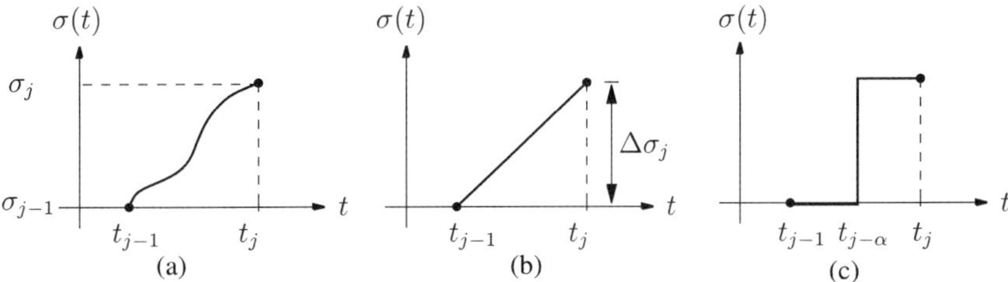


Figure 4: Stress increase within time increment $j$: (a) real situation and (b,c) possible approximations for numerical implementation

For the case of the considered PL model, with $J(t-\tau) = 1/E + J_a[(t-\tau)/\bar{\tau}]^k$, this leads to

$$\Delta\varepsilon_j = \frac{\Delta\sigma_j}{\Delta t_j}\int_{t_{j-1}}^{t_j}\left[\frac{1}{E} + J_a\left(\frac{t_{n+1}-\tau}{\bar{\tau}}\right)^k\right]d\tau = \Delta\sigma_j J_j \tag{7}$$

where

$$J_j = \frac{1}{E} - \frac{J_a}{\bar{\tau}^k\,\Delta t_j}\frac{(t_{n+1}-t_j)^{k+1} - (t_{n+1}-t_{j-1})^{k+1}}{k+1} \tag{8}$$

2. By a step function, with a stress increase of $\Delta\sigma_j$ at $t_{j-\alpha}$, with $0 \leq \alpha \leq 1$, reading

$$\sigma(t) = \sigma_{j-1} + H(t_{j-\alpha})\,\Delta\sigma_j \tag{9}$$

where $H(\bullet)$ is the Heavyside step function. The contribution to the viscoelastic strain $\varepsilon_{n+1}$ is obtained as

$$\Delta\varepsilon_j = \Delta\sigma_j\int_{t_{j-1}}^{t_j} J(t_{n+1}-\tau)\delta(t_{j-\alpha})d\tau = \Delta\sigma_j J_j \tag{10}$$

with $J_j = J(t_{n+1} - t_{j-\alpha})$.

Based on Equations (7) and (10), the strain at the end of the time increment $(n+1)$ is obtained for both presented modes of stress update as

$$\varepsilon_{n+1} = \sum_{j=1}^{n+1}\Delta\varepsilon_j = \sum_{j=1}^{n} J_j\Delta\sigma_j + J_{n+1}\Delta\sigma_{n+1} \tag{11}$$

finally giving access to the stress at the end of the time increment $(n+1)$ as

$$\sigma_{n+1} = \sigma_n + \Delta\sigma_{n+1} = \sigma_n + \frac{1}{J_{n+1}}\left(\varepsilon_{n+1} - \sum_{j=1}^{n} J_j\Delta\sigma_j\right) \tag{12}$$

Figure 5 shows a comparison of numerical results obtained from three modes of

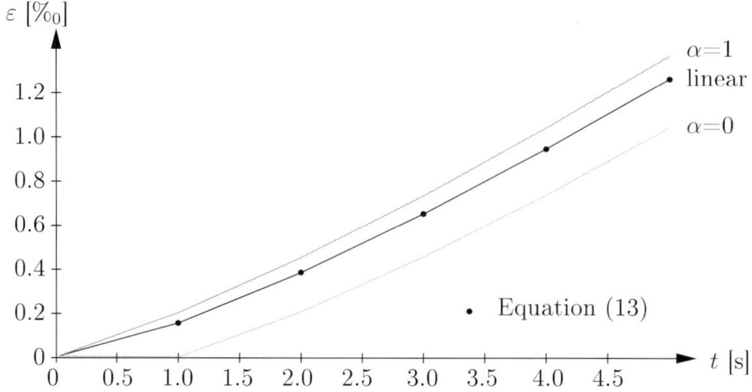


Figure 5: Strain history for linearly increasing stress calculated with the PL model: comparison of different modes of stress update ($\dot{\sigma}$=0.1 MPa/s, $\Delta t = 1$ s, $E = 28000$ MPa, $J_a = 0.002$ MPa$^{-1}$, $k = 0.3$, and $\bar{\tau} = 1$ s)

stress update considering a linearly increasing stress history with $\sigma(t) = \dot{\sigma}t$ where $\dot{\sigma}$=constant

$$\varepsilon(t) = \int_0^t J(t-\tau)d\sigma(\tau) = \left[\frac{t}{E} + \frac{J_a}{\bar{\tau}^k}\frac{t^{k+1}}{k+1}\right]\dot{\sigma} \tag{13}$$

Whereas consideration of step-wise increase of stress at the beginning of the time increment ($\alpha = 1$) overestimates and a step-wise increase of stress at the end of the time increment ($\alpha = 0$) underestimates the viscoelastic deformations, the stress update considering a linear increase reproduces the analytical solution.

## 1.2 Consideration of temperature changes

In consequence of temperature changes, the characteristic time $\bar{\tau}$ appearing in Equation (1) changes with time and influences the strain evolution. Accordingly, time integration of Equation (1), gives

$$J(t-\tau) = \frac{1}{E} + H\int_\tau^t \left(\frac{\hat{t}-\tau}{\bar{\tau}[T(\hat{t})]}\right)^p d\hat{t} \tag{14}$$

Considering Equation (14) within the convolution integral for the viscoelastic strain, one gets

$$\varepsilon(t) = \int_0^t J(t-\tau)d\sigma = \frac{\sigma(t)}{E} + \int_0^t \left\{\int_\tau^t k\frac{J_a}{\bar{\tau}^k}(\hat{t})(\hat{t}-\tau)^{k-1}\,d\hat{t}\right\}\frac{\partial\sigma}{\partial\tau}d\tau \tag{15}$$

In the presented model implementation, the temperature history is considered following the employed mode of stress update. Accordingly, for the linear stress update

(Equation (5)) a linear change of $J_a/\bar{\tau}^k$ in Equation (15) is considered, reading for the $j$-th increment:

$$\frac{J_a}{\bar{\tau}^k}(\hat{t}) = \frac{J_a}{\bar{\tau}^k}(T_{j-1}) + \frac{\hat{t} - t_{j-1}}{\Delta t_j}\left[\frac{J_a}{\bar{\tau}^k}(T_j) - \frac{J_a}{\bar{\tau}^k}(T_{j-1})\right] \tag{16}$$

Equation (16) together with the linear stress update are considered in the integral of Equation (15), allowing us the analytical integration of the so-obtained expressions. The respective results were implemented into the finite element (FE) program FEAP. Figures 6 and 7 show a comparison of the obtained strain history calculated for the uniaxial case for different prescribed stress and temperature histories considering the linear stress/temperature update. Figure 6 illustrates the strain evolution for a constant stress in case of (a) a constant temperature and (b) a temperature increase after 1 hour. The temperature increase results in a change of the characteristic time $\bar{\tau}$ and, thus, in a sharp bend in the strain evolution. Figure 7 illustrates the strain evolution for a linear increasing temperature history for the case of (a) a constant stress and (b) a linear stress increase. The change of material parameters with increasing temperature (and stress) results in a nonlinear increase of strain. For the case of linearly increasing stress/temperature, the numerical solution corresponds to the respective analytical solution. In case the stress history is considered by a step function during integration of the convolution integral, the temperature is set constant within each time increment, with $T = T(t_{j-\alpha})$=constant. Accordingly, the strain at the end of the time increment $(n+1)$ is given by

$$\varepsilon_{n+1} = \sum_{j=1}^{n+1} \Delta\varepsilon_j = \sum_{j=1}^{n+1} J_j \Delta\sigma_j \tag{17}$$

with

$$\begin{aligned} J_j = &\left[\frac{1}{E} + \frac{J_a}{\bar{\tau}^k}[T(t_{j-\alpha})]\,(t_j - t_{j-\alpha})\right] + \\ &+ \sum_{i=j+1}^{n+1} \frac{J_a}{\bar{\tau}^k}[T(t_{j-\alpha})]\left[(t_i - t_{j-\alpha})^k - (t_{i-1} - t_{j-\alpha})^k\right] \end{aligned} \tag{18}$$

## 2 Multiscale determination of model parameters

Recent developments in the finer-scale characterization of materials provide the basis for the development of so-called "bottom-up" multiscale models by relating the macroscopic model parameters to finer-scale properties of the material. After identification of the material properties at the finer scales, upscaling techniques are used to shift the finer-scale information from an observation scale to the next-higher one, considering the respective material morphology.

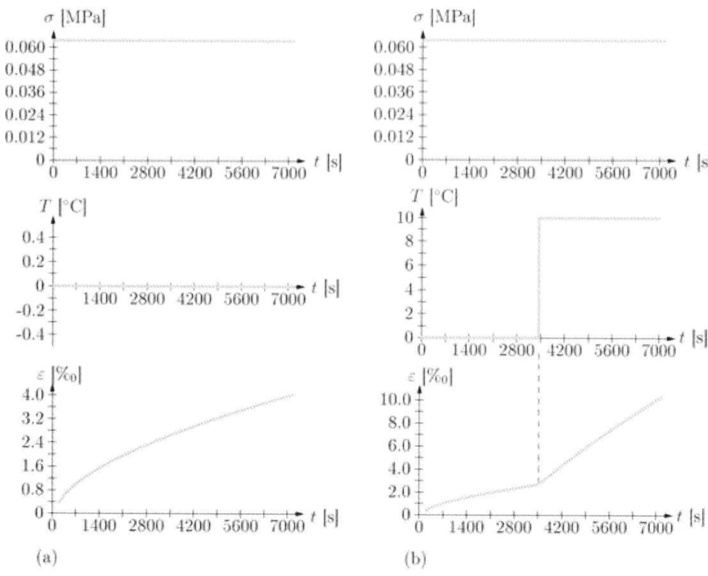


Figure 6: Illustration of effect of temperature changes on strain evolution: constant stress with (a) constant temperature and (b) temperature increase at $t = 3600$ s considering linear stress/temperature update

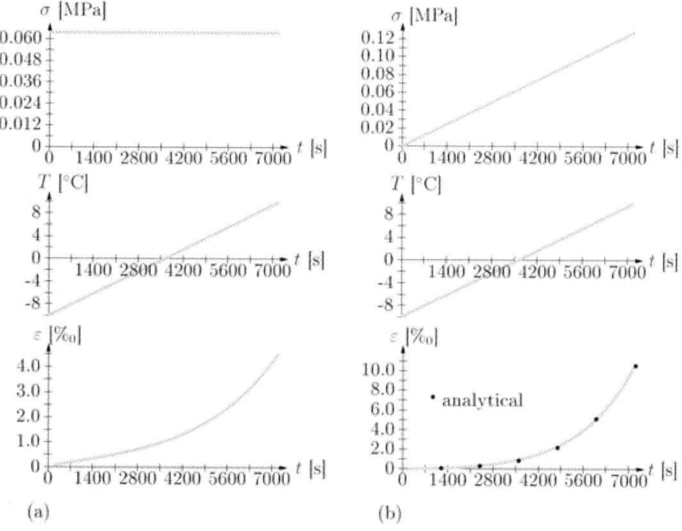


Figure 7: Illustration of effect of temperature changes on strain evolution: linear temperature increase with (a) constant stress and (b) linear stress increase considering linear stress/temperature update

## 2.1 Homogenization of elastic properties – continuum micromechanics

Continuum micromechanics is an analytical technique for determination of the effective behavior of composites, taking the arrangement and properties of the material phases into account. Within continuum micromechanics, the effective material tensor $\mathbb{C}_{eff}$ is given as

$$\mathbb{C}_{eff} = \langle \mathbb{c} : \mathbb{A} \rangle_V \tag{19}$$

When considering an idealized microstructure, the localization tensor $\mathbb{A}$ can be estimated. For a microstructure showing a clear matrix-inclusion morphology, the Mori-Tanaka scheme [1] is used. Hereby, the localization tensor for $I$-th inclusion is given by

$$\mathbb{A}_I = \left[\mathbb{I} + \mathbb{S}_I : \left(\mathbb{c}_M^{-1} : \mathbb{c}_I - \mathbb{I}\right)\right]^{-1} : \left\{ f_M \mathbb{I} + f_I \left[\mathbb{I} + \mathbb{S}_I : \left(\mathbb{c}_M^{-1} : \mathbb{c}_I - \mathbb{I}\right)\right]^{-1} \right\}^{-1} \tag{20}$$

where the indices "$I$" and "$M$" refer to the inclusion and matrix phase, respectively. In Equation (20), $f_I$ and $f_M$ represent the volume fraction of the inclusions and the matrix, respectively, and $\mathbb{S}$ denotes the so-called fourth-order Eshelby tensor. From $\langle \mathbb{A} \rangle_V = \mathbb{I}$, $\langle \mathbb{A} \rangle_{V_M}$ is obtained as

$$\langle \mathbb{A} \rangle_{V_M} = \frac{1}{f_M} \left(\mathbb{I} - f_I \mathbb{A}_I\right) \tag{21}$$

Considering Equations (20) and (21) in Equation (19), the effective shear modulus $\mu_{eff}$ is obtained for the case of spherical inclusions as:

$$\mu_{eff} = \frac{f_M \mu_M + f_I \mu_I \left[1 + \beta\left(\mu_I/\mu_M - 1\right)\right]^{-1}}{f_M + f_I \left[1 + \beta\left(\mu_I/\mu_M - 1\right)\right]^{-1}} \tag{22}$$

where $\mu_I$ and $\mu_M$ denote the shear moduli of the inclusions and the matrix, respectively. $\beta = 6(k_M + 2\mu_M)/[5(3k_M + 4\mu_M)]$, with $k_M$ as the bulk modulus of the matrix material.

## 2.2 Homogenization of viscoelastic properties: the elastic-viscoelastic correspondence principle

Applying the Laplace-Carson transformation

$$\mathcal{LC}\left[f(t)\right] = f^*(p) = p \int_0^\infty f(t) e^{-pt} dt \tag{23}$$

to Equation (3), giving

$$\varepsilon^*(p) = J^*(p)\sigma^*(p) \tag{24}$$

reveals on analogy to the linear-elastic constitutive law, referred to as elastic-viscoelastic correspondence principle (see, *e.g.*, [2]). Accordingly, material parameters in the solution of the respective elastic problem may be replaced by the respective Laplace-Carson transformed viscoelastic parameters. With the elastic solution for the effective properties of composite materials at hand (Equation (22)), the correspondence principle gives access to the effective (homogenized) deviatoric creep compliance of a matrix/inclusion-composite as

$$J_{eff}^{dev*}(p) = \frac{f_M + f_I\left[1 + \beta^*\left(J_M^*/J_I^* - 1\right)\right]^{-1}}{f_M/J_M^* + f_I/J_I^*\left[1 + \beta^*\left(J_M^*/J_I^* - 1\right)\right]^{-1}} \tag{25}$$

with $\beta^* = 6(k_M^* + 2/J_M^*)/[5(3k_M^* + 4/J_M^*)]$. For the case of deviatoric creep of the matrix and elastic behavior of the inclusions, $k_M^* = k_M$ and $1/J_I^* = \mu_I$. Application of the inverse Laplace-Carson transformation,

$$J_{eff}^{dev}(t) = \mathcal{LC}^{-1}\left[J_{eff}^*(p)\right] \tag{26}$$

leads to the effective deviatoric creep compliance of the composite material in the time domain.

# 3 Application to asphalt

For implementation of the presented multiscale model within the numerical simulation of asphalt pavements, three steps are required:

### Step 1: Parameter identification for binder material

For the identification of the mechanical behavior of the binder material, two types of experiments, the bending-beam rheometer (BBR) test in the low temperature regime (-30 $< T <$ 0°C) and the dynamic-shear rheometer (DSR) test in the elevated temperature regime (0 $< T <$ 80°C) are conducted:

*Bending beam rheometer (BBR):* Based on the monitored displacement history $u(t)$, the creep compliance of the BBR $J^{BBR}$ [MPa$^{-1}$] is obtained from beam theory as

$$J^{BBR}(t) = \frac{\varepsilon(t)}{\sigma} = \frac{4bh^3}{Fl^3}u(t) \tag{27}$$

When considering deviatoric creep only, the deviatoric creep compliance $J^{dev}$ is given by

$$J^{dev}(t) = 3J^{BBR} = \frac{12bh^3}{Fl^3}u(t) \tag{28}$$

The experimental results are well described by the aforementioned PL model, with the deviatoric creep compliance $J^{dev}$ and the deviatoric creep-compliance rate $dJ^{dev}/dt$ (see Figure 8(b)) reading

$$J^{dev} = \frac{1}{\mu_0} + J_a\left(\frac{t}{\tau}\right)^k \quad \text{and} \quad dJ^{dev}/dt = \frac{J_a\,k}{\tau}\left(\frac{t}{\tau}\right)^{k-1} = Ht^p \tag{29}$$

where $H$ [MPa$^{-1}$ s$^{-1}$] represents the creep-compliance rate at $t = \bar{\tau}$ and $p$ [-] is the slope of the linear model response in the $\log dJ/dt - \log t$ diagram. BBR experiments on pure bitumen were carried out and the model parameters $p$ and $H$ were extracted from the creep-compliance rate. Since BBR tests are performed at different temperatures, the temperature-dependence of the model parameter $p$ and $H$ are obtained as

$$p(T) = p_0 + c(T - \bar{T}) \tag{30}$$

and

$$H(T) = H_0 \exp\left[-\frac{E_a}{R}\left(\frac{1}{T} - \frac{1}{\bar{T}}\right)\right] \tag{31}$$

with $T$ as the actual temperature and $\bar{T}$ as the reference temperature. $H_0$ is the creep-compliance rate at $t = \bar{\tau}$ for $T = \bar{T}$, $E_a$ [J/mol] is the activation energy, and $R$ is the gas constant.

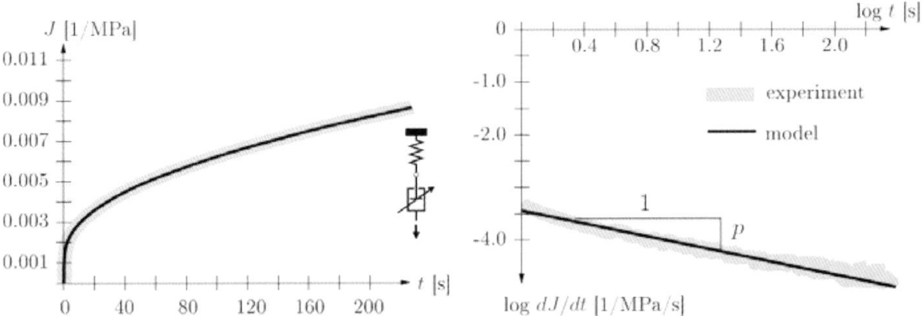


Figure 8: Comparison of PL-model response with experimental data from BBR: (a) creep compliance and (b) creep-compliance rate

*Dynamic shear rheometer (DSR):* DSR tests are conducted under cyclic loading undergoing a temperature and frequency sweep from -20 to +46°C and 0.1 to 50 Hz, respectively. The DSR delivers the phase angle $\varphi$ and the complex shear modulus $\mu^*$, giving the storage and loss modulus as $\mu' = \mathrm{Re}(\mu^*) = \mu^* \cos\varphi$ and $\mu'' = \mathrm{Im}(\mu^*) = \mu^* \sin\varphi$. The storage and loss modulus in viscoelastic solids measure the stored energy, representing the elastic portion, and the energy dissipated as heat, representing the viscous portion. The experimentally-obtained values for $\mu'$ and $\mu''$ obtained for two types of bitumen are plotted in the so-called Cole-Cole diagram (see Figure 9). The parameters of the PL model describing the material response of pure bitumen are obtained from curve fitting, aiming at the best fit between the response corresponding to the PL model and the experimental data in the Cole-Cole diagram. This approach is illustrated in Figure 9 for bitumen B50/70 and a polymer-modified bitumen pmB45/80-65.

**Step 2: Upscaling of viscoelastic properties and validation**

In order to validate the model parameters for asphalt obtained from upscaling using

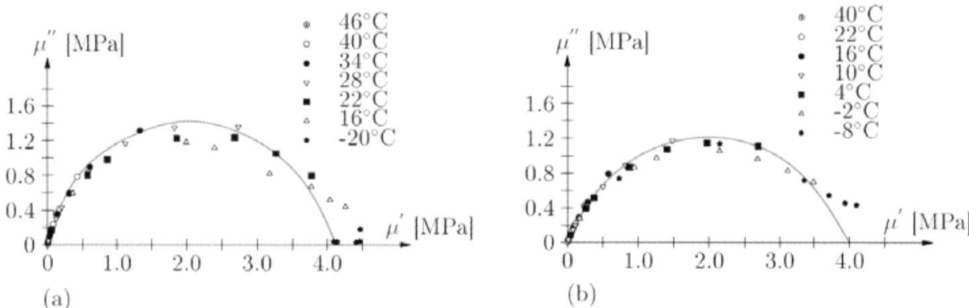


Figure 9: DSR results and PL-model response for (a) bitumen B50/70 (model parameters: $1/\mu_0 = 0.24$ MPa$^{-1}$, $J_a^{dev} = 1.2$ MPa$^{-1}$, $k = 0.8$, $\bar{\tau} = 1$ s) and (b) polymer-modified bitumen pmB45/80-65 (model parameters: $1/\mu_0 = 0.25$ MPa$^{-1}$, $J_a^{dev} = 1.0$ MPa$^{-1}$, $k = 0.7$, $\bar{\tau} = 1$ s)

continuum micromechanics, static creep tests are used for determination of parameters representing the long-term response and cyclic tests are performed at different frequency and temperature regimes in order to specify elastic and short-term viscous properties. The presented upscaling scheme was applied to four types of asphalt, with the mix design given in Table 3. Hereby, limestone dust was used as filler for all considered types of asphalt. Two types of aggregates were used: Diabas for SMA11 and Hollitzer for BT22.

| asphalt type | bitumen [vol%] | filler [vol%] | aggregate 0-4mm [vol%] | aggregate 4-22mm [vol%] | air [vol%] |
|---|---|---|---|---|---|
| SMA11-B70/100-D | 15.7 | 3.7 | 23.1 | 54.4 | 3.1 |
| SMA11-pmB45/80-65-D | 15.5 | 3.7 | 22.9 | 54.1 | 3.7 |
| BT22-B50/70-H | 11.1 | 0.4 | 33.3 | 49.9 | 5.2 |
| BT22-pmB45/80-65-H | 11.1 | 0.4 | 33.4 | 50.2 | 4.8 |

Table 1: Composition of considered types of asphalt

*Static tests:* Both $J_{a,eff}^{dev}$ and $k_{eff}$ obtained from upscaling, describing the long-term response of the material, were validated by static uniaxial creep tests using prismatic asphalt specimens with cross sections of 50 mm × 50 mm for testing of SMA11 and 60 mm × 60 mm for testing BT22. Within the conducted experimental program, the load level is adapted to the tensile strength of asphalt at the respective testing temperature. With the input parameters for bitumen given in Table 3 at hand, Equation (25) is used to compute the effective deviatoric creep parameters for the considered types of asphalt. Hereby, upscaling is performed in a step-wise manner, following the multiscale model. Thus, only bitumen and filler are considered in the first step. In the second step, this bitumen-filler composite (mastic) becomes the new matrix

| bitumen | $H_0$ [mm/m/MPa/s] | $p_0$ [-] | $c$ [-] | $E_a/R$ [K] |
|---|---|---|---|---|
| B70/100 | 0.376 | -0.41 | 0.01 | 9500 |
| pmB45/80-65 | 0.611 | -0.46 | 0.011 | 8600 |
| B50/70 | 0.254 | -0.51 | 0.013 | 10600 |

Table 2: Model parameters for considered types of bitumen obtained from BBR and DSR tests ($\bar{T}$= -12°C)

material where additional aggregates, ranging from 0 to 4 mm, are added. In a third step, aggregates, ranging from 4 to 22 mm, and air voids are added to this new matrix material. When considering two inclusion phases (*e.g.*, aggregate and air within the third step), Equation (25) is extended to

$$J_{eff}^{dev*} = \frac{f_M + f_s\left[1+\beta^*\left(\frac{J_M^*}{J_s^*}-1\right)\right]^{-1} + f_a\left[1+\beta^*\left(\frac{J_M^*}{J_a^*}-1\right)\right]^{-1}}{f_M/J_M^* + f_s/J_s^*\left[1+\beta^*\left(\frac{J_M^*}{J_s^*}-1\right)\right]^{-1} + f_a/J_a^*\left[1+\beta^*\left(\frac{J_M^*}{J_a^*}-1\right)\right]^{-1}} \tag{32}$$

with $\beta^* = 6(k_M^* + 2/J_M^*)/[5(3k_M^* + 4/J_M^*)]$ and the indices $M$, $s$, and $a$ referring to matrix, stone, and air phase, respectively.

For all three upscaling steps, the matrix material is associated with viscoelastic behavior described by the PL model, whereas elastic behavior is assigned to the inclusions. Under hydrostatic loading, the matrix is assumed to behave elastically, giving $k_M^* = k_M$. Inserting the Laplace-Carson transformed creep compliance, $k_M^* = k_M$, $1/J_s^* = \mu_s = 2 \times 10^5$ MPa$^{-1}$ and $1/J_a^* = \mu_a = 0$ into Equations (25) and (32), respectively, and performing the inverse Laplace-Carson transformation gives access to the effective creep compliance $J_{eff}^{dev}(t)$ and, thus, to the effective creep parameters $H_{eff}$ and $p_{eff}$ of asphalt concrete. The results obtained from upscaling from the bitumen-scale to the macroscale are depicted in Figures 10 to 13 and compared with the respective values obtained from the uniaxial creep experiments. The slope of the ln($H$)-1/$T$ curve obtained from upscaling gives access to the activation energy which remains unchanged during scale transition from the bitumen scale to the macroscale. The so-obtained temperature dependence of the creep parameters $H$ and $p$, which was already introduced at the bitumen-scale, captures the experimentally-obtained results for asphalt well.

*Cyclic tests:* Cyclic tests are performed at different temperatures (ranging from, *e.g.*, -20 to 45 °C) and different frequencies, ranging from 0.1 to 20 Hz, focusing on the short-term material response of asphalt. Hereby, either a uniaxial loading condition using prismatic specimens or a bending-beam experimental setup is employed. From the measured load (stress) and displacement (strain) history, the complex modulus $E^*$ and the phase angle $\varphi$ are determined and plotted in the Cole-Cole diagram (see Figures 14 and 15). The experimental results are compared with the material response of asphalt predicted by the multiscale model. This comparison is shown in

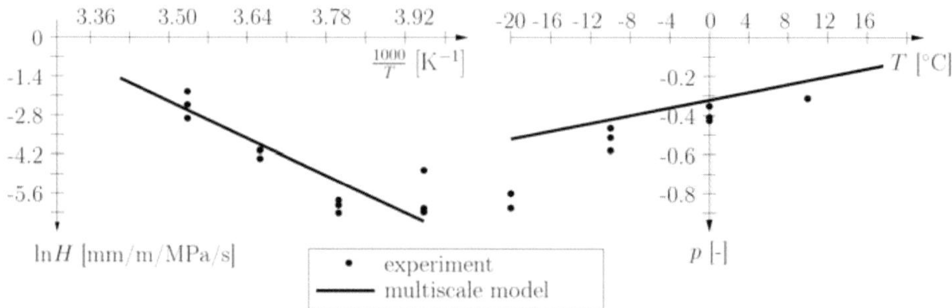


Figure 10: Experimental results for $H$ and $p$ and multiscale predictions for SMA11-B70/100-D

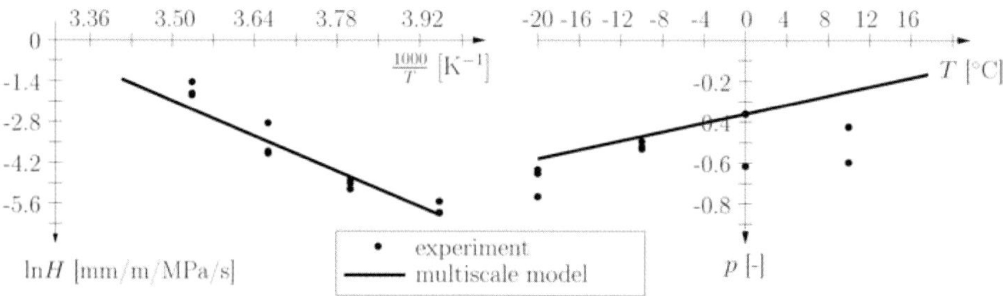


Figure 11: Experimental results for $H$ and $p$ and multiscale predictions for SMA11-pmB45/80-65-D

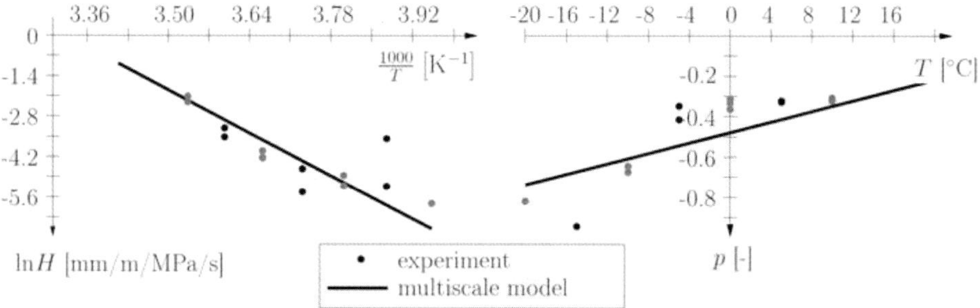


Figure 12: Experimental results for $H$ and $p$ and multiscale predictions for BT22-B50/70-H

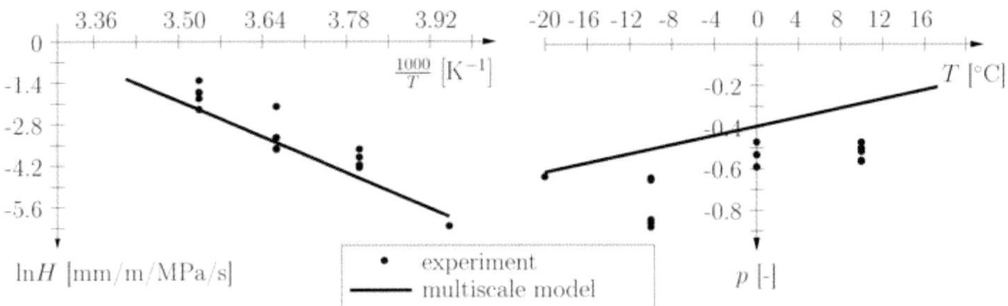


Figure 13: Experimental results for $H$ and $p$ and multiscale predictions for BT22-pmB45/80-65-H

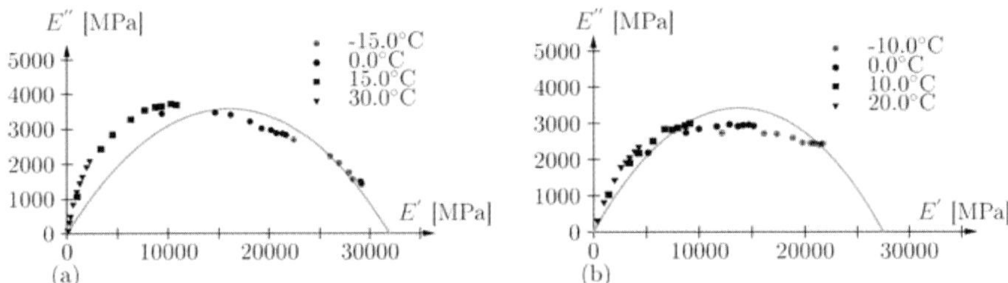


Figure 14: Experimental results from cyclic tests and multiscale predictions for (a) SMA11-70/100-D (model parameters from upscaling: $1/\mu_{0,eff} = 3.0\ 10^{-5}$ MPa$^{-1}$, $J^{dev}_{a,eff} = 1.7\ 10^{-5}$ MPa$^{-1}$, $k_{eff} = 0.3$, $\bar{\tau} = 1$ s) and (b) SMA11-pmB45/80-65-D (model parameters from upscaling: $1/\mu_{0,eff} = 3.03\ 10^{-5}$ MPa$^{-1}$, $J^{dev}_{a,eff} = 3.5\ 10^{-5}$ MPa$^{-1}$, $k_{eff} = 0.3$, $\bar{\tau} = 1$ s)

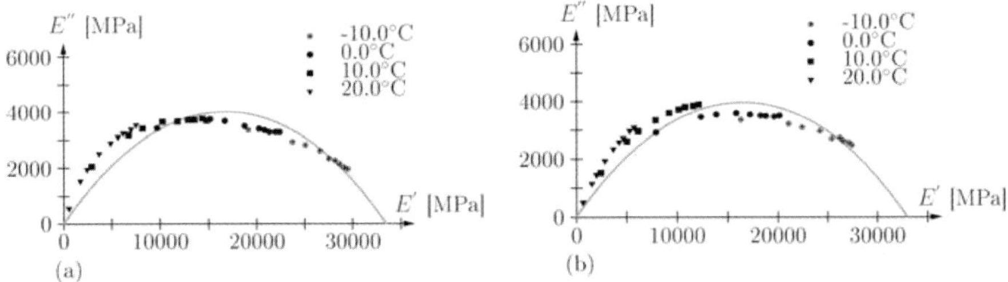


Figure 15: Experimental results from cyclic tests and multiscale predictions for (a) BT22-50/70-H (model parameters from upscaling: $1/\mu_{0,eff} = 3.125\ 10^{-5}$ MPa$^{-1}$, $J^{dev}_{a,eff} = 2.93\ 10^{-5}$ MPa$^{-1}$, $k_{eff} = 0.28$, $\bar{\tau} = 1$ s) and (b) BT22-pmB45/80-65-H (model parameters from upscaling: $1/\mu_{0,eff} = 3.57\ 10^{-5}$ MPa$^{-1}$, $J^{dev}_{a,eff} = 5.4\ 10^{-5}$ MPa$^{-1}$, $k_{eff} = 0.3$, $\bar{\tau} = 1$ s)

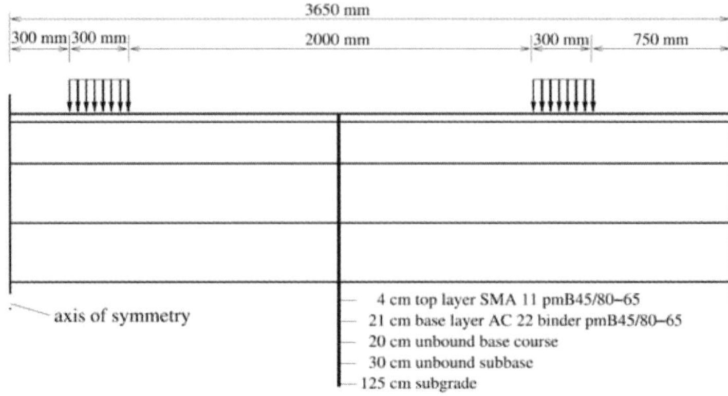


Figure 16: Geometric dimensions of the considered pavement structure

Figures 14 and 15 for the four types of asphalt listed in Table 3. The model predictions fit the experimentally-obtained results in the low-temperature regime well. In fact, the PL model will be used in the following section for assessment of the risk of low-temperature cracking in flexible pavements only.

**Step 3: Application to low-temperature assessment of flexible pavements**

Surface-initiated top-down cracking in consequence of stresses caused by a temperature drop superposed by traffic-initiated stresses is a major mode of deterioration of asphalt pavements. In order to assess the risk of damage of flexible pavements, upscaling of viscoelastic model parameters outlined in Section 2 is combined with a numerical analysis tool.

The pavement structure considered in the numerical analysis (see Figure 16) consists of a 40 mm bituminous top layer of stone-mastic asphalt SMA 11 PmB45/80-65, a 210 mm base layer of high modulus asphalt concrete AC 22 PmB45/80-65, a 600 mm unbound base course/sub-base of coarse gravel, and a sub-grade. The behavior of the unbound layers is described by an elastic material model, with $E_{basecourse} = 266$ MPa, $E_{subbase} = 140$ MPa, $E_{subgrade} = 70$ MPa, and $\nu = 0.3$. The time span assigned to loading in consequence of traffic is set to 0.02 s, corresponding to a vehicle velocity of 80 km/h. The axle load $P$ [kN] is distributed over a circular contact area for each tire, given by the radius

$$a = \sqrt{\frac{P}{2p\pi}} \tag{33}$$

where $p$ is the internal tire pressure (0.7 MPa in the present study).

The parameters of the PL model describing the mechanical behavior of the top and the base layers were determined from upscaling of viscoelastic model parameters from the bitumen scale towards the asphalt scale following the procedure outlined in Section 2

| | $E$ [MPa] | $J_a/\bar{\tau}_0^k$ [mm/m/MPa/s$^k$] | $k$ [-] | $E_a/R$ [K] |
|---|---|---|---|---|
| SMA 11 PmB45/80-65 | 28570 | 0.01752 | 0.53 | 22800 |
| AC 22 PmB45/80-65 | 28985 | 0.02011 | 0.49 | 22800 |

Table 3: Determined model parameters for SMA 11 PmB45/80-65 and AC 22 PmB45/80-65 using $T_0 = -12\,°\mathrm{C}$

(see Table 3).

The solution of the coupled thermo-mechanical problem is performed in two steps: First, the temperature distribution in the road section is determined on the basis of the prescribed temperature cooling scenario. The so-obtained temperature profiles serve as input for the second step, the mechanical analysis, considering thermal shrinkage, the change of material parameters with temperature and, finally, the traffic load. This analysis is performed by using the finite element program FEAP [11] using the linear stress/temperature update presented in Section 1.

Figure 17 shows the distribution of the surface stress perpendicular to the road axis for different time instants. After cooling was simulated in the analysis, the traffic load is applied, resulting in an increase of tensile stresses between the two tires.

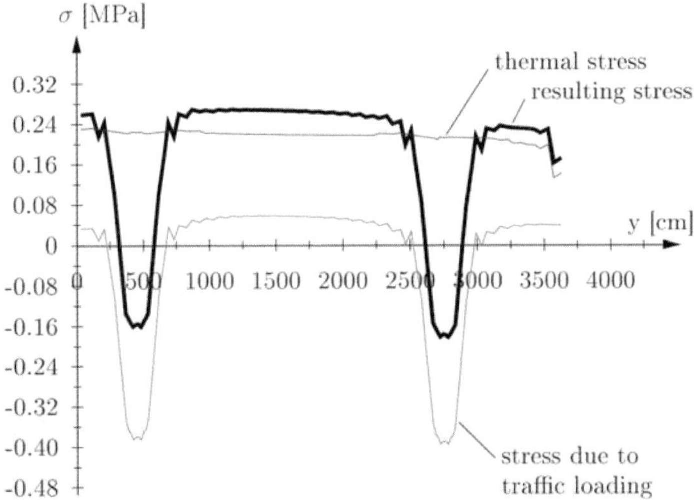


Figure 17: Distribution of surface stress perpendicular to the road axis in consequence of temperature changes and additional traffic load

In order to assess the influence of variation in filler, sand, and stone content on the performance of the pavement structure, their volume contents were varied, with the multiscale model giving the new, modified model parameters. The impact on the resulting stresses obtained from the 3D analyis is demonstrated in Figure 18.

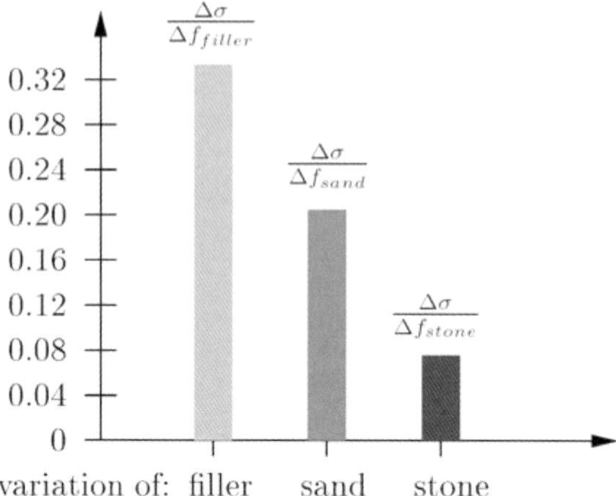


Figure 18: Influence of variation in filler, sand, and stone content in upscaling scheme on resulting stresses between the two tires

# 4 Conclusions

In this paper, a method for the risk assessment of low-temperature cracking of flexible pavements by combining multiscale modeling of asphalt with the computational analysis of pavement structures was presented.

From the presented methods and the obtained numerical results, the following conclusions can be drawn:

- Concerning the implementation of the power-law model into the finite element program FEAP, the accuracy of the integration of the convolution integrals was improved by replacing the step-wise stress history by a linear change of stress and temperature within each time increment.

- Within the multiscale model, the thermorheological properties were assigned to the bitumen only. Via upscaling, the viscoelastic properties of asphalt were determined. For upscaling, continuum micromechanics was employed, using the so-called "correspondence principle", taking the typical matrix-inclusion morphology of asphalt (Mori-Tanaka scheme). The viscous behavior of bitumen was described by the Power-law model, with the respective model parameters serving as input for the multiscale model. The predictive capabilities of the multiscale model were validated by static and cyclic creep tests performed on four types of asphalt, showing good agreement between the experimental results and the multiscale model prediction.

Thus, based on the presented multiscale analysis framework, the stress state in the pavement structure can be computed for various asphalt mix designs, cooling scenarios, and bearing capacities of the unbound layers. The risk of low-temperature top-down cracking is estimated by comparing the obtained surface stresses with the respective material strength, *i.e.*, the tensile strength of asphalt at the respective temperature. Finally, the presented multiscale model allows the goal-oriented optimization of the mix design of asphalt. By adapting the mix design, serving as input for the multiscale model, the sensitivity of the viscoelastic response of asphalt with respect to certain modifications in the mix design can be determined. A study considering a variation of the filler, sand, and stone content in the upscaling procedure showed that a rise in filler content has the greatest influence on the stress state in the pavements.

## Acknowledgements

The first author is grateful for financial support by the Austrian Academy of Sciences via the DOC-FFORTE program. Financial support by the Christian Doppler Gesellschaft (Vienna, Austria) is gratefully acknowledged.

## References

[1] T. Mori, K. Tanaka, "Average stress in a matrix and average elastic energy of materials misfitting inclusions", Acta Metallurgica, 21:571-574, 1973.

[2] J. Mandel, "Méchanique des milieux continues [Mecanics of continuous media]", Gauthier, Paris, 1966. In French.

[3] E.G. Aigner, R. Lackner, Ch. Pichler, "Micromechanics-Based Determination of Viscoelastic Properties of Asphalt Concrete" Journal of Materials in Civil Engineering (ASCE), 2007. Submitted for publication.

[4] M. Wistuba, R. Lackner, M. Spiegl, R. Blab, "Low Temperature Performance Prediction of Asphalt Mixtures - New Approach Based on Fundamental Test Methods and Numerical Modeling", International Journal of Pavement Engineering, 7(2): 121-132, 2006.

[5] R. Lackner, R. Blab, A. Jger, M. Spiegl, K. Kappl, M. Wistuba, B. Gagliano, J. Eberhardsteiner, "Multiscale modeling as the basis for reliable predictions of the behavior of multi-composed Materials", In Progress in Engineering Computational Technology, B.H.V. Topping and C. A. Mota Soares (ed.). Saxe-Coburg Publications, Stirling, Chapter 8: 153-187, 2004.

[6] C. Huet, "Étude par une méthode d'impédance du comportement viscoélastique des matériaux hydrocarbonés [Study of the viscoelastic behavior of bituminous mixtures by method of impedance]", PhD thesis, Faculte des Sciences de Paris, Paris, 1963.

[7] G. Sayegh, "Contribution à l'étude des propriétés viscoélastique des bitumes purs et des bétons bitumineux [Contribution of viscoelastic properties of pure bitumen on asphalt concrete]", PhD thesis, Sorbonne, Paris, 1965.

[8] F. Olard, "Comportement thermoméchanique des Enrobés bitumineux à basses températures. Relations entre les propriétés du liant et de l'enrobé [Thermome-

chanical behavior of bituminous mixtures at low temperatures. Relations between characteristics of binder and properties of bituminous mixtures]", PhD thesis, Ecole Nationale des TPE, Lyon, 2003.

[9] W. Findley, J. Lai, K. Onaran, "Creep and relaxation of nonlinear viscoelastic materials", Dover Publications, New York, 1988.

[10] H. Stehfest, "Algorithm 368: Numerical inversion of Laplace transforms", Communications of the ACM, 13, 47-49, 1970.

[11] R. Taylor, "FEAP - A Finite Element Analysis Program (Version 7.5 User Manual)", Department of Civil and Environmental Engineering, University of California at Berkley, California, USA, 2004.

©Saxe-Coburg Publications, 2008.
Trends in Engineering Computational Technology
B.H.V. Topping and M. Papadrakakis, (Editors)
Saxe-Coburg Publications, Stirlingshire, Scotland, 229-245.

# Chapter 12

# Advances in the Meccano Technique for Adaptive Tetrahedral Mesh Generation

**R. Montenegro[1], J.M. Cascón[2], E. Rodríguez[1], G. Cascón[2] and J.M. Escobar[1]**
**[1] University Institute for Intelligent Systems and Numerical Applications in Engineering University of Las Palmas de Gran Canaria, Spain**
**[2] Department of Mathematics, Faculty of Sciences University of Salamanca, Spain**

## Abstract

This paper introduces a new automatic strategy for adaptive tetrahedral mesh generation. A local refinement/derefinement algorithm for nested triangulations and a simultaneous untangling and smoothing procedure are the main techniques involved. The mesh generator is applied to 3-D complex domains whose boundaries are projectable on external faces of a *meccano* approximation composed of cuboids. The domain surfaces must be given by a mapping between *meccano* surfaces and object boundary.

**Keywords:** tetrahedral mesh generation, adaptive refinement/derefinement, nested meshes, mesh smoothing, mesh untangling.

# 1 Introduction

In finite element simulation in engineering problems, it is crucial to automatically adapt the three-dimensional discretization to geometry and to solution. In the past, many authors have devoted great effort to solving this problem in different ways [1–4], but automatic 3-D mesh generation is still an open problem. Generally, as the complexity of the problem increases (domain geometry and model), the methods for approximating the solution become more complicated. At present, it is well known that most mesh generators are based on Delaunay triangulation and the advancing front technique. On the other hand, local adaptive refinement strategies are employed to adapt the mesh to singularities of numerical solution. These adaptive methods usually involve remeshing or nested refinement [5–8]. Another interesting idea is to adapt simultaneously the model and the discretization in different regions of the domain. A perspective on adaptive modeling and meshing is studied in [9]. The main objective of all these adaptive techniques is to achieve a good approximation of the *real* solution with minimal user intervention and low computational cost. For this purpose, the mesh element quality is also an essential aspect for the efficiency and numerical behavior of finite element method. The element quality measure should be understood depending on the isotropic or anisotropic character of the numerical solution.

In this paper we present new ideas and applications of an innovative tetrahedral mesh generator which was introduced in [10–12]. This automatic mesh generation strategy uses no Delaunay triangulation, nor advancing front technique, and it simplifies the geometric discretization problem for 3-D complex domains, whose surfaces can be mapped from a *meccano* face to object boundary. The main idea of the new mesh generator is to combine a local refinement/derefinement algorithm for 3-D nested triangulations [6] and a simultaneous untangling and smoothing procedure [13]. The resulting adaptive meshes have an appropriate quality for finite element applications.

The mesh generator starts building a *meccano* approximation formed by cuboids. Then, a coarse and valid hexahedral mesh of the *meccano* approximation is generated. The automatic subdivision of each hexahedron into six tetrahedra produces an initial tetrahedral mesh of the *meccano* approximation. The main idea is to construct a sequence of nested meshes by refining only those tetrahedra with a face on the *meccano* boundary. The virtual projection of the *meccano* external faces defines a valid triangulation on the domain boundary. Then a 3-D local refinement/derefinement is carried out so that the approximation of domain surfaces verifies a given precision. Once this objective is reached, those nodes placed on the *meccano* boundary are really projected on their corresponding true boundary, and inner nodes are relocated using a suitable mapping. As the mesh topology is kept during node movement, poor quality or even inverted elements could appear in the resulting mesh; therefore, we finally apply a mesh optimization procedure.

At present, the refinement/derefinement module is implemented in ALBERTA code [14, 15]. This software can be used for solving several types of 1-D, 2-D or 3-D problems with adaptive finite elements, but we only use the local refinement/derefinement algorithm. It is based on local bisection. Actually, ALBERTA has implemented an efficient data structure and adaption for 3-D domains which can be decomposed into hexahedral elements as regular as possible. Each hexahedron is subdivided into six tetrahedra by constructing a main diagonal and its projections on its faces, see Figure 2(a). The local bisection of the resulting tetrahedra is recursively carried out by using general ideas of the longest edge [17] and the newest vertex bisection methods. The refinement of a given triangulation is performed by a recursive algorithm. In order to guarantee that this procedure terminates in a finite number of iterations, the algorithm requires that the refinement edge of an element in the initial mesh is the same for all elements that share this edge. Details about the local refinement technique implemented in ALBERTA and restrictions on initial mesh are analyzed in [6, 14, 16]. This strategy works very efficiently for initial meshes obtained by subdivision of regular quadrilateral or hexahedral elements. In these cases, the degeneration of the resulting 2-D or 3-D triangulations after successive refinements is avoided. The restriction on the initial element shapes and mesh connectivities makes it necessary to develop a particular mesh generator for ALBERTA. The presented algorithm produces compatible meshes with ALBERTA. Obviously, our mesh generation technique could be developed for other types of local refinement/derefinement algorithms for tetrahedral meshes [5,7,8].

The *meccano* technique presents several advantages with respect to more traditional approaches, such as Delaunay triangulation or the advancing front technique [1–4]. Delaunay triangulation requires a control in order to avoid *slivers*. Furthermore, the mesh conformity with the object boundary is not a trivial problem for complex geometry. On the other hand, advancing front technique requires a suitable surface triangulation. In addition, an appropriate definition of element sizes is demanded for obtaining good quality tetrahedra. Other technical difficulties appear when these two methods are applied to objects comprising different materials.

Our approach is based on the combination of several former procedures (refinement, derefinement, projection, untangling and smoothing) which are not in themselves new, but the overall integration is an original contribution. Authors have used them in different ways. Triangulations for convex domains can be constructed from a coarse mesh by using refinement/projection [15]. Adaptive nested meshes have been constructed with refinement and derefinement algorithms for evolution problems [18]. Large domain deformations can lead to severe mesh distortions, especially in 3-D. Mesh optimization is thus key for keeping mesh shape regularity and for avoiding a costly remeshing [19,20]. In traditional mesh optimization, mesh moving is guided by the minimization of certain overall functions, but it is usually done in a local fashion. In general, this procedure involves two steps [21,22]: the first is for mesh untangling and the second one for mesh smoothing. Each step leads to a different objective function. In this paper, we use the improvement proposed by [13], where a simultaneous untangling and smoothing guided by the same objective function is introduced.

Some advantages of our technique are that: surface triangulation is automatically constructed, the final 3-D triangulation is conforming with the object boundary, inner surfaces are automatically preserved (for example, interface between several materials), node distribution is adapted in accordance with the object geometry, and parallel computations can easily be developed for meshing the *meccano* pieces. Nevertheless an admissible mapping between the *meccano* boundary and the object surface must be defined. Some effort should be made in that respect in the future.

In the following section we present a description of the main stages of the new mesh generation procedure. In Section 3 we show a test problem which illustrates the efficiency of this strategy. Finally, conclusions and future research are presented in Section 4.

# 2 Description of the Mesh Generator

In this section, we present the main ideas that have been introduced in the mesh generation procedure. The following algorithm describes the whole *mesh generation approach*

Mesh generation

1. Construct a *meccano* approximation formed by cuboids.
2. Define an admissible mapping between the *meccano* approximation and the object boundaries.
3. Construct a valid hexahedral mesh of the *meccano* approximation.
4. Construct a coarse tetrahedral mesh from the previous hexahedral mesh.
5. Generate a local refined tetrahedral mesh of the *meccano* for a given precision.
6. Move the boundary nodes of the *meccano* to the object surface according to the mapping defined in 2.
7. Relocate the inner nodes of the *meccano*.
8. Optimize the actual tetrahedral mesh applying the simultaneous untangling and smoothing procedure.

In Sections 2.1 and 2.2, we start with the definition of the domain and its subdivision in an initial 3-D triangulation that verifies the restrictions imposed by ALBERTA. In Section 2.3, we continue with the presentation of different strategies to obtain an adapted mesh which can approximate the boundaries of the domain within a given precision. We construct a mesh of the domain by projecting the boundary nodes from a *meccano* plane face to the true boundary surface and by relocating the inner nodes. These two steps are summarized in Sections 2.4 and 2.5, respectively. Finally, in Section 2.6 we present a procedure to optimize the resulting mesh.

## 2.1 Object *Meccano*

The first step of the procedure is to construct a *meccano* approximation. We have developed a simple CAD application that allows the user to generate a meccano approximation by connecting simple cuboids. This toolkit, called *MECCANO*, has the most common options of a graphic design application. The user can move, resize, rotate, delete, undelete, copy or paste cuboids, and edit the properties of each piece of the meccano approximation. Our application also verifies if the restrictions imposed on the topology and structure are correct each time the user makes some changes in the scene. MECCANO is based on library Coin3D [23]. Coin3D is a high level toolkit for developing 3D simulation and visualization applications. It is built on OpenGL and

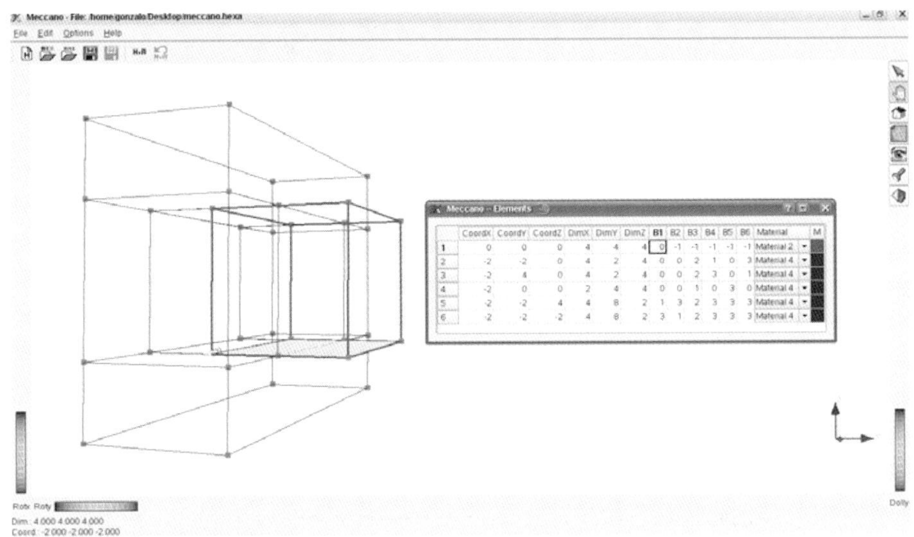


Figure 1: View of the main window of toolkit *MECCANO*. The *meccano* approximation in picture corresponds to the example of Section 3

uses scene graph data structures to render 3D graphics in real-time. In Figure 1 a view of the main window of *MECCANO* is shown.

Note that, in general the union of cuboids is non-valid hexahedral mesh. The general idea of the *meccano* technique could be understood as the connection of different polyhedral pieces. The use of cuboid pieces is an initial particular case.

Once the *meccano* approximation is fixed, we have to define an *admissible* mapping between the boundary faces of the *meccano* and the boundary of the object. We now introduce this concept. Let $\Sigma_0$ be the boundary of the *meccano* and $\Sigma$ the boundary of the object. We denote $\Sigma_0^i$ the *i-th* face of the *meccano* boundary, such that $\Sigma_0 = \bigcup_{i=1}^{n} \Sigma_0^i$ where $n$ is the number of *meccano* boundary faces. We define $\Pi : \Sigma_0 \to \Sigma$ as a piecewise function, such that $\Pi_{|\Sigma_0^i} = \Pi^i$ where $\Pi^i : \Sigma_0^i \to \Pi^i(\Sigma_0^i) \subset \Sigma$. Then, $\Pi$ is called an *admissible* mapping if it satisfies:

1. Functions $\{\Pi^i\}_{i=1}^n$ are compatible on $\Sigma_0$. That is $\Pi^i_{|\Sigma_0^i \cap \Sigma_0^j} = \Pi^j_{|\Sigma_0^j \cap \Sigma_0^i}$, $\forall i, j = 1, \ldots, n$, with $i \neq j$ and $\Sigma_0^i \cap \Sigma_0^j \neq \emptyset$.
2. Global mapping $\Pi$ is continuous and biyective between $\Sigma_0$ and $\Sigma$.
3. Functions $\Pi^i$ are differentiable on $\Sigma_0^i$.

We note that, if the mesh size of the *meccano* boundary is not small enough, the resulting surface mesh could be non-valid. The appropriate element size depends on gradient of $\Pi^i$. We also note that admissible mapping is not unique. Obviously, the quality of the resultant surface mesh depends on the chosen mapping.

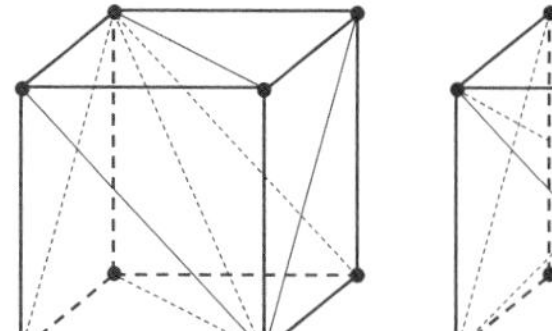
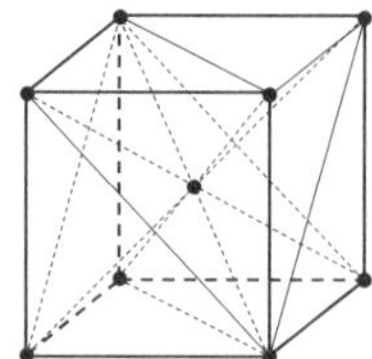
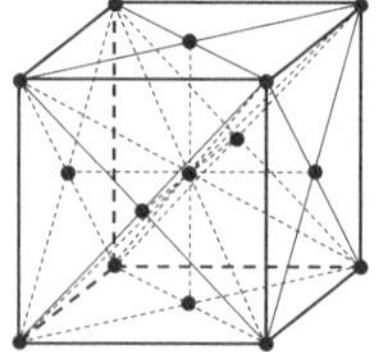
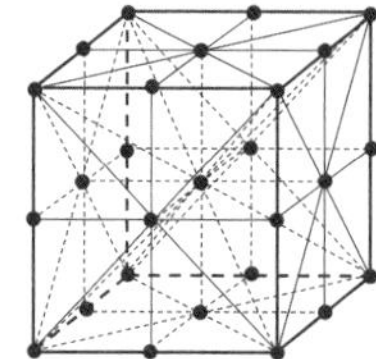

Figure 2: Refinement of a cube by using Kossaczky's algorithm: (a) cube subdivision into six tetrahedra, (b) bisection of all tetrahedra by inserting a new node in the cube main diagonal, (c) new nodes in diagonals of cube faces and (d) global refinement with new nodes in cube edges

## 2.2 Coarse Tetrahedral Mesh of the *Meccano*

The *meccano* is now decomposed into a coarse and valid hexahedral mesh by an appropriate subdivision of initial cuboids. Then, we build a coarse tetrahedral mesh by splitting each hexahedron into six tetrahedra [6]. For this purpose, it is necessary to define a main diagonal on each hexahedron and corresponding diagonal on its faces. For an example of the subdivision of a cube, see Figure 2(a). In order to get a conforming tetrahedral mesh, all hexahedra are subdivided in the same way, maintaining compatibility between the diagonal of their faces. The resulting initial mesh $\tau_1$ can be introduced in ALBERTA since it verifies the imposed restrictions about topology and structure. The user can introduce in the code the necessary number of recursive global bisections [6] for fixing a quasi-uniform element size in the whole initial mesh. Three consecutive global bisections for a cube are presented in Figures 2 (b), (c) and (d). The resulting mesh of Figure 2(d) contains 8 cubes similar to the one shown in Figure 2(a). Obviously, the quality of the resulting tetrahedral mesh is directly related to the quality of the previous hexahedral mesh. Therefore, although the ideal case is the subdivision of the cuboids into cubes, it is not always possible.

## 2.3 Local Refined Mesh of the *Meccano*

The next step in the mesh generator includes a recursive adaptive local refinement strategy of those tetrahedra with a face placed on a boundary face of the initial coarse mesh. The refinement process is done in such a way that the true surfaces are approximated by a linear piecewise interpolation within a given precision. That is, we seek an adaptive triangulation on the *meccano* boundary faces, so that the resulting triangulation after node mapping on the object true boundary is a good approximation of this boundary. The user has to introduce as input data a parameter $\varepsilon$, which is a tolerance to measure the separation allowed between the linear piecewise interpolation and the true surface. At present, we have considered two criteria: the first related to the Euclidean distance between both surfaces and the second attending to the difference in terms of volume.

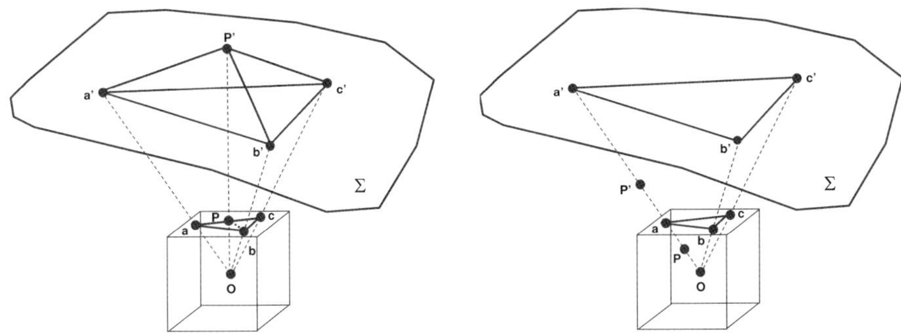


Figure 3: Node mapping from *meccano* to real domain: (a) transformation of an external node P and (b) of an inner node P

To illustrate these criteria, let $abc$ be a triangle placed on the *meccano* boundary, and $a'b'c'$ the resulting triangle after projecting the nodes $a$, $b$ and $c$ on surface $\Sigma$, see Figure 3. We define two different criteria to decide whether it is necessary to refine the triangle (and consequently the tetrahedra containing it) in order to improve the approximation.

For any point $Q$ in the triangle $abc$ we define $d_1^Q$ as the euclidean distance between the mapping of $Q$ on $\Sigma$, $Q'$, and the plane defined by $a'b'c'$. This definition is an estimate of the distance between the surface of the object and the current piecewise approximation.

We also introduce a measure in terms of volume, then, for any $Q$ in the triangle $abc$ we define $d_2^Q$ as the volume of the *virtual* tetrahedron $a'b'c'Q'$. In this case, $d_2^Q$ is an estimate of the lost volume in the linear approximation by the face $a'b'c'$ of the true surface.

The threshold of whether to refine the triangle or not is given by a tolerance $\varepsilon_i$ fixed by the user. With the previous definition, $d_1^Q < \varepsilon_1$ for all $Q$ in the boundary on the *meccano* implies that the distance between the surface of the object and its piecewise linear approximation is less than $\varepsilon_1$. On the other hand, $d_2^Q < \varepsilon_2$ for all $Q$ in the boundary on the *meccano* would mean that the lost volume per boundary face is bounded by $\varepsilon_2$. Alternatively, $\varepsilon_2$ could be defined as the allowed difference of volumes and we could use an equidistribution strategy as is usual in *a-posteriori* error estimates. Nevertheless, here we prefer to use a local version of $\varepsilon_2$, so the difference of volumes is estimated by multiplying $\varepsilon_2$ by the number of boundary faces of the final approximation.

Obviously, other measures could be introduced in line with the desired approximation type (curvature, points properties, *etc.*). What is more, the user could consider the combination of several measures simultaneously.

Once we have defined separation measures $d_i$ and tolerances $\varepsilon_i$, we propose two different strategies for reaching our objective in the following subsections.

### 2.3.1 Sequence of Refinements and Derefinement

The first strategy consists of a simple method. It combines a sequence of refinement steps with one derefinement step.

Simple refinement step. We construct a sequence of tetrahedral nested meshes by recursive bisection of all tetrahedra that contain a face located on the *meccano* boundary faces; see Figure 2. The number of bisections $n_b$ is determined by the user as a function of the desired resolution. At this point, we identify the true surface with the linear approximation obtained with this resolution. So, we have a uniform distribution of nodes on these *meccano* faces and we can consider their *virtual* mapping on the object boundary.

Derefinement step. We apply a derefinement procedure, which is a generalization of the strategy developed in [18]. This criterion fixes which tetrahedra introduced in the refinement sequence can be eliminated without damaging the approximation for the prescribed tolerance $\varepsilon_i$. The derefinement is applied iteratively to the current mesh and concludes either when there are no elements to remove or when the coarse mesh is reached.

Note that each tetrahedron $T$ (generated by bisection of its *father*) has a so-called *newest* node $P$. This node was introduced at the middle point of a prescribed edge $ac$ of the father of $T$ to generate its two sons, see Figure 3(a). The derefinement criterion is efficient because it only computes the distance $d_i^P$ relative to this *newest* node to decide if the tetrahedron $T$ could be removed. This distance is related to the face $abc$ (or its virtual mapping $a'b'c'$) of the father of $T$.

Then, the *derefinement criterion*, associated to a tetrahedron $T$ of the sequence of nested meshes, can be introduced as:

> Derefinement criterion. Tetrahedron $T$ is marked to be derefined, if it satisfies **one** of the following conditions:
>
> 1. The *newest* node $P$ of $T$ is interior.
> 2. The *newest* node $P$ of $T$ is placed on the boundary of the *meccano* and $d_i^P < \varepsilon_i$.

A marked tetrahedron $T$ will be removed only if all the elements generated by the bisection of the edge $ac$ of its father are also marked. So, the *refinement/derefinement procedure* to construct a local refined tetrahedral mesh of the *meccano* is summarized in the following algorithm:

> Refinement/derefinement procedure
>
> 1. Set $n_b$ and $\varepsilon_i$.
> 2. Construct the coarse tetrahedral mesh of the *meccano*.
> 3. Refine $n_b$ times all tetrahedra with at least one face placed on the *meccano* boundary.

4. Mark for derefinement all tetrahedra that satisfy the *derefinement criterion* for a distance $d_i$ and a tolerance $\varepsilon_i$.
5. Derefine the mesh.
6. If the mesh was modified, go to step 4.

### 2.3.2 Sequence of Local Refinements

The second strategy also starts with the coarse mesh of the *meccano*, but it only applies local refinement to obtain the fine one. In this case the *refinement criterion* for tetrahedron $T$ is:

Refinement criterion. Tetrahedron $T$ is marked to be refined, if it satisfies the following **two** conditions:

1. $T$ has a face $F$ on the boundary of the *meccano*.
2. $d_i^Q \geq \varepsilon_i$ for some point $Q$ located on face $F$ of $T$.

From a numerical point of view, the number of points $Q$ (analyzed in this strategy) is reduced to a set of points on a uniform mesh of a given resolution, or a set of points of quadrature. Finally, the *refinement strategy* for constructing a local refined tetrahedral mesh of the *meccano* is summarized in the following algorithm:

Refinement procedure

1. Set $n_b$ and $\varepsilon_i$.
2. Construct the coarse tetrahedral mesh of the *meccano*.
3. Mark for refinement all tetrahedra which satisfy the *refinement criterion* for a distance $d_i$ and a tolerance $\varepsilon_i$.
4. Refine the mesh.
5. If the number of refinement steps is less than $n_b$ and the mesh was modified, go to step 3.

While the first strategy is simpler, it could lead to problems with memory requirements if the number of tetrahedra is very high before applying the derefinement algorithm. For example, this situation can occur when there are surfaces defined by very high resolution functions. Nevertheless, the user could control the number of recursive bisections $n_b$ and the tolerance $\varepsilon_i$.

On the other hand, the problem of the second strategy is to determine whether a face placed on *meccano* boundary must be subdivided to achieve the desired approximation of the true surface. This analysis must be done every time that a boundary face is subdivided into its two *son* faces. Suppose, for example, that the true surface is given by a discrete function. Then, the subdivision criterion should stop for a particular face when all the surface discretization points, defined on this face, have been analyzed and all of them verify the approximation criterion. So, this second strategy has the inconvenience that each surface discretization point could be studied many times and, therefore, it generally involves a higher computational cost than the first strategy. Nevertheless, both of those strategies could be faster depending on the geometry of the object surface and the parameters fixed by the user.

## 2.4 External Node Mapping on Object Boundary

Although ALBERTA has already implemented a node projection on a given boundary surface during the bisection process, it has two important restrictions: nodes belonging to the initial mesh are not projected and inverted elements could appear in the case of projecting new nodes on complex surfaces (*i.e.* non-convex object). In the latter case, the code does not work properly since it is only prepared to manage *valid* meshes.

Therefore, a new strategy must be developed in the mesh generator. The projection (or mapping) is really done once we have defined the local refined mesh by using one of the two methods proposed in the previous section. Then, the nodes placed on the *meccano* faces are projected (or mapped) on their corresponding true surfaces, maintaining the position of the inner nodes of the *meccano* triangulation.

After this process, we obtain a valid triangulation of the domain boundary, but a tangled tetrahedral mesh could appear. Inner nodes of the *meccano* could now be located even outside the domain. Thus, an optimization of the mesh is necessary. Although the final optimized mesh does not depend on the initial position of the inner nodes, it is better for the optimization algorithm to start from a mesh with as good a quality as possible. Therefore, we propose to relocate the inner nodes of the *meccano* in a reasonable position before the mesh optimization.

## 2.5 Relocation of Inner Nodes

There would be several strategies for defining an appropriate position for each inner node of the domain. An acceptable procedure is to modify their relative position as a function of the distance between boundary surfaces before and after their projections. This relocation is done relative to proportional criteria along the corresponding projection line. For example, relocation of inner node $P$ in its new position $P'$, such that $OP' = OP \times Oa' \,/\, Oa$, is represented in Figure 3(b).

Although this node movement does not solve the tangle mesh problem, it normally lessens it. In other words, the resulting number of inverted elements is lower and the mean quality of valid elements is greater.

## 2.6 Object Mesh Optimization: Untangling and Smoothing

An efficient procedure is necessary to optimize the current mesh. This process must be able to smooth and untangle the mesh and is crucial in the proposed mesh generator.

The most usual techniques to improve the quality of a *valid* mesh, that is, a mesh with no inverted elements, are based upon local smoothing. In short, these techniques consist of finding the new positions that the mesh nodes must hold, in such a way that they optimize an objective function. Such a function is based on a certain measurement of the quality of the *local submesh,* $N(v)$, formed by the set of elements connected to the *free node* $v$ whose coordinates are given by $\mathbf{x}$. We have considered the following objective function (1) derived from an *algebraic mesh quality metric* studied in [20], but it would also be possible to use other objective functions that have barriers like those presented in [19]:

$$K(\mathbf{x}) = \left[\sum_{m=1}^{M} \left(\frac{1}{q_{\eta_m}}\right)^p (\mathbf{x})\right]^{\frac{1}{p}}, \tag{1}$$

where $M$ is the number of elements in $N(v)$, $q_{\eta_m}$ is an algebraic quality measure of the $m$-th element of $N(v)$ and p is usually chosen as $1$ or $2$. Specifically, we have considered the mean ratio quality measure, which for a tetrahedron is $q_\eta = \frac{3\sigma^{\frac{2}{3}}}{|S|^2}$ and for a triangle is $q_\eta = \frac{2\sigma}{|S|^2}$, being $|S|$ the Frobenius norm of matrix $S$ associated with the affine map from the *ideal* element (usually equilateral tetrahedron or triangle) to the physical one, and $\sigma = \det(S)$. Other algebraic quality measures can be used as, for example, the metrics based on the condition number of matrix $S$, $q_\kappa = \frac{\rho}{|S||S^{-1}|}$, where $\rho = 2$ for triangles and $\rho = 3$ for tetrahedra.

As it is a local optimization process, we cannot guarantee that the final mesh is globally optimum. Nevertheless, after repeating this process several times for all the nodes of the current mesh, quite satisfactory results can be achieved. Objective functions are usually appropriate to improve the quality of a valid mesh, but they do not work properly when there are inverted elements. This is because they present singularities (barriers) when any tetrahedron of $N(v)$ changes the sign of its Jacobian determinant. To avoid this problem it is possible to proceed as Freitag *et al.* in [21, 22], where an optimization method consisting of two stages is proposed. In the first, the possible inverted elements are untangled by an algorithm that maximizes their negative Jacobian determinants [21] while, in the second, the resulting mesh from the first stage is smoothed using another objective function based on a quality metric of the tetrahedra of $N(v)$ [22]. After the untangling procedure, the mesh has a very poor quality because the technique has no motivation to create good-quality elements. As remarked in [22], it is not possible to apply a gradient-based algorithm to optimize the objective function because it is not continuous all over $\mathbb{R}^3$, making it necessary to use other non-standard approaches.

We have proposed an alternative to this procedure [13], so the untangling and smoothing are carried out at the same stage. For this purpose, we use a suitable modification of the objective function such that it is regular all over $\mathbb{R}^3$. It consists of substituting the term $\sigma$ in the quality metrics with the positive and increasing function $h(\sigma) = \frac{1}{2}(\sigma + \sqrt{\sigma^2 + 4\delta^2})$. When a feasible region (subset of $\mathbb{R}^3$ where $v$ could be placed, $N(v)$ being a valid submesh) exists, the minima of the original and modified objective functions are very close and, when this region does not exist, the minimum of the modified objective function is located in such a way that it tends to untangle $N(v)$. The latter occurs, for example, when the fixed boundary of $N(v)$ is tangled. With this approach, we can use any standard and efficient unconstrained optimization method [24] to find the minimum of the modified objective function.

In addition, a smoothing of the boundary surface triangulation could be applied before the movement of inner nodes of the domain by using the new procedure presented in [25] and [26]. This surface triangulation smoothing technique is also based on a vertex repositioning defined by the minimization of a suitable objective function. The original problem on the surface is transformed into a two-dimensional one on the *parametric space*. In our case, the parametric space is a plane, chosen in terms of the local mesh, in such a way that this mesh can be optimally projected performing a *valid* mesh, that is, without *inverted* elements.

# 3 Test Example

The performance of our new mesh generator is shown in the following application. We use our strategy to obtain a 3D mesh of the $25^{\text{th}}$ Civil-Comp Conference logo. A few more examples can be found at [10–12], where we analyze different issues of our algorithm: refinement strategies, derefinement parameter, surface deviation of the resulting mesh, applications to complex geometries, etc.

We are going to stamp the $25^{\text{th}}$ Civil-Comp Conferences logo on a *button*. This object is the union of a sphere with ratio 2, and semi-sphere with ratio 4, both with the same center point. We stamp the number “25” on the uncovered surface of the sphere, and the legend “- Civil-Comp Conferences - 1983 - 2008” on the uncovered flat place of the semi-sphere.

The input data can be easily generated with toolkit MECCANO. It consists of five cuboids for the semi-sphere and one cuboid for the small sphere, see Figure 1 or Figure 4(a). This *meccano* approximation is included in a parallelepiped whose dimensions are $8 \times 6 \times 8$. A one-to-one projection between *meccano* and the object is defined by a sphere projection.

The *meccano* is first split into 288 cubes, see Figure 4(b), and then into 3-D triangulation of 1728 tetrahedra and 455 nodes, see Figure 4(c). We apply $9$ recursive bisections on all tetrahedra which have a face placed on the *meccano* boundary or on the sphere interface, and an additional $9$ recursive bisections on the area where the legend will be stamped. This mesh contains 469672 nodes and 2161960 tetrahedra. The derefinement criterion is applied to capture the sphere curvature and the legend, according to Section 2.3. We use here $\varepsilon_2 = 0.0001$ for the curvature, and decide to keep in the surface triangulation all triangles in the interface of the legend. The resulting mesh contains 187750 nodes and 43710, see Figure 4(d).

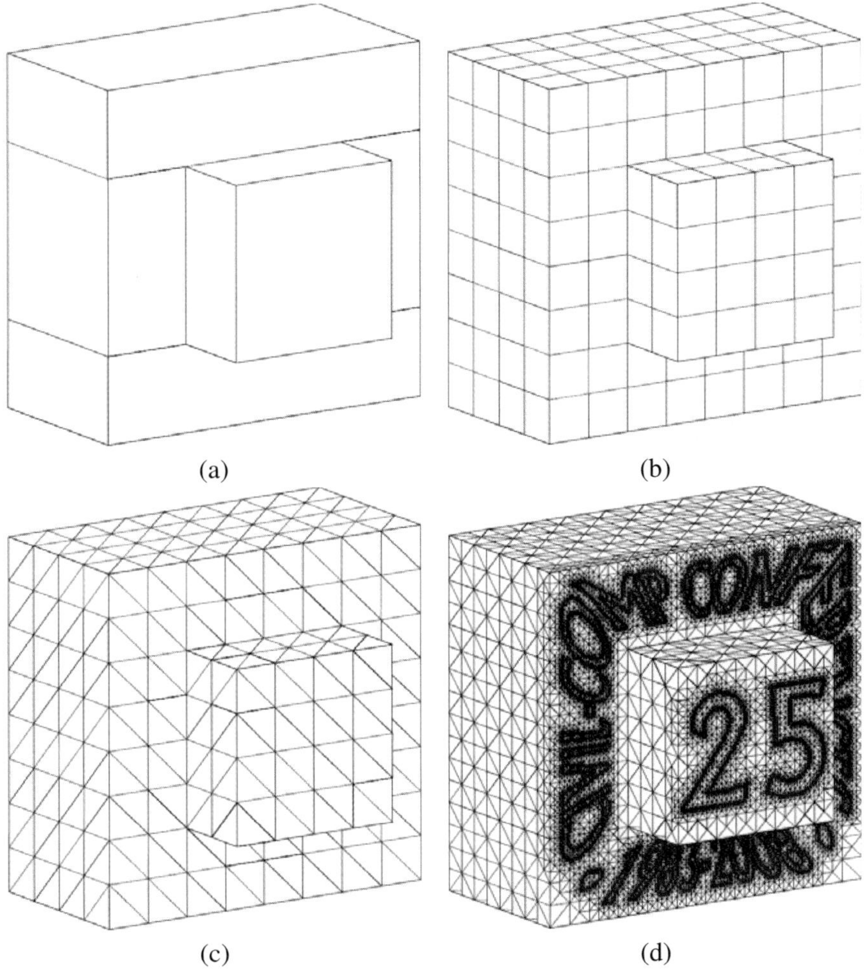


Figure 4: Main stages of the mesh generator: (a) *meccano*, (b) valid mesh of cubes, (c) coarse tetrahedral mesh, (d) mesh adaption after applying the refinement/derefinement procedure

The projection of this *meccano* surface triangulation on the true surface produces a 3-D tangled mesh with 19682 inverted elements, see Figure 5(a). The relocation of inner nodes by using a proportional criterion reduces the number of inverted tetrahedra to 88, Figure 5(b). After 9 iterations, the mesh optimization algorithm of Section 2.6 converts the tangled mesh into the one presented in Figure 5(c).

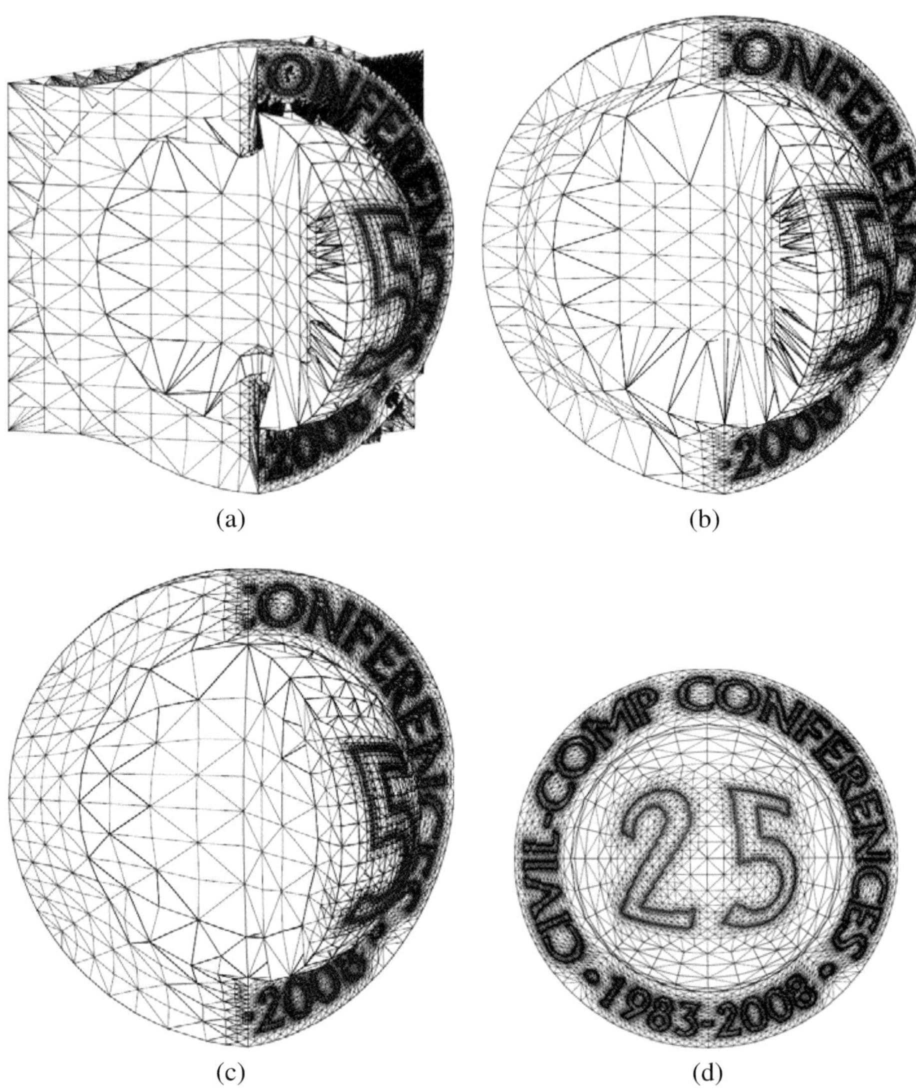


Figure 5: Main stages of the mesh generator (continued): (a) cross section of tangled mesh after the projection, (b) cross section of tangled mesh after relocation, (c) cross section of resulting mesh after mesh optimization process. (d) fontral view of the final mesh

The mesh quality is improved to a minimum value of $0.32$ and an average $\overline{q}_{\kappa} = 0.69$. The quality curves for the initial, intermediate and final triangulations are shown in Figure 6.

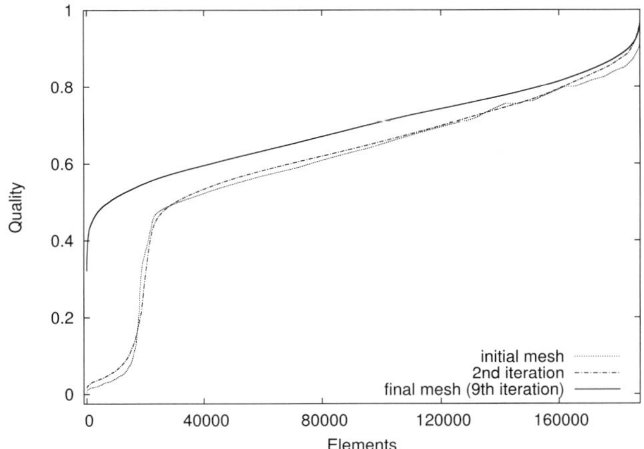


Figure 6: Quality curves, using $q_{\kappa} = \frac{3}{|S||S^{-1}|}$, for the initial and optimized meshes after two and nine (final mesh) iterations

The CPU time for constructing the initial mesh (refinement/derefinement, projection and inner relocation) is approximately $37$ seconds and for its optimization (untangling and smoothing) is $120$ seconds on a Intel Xeon processor, $3.06$ $GHz$ and $4$ $Gb$ RAM memory.

# 4 Conclusions and Future Research

The proposed mesh generator is an efficient method for creating tetrahedral meshes on domains with boundary faces projectable on a *meccano* boundary. We remark that it requires minimum user intervention and has a low computational cost.

Although this procedure is at present limited in applicability for highly complex geometries, it results in a very efficient approach to the problems that fall within the mentioned class. At present, the user has to define the *meccano* associated to the object and mapping between *meccano* and object surfaces. Once these aspects are fixed, the mesh generation procedure is fully automatic. A simple CAD application has been developed to design the *meccano* approximation. In future works, we will increase the functionality of our CAD application, including new types of pieces in the construction of the *meccano*, and adding an automatic mapping between the boundary faces of the *meccano* and the boundary of the object. Specifically, object surface patches should be defined using *meccano* surfaces as parametric spaces.

The mesh generation technique is based on sub-processes (subdivision, projection, optimization) which are not in themselves new, but the overall integration using a simple shape as a starting point is an original contribution of this paper and it has some obvious performance advantages. We have also introduced a generalized derefinement condition for a simple approximation of surfaces. Finally, another interesting property of the new mesh generation strategy is that it automatically fixes the boundary between materials and achieves a good mesh adaption to the geometrical characteristics of the domain.

## Acknowledgments

This work has been supported by the Spanish Government, "Secretaría de Estado de Universidades e Investigación", "Ministerio de Educación y Ciencia", and FEDER, grant contracts: CGL2004-06171-C03-02-03 and CGL2007-65680-C03-01-03.

## References

[1] G.F. Carey, "Computational Grids: Generation, Adaptation, and Solution Strategies", Taylor & Francis, Washington, 1997.

[2] P.J. Frey, P.L. George, "Mesh Generation", Hermes Science Publishing, Oxford, 2000.

[3] P.L. George, H. Borouchaki, "Delaunay Triangulation and Meshing: Application to Finite Elements", Editions Hermes, Paris, 1998.

[4] J.F. Thompson, B. Soni, N. Weatherill, "Handbook of Grid Generation", CRC Press, London, 1999.

[5] J.M. González-Yuste, R. Montenegro, J.M. Escobar, G. Montero, E. Rodríguez, "Local Refinement of 3-D Triangulations Using Object-oriented Methods", Adv. Eng. Soft., 35, 693-702, 2004.

[6] I. Kossaczky, "A Recursive Approach to Local Mesh Refinement in Two and Three Dimensions", J. Comput. Appl. Math., 55, 275-288, 1994.

[7] R. Löhner, J.D. Baum, "Adaptive H-Refinement on 3-D Unstructured Grids for Transient Problems", Int. J. Num. Meth. Fluids, 14, 1407-1419, 1992.

[8] M.C. Rivara, C. Levin, "A 3-D Refinement Algorithm Suitable for Adaptive Multigrid Techniques", J. Comm. Appl. Numer. Meth., 8, 281-290, 1992.

[9] G.F. Carey, "A Perspective on Adaptive Modeling and Meshing (AM&M)", Comput. Meth. Appl. Mech. Eng., 195, 214-235, 2006.

[10] R. Montenegro, J.M. Cascón, J.M. Escobar, E. Rodríguez, G. Montero, "Implementation in ALBERTA of an automatic tetrahedral mesh generator", in: "Proceeding 15th International Meshing Roundtable", Springer, Berlin, 325-338, 2006.

[11] J.M. Cascón, R. Montenegro, J.M. Escobar, E. Rodríguez, G. Montero, "A new *meccano* technique for adaptive 3-D triangulation" in: "Proceedings of 16th International Meshing Roundtable", Springer, Berlin, 103-120, 2007.

[12] R. Montenegro, J.M. Cascón, J.M. Escobar, E. Rodríguez, G. Montero, "An automatic strategy for adaptive tetrahedral mesh generation", Applied Numerical Mathematics (to appear).

[13] J.M. Escobar, E. Rodríguez, R. Montenegro, G. Montero, J.M. González-Yuste, "Simultaneous Untangling and Smoothing of Tetrahedral Meshes", Comput. Meth. Appl. Mech. Eng., 192, 2775-2787, 2003.

[14] "ALBERTA - An Adaptive Hierarchical Finite Element Toolbox", http://www.alberta-fem.de/.

[15] A. Schmidt, K.G. Siebert, Design of Adaptive Finite Element Software: The Finite Element Toolbox ALBERTA, "Lecture Notes in Computer Science and Engineering", Vol. 42, Springer, Berlin, 2005.

[16] W.F. Mitchell, "A Comparison of Adaptive Refinement Techniques for Elliptic Problems", ACM Trans. Math. Soft., 15, 326-347, 1989.

[17] M.C. Rivara, "A Grid Generator Based on 4-Triangles Conforming. Mesh-refinement Algorithms", Int. J. Num. Meth. Eng., 24, 1343-1354, 1987.

[18] L. Ferragut, R. Montenegro, A. Plaza, "Efficient Refinement/Derefinement Algorithm of Nested Meshes to Solve Evolution Problems", Comm. Num. Meth. Eng., 10, 403-412, 1994.

[19] P.M. Knupp, "Achieving Finite Element Mesh Quality Via Optimization of the Jacobian Matrix Norm and Associated Quantities. Part II-A Frame Work for Volume Mesh Optimization and the Condition Number of the Jacobian Matrix", Int. J. Num. Meth. Eng., 48, 1165-1185, 2000.

[20] P.M. Knupp, "Algebraic Mesh Quality Metrics", SIAM J. Sci. Comput., 23, 193-218, 2001.

[21] L.A. Freitag, P. Plassmann, "Local Optimization-based Simplicial Mesh Untangling and Improvement", Int. J. Num. Meth. Eng. 49, 109-125, 2000.

[22] L.A. Freitag, P.M. Knupp, "Tetrahedral Mesh Improvement Via Optimization of the Element Condition Number", Int. J. Num. Meth. Eng., 53, 1377-1391, 2002.

[23] "Coin 3D-3D Graphics Development Tools", http://www.coin3d.org/.

[24] M.S. Bazaraa, H.D. Sherali, C.M. Shetty, "Nonlinear Programing: Theory and Algorithms", John Wiley and Sons Inc., New York, 1993.

[25] J.M. Escobar, G. Montero, R. Montenegro, E. Rodríguez, "An Algebraic Method for Smoothing Surface Triangulations on a Local Parametric Space", Int. J. Num. Meth. Eng., 66, 740-760, 2006.

[26] R. Montenegro, J.M. Escobar, G. Montero, E. Rodríguez, "Quality improvement of surface triangulations", in: "Proceeding 14th International Meshing Roundtable", Springer, Berlin, 469-484, 2005.

©Saxe-Coburg Publications, 2008.
Trends in Engineering Computational Technology
B.H.V. Topping and M. Papadrakakis, (Editors)
Saxe-Coburg Publications, Stirlingshire, Scotland, 247-269.

# Chapter 13

# Atoms, Molecules and Flows: Recent Advances and New Challenges in their Multi-Scale Numerical Modeling at the Beginning of the Third Millenium

**F. Chinesta[1], A. Ammar[2], H. Lamari[2] and N. Ranc[1]**
**[1] Laboratoire de Mécanique des Systèmes et des Procédés**
**UMR 8106 CNRS-ENSAM, Paris, France**
**[2] Laboratoire de Rhéologie**
**INPG, UJF, CNRS (UMR 5520), Grenoble, France**

## Abstract

Nano-science and nano-technology as well as the fine description of the structure and mechanics of materials from the nanometric to the micrometric scales need descriptions ranging from the quantum mechanics to the kinetic theory descriptions characteristic of statistical mechanics. This paper explores the modelling at these scales and points out the main challenges related to the numerical solution of such models that sometimes are discrete but involve an extremely large number of particles (as in the case of molecular dynamics simulations or coarse-grained molecular dynamics) and other times are continuous but they are defined in highly multidimensional spaces leading to the well known curse of dimensionality issues.

**Keywords:** multi-scale modelling, quantum mechanics, molecular dynamics, Brownian dynamics, kinetic theory, statistical mechanics, model reduction, curse of dimensionality, separated representation, finite sums decompositions.

## 1 Introduction

The fine description of the mechanics and structure of materials at the micro, nano and sub-nanometric scales introduces some specific challenges related to the impressive number of degrees of freedom required or the highly dimensional spaces in which those models are defined. Despite the fact that spectacular progress has been accomplished in the context of computational mechanics in the last decade, the treatment of those models, as we describe in the present work, needs further development.

The brute force approach cannot be considered as a possibility for treating this kind of models. Thus, some specialists such as the Nobel Prize R.G. Laughlin, affirmed that no computer existing, or that will ever exist, can break the barriers found in quantum mechanics because it is a catastrophe of dimension [1].

We can understand the catastrophe of dimension by assuming a model defined in a hyper-cube in a space of dimension $D$, $\Omega = ]-L, L[^D$. Now, if we define a grid to discretize the model, as it is usually performed in the vast majority of numerical methods (finite differences, finite elements, finite volumes, spectral methods *etc.*), consisting of $N$ nodes on each direction, the total number of nodes will be $N^D$. If we assume that for example $N \approx 10$ (an extremely coarse description) and $D \approx 80$ (much lower than the usual dimensions required in quantum or statistical mechanics), the number of nodes involved in the discrete model reaches the astronomical value of $10^{80}$ that represents the presumed number of elementary particles in the universe! We shall come back to the analysis of these systems later.

Thus, progress in this field needs further developments on the physical modelling as well as the introduction of new ideas and methods in the context of computational physics. In this work we are exploring different modelling scales, starting from the finest one, the quantum mechanics, to derive molecular dynamics, Brownian dynamics and finally kinetic theory models. The main particulars of such models, the recent advances and the new challenges will be emphasized.

# 2 From quantum mechanics to statistical mechanics: a walk on the frontier of the simulable world

## 2.1 The finest description: the quantum approach

### 2.1.1 The Schrödinger equation

The quantum state of a given electronic distribution could be determined by solving the Schrödinger equation. This equation has been for a long time considered as one of the finest descriptions of the world. However, before focusing on the challenges of its numerical solution, we would like recall that this equation is not relativistic and that it fails when it is applied to describe heavy atoms. Moreover the Pauli's principle constraint was introduced in the Schrödinger formalism in an *ad hoc* way, being the reason for the main numerical difficulties.

Some simplified hypotheses are usually introduced, as for example the Born-Oppenheimer that states that the nuclei can be in a first approximation assumed as classical point-like particles, that the state of electrons only depends on the nuclei positions and that the electronic ground state corresponds to the one that minimizes the electronic energy for a nuclear configuration. This equation defines a multidimensional problem whose dimension increases linearly with the number of the electrons in the system.

Thus, the knowledge of a quantum system reduces to the determination of the wavefunction $\Psi\left(\mathbf{x}_1, \mathbf{x}_2, \cdots, \mathbf{x}_N, t; \mathbf{X}_1, \cdots, \mathbf{X}_M\right)$ (that establishes that the electronic wavefunction depends parametrically on the nuclei positions) whose evolution is governed by the Schrödinger equation:

$$i\hbar\frac{\partial\Psi}{\partial t} = -\frac{\hbar^2}{2m_e}\sum_{e=1}^{e=N}\nabla_e^2\Psi + \sum_{e=1}^{e=N-1}\sum_{e'=e+1}^{e'=N}V_{ee'}\Psi + \sum_{e=1}^{e=N}\sum_{n=1}^{n=M}V_{en}\Psi \tag{1}$$

where $N$ is the number of electrons and $M$ the number of nuclei, the last ones assumed located and fixed at positions $\mathbf{X}_j$. Each electron is defined in the whole physical space $\mathbf{x}_j \in \mathbf{R}^3$, $i = \sqrt{-1}$, $\hbar$ represents the Planck's constant divided by $2\pi$ and $m_e$ is the electron mass.

The differential operator $\nabla_e^2$ is defined in the conformation space of each particle, *i.e.*: $\nabla_e^2 \equiv \partial^2/\partial x_e^2 + \partial^2/\partial y_e^2 + \partial^2/\partial z_e^2$. The Coulomb's potentials accounting for the electron-electron and electron-nuclei interactions are written as:

$$V_{ee'} = \frac{(q_e)^2}{\|\mathbf{x}_e - \mathbf{x}_{e'}\|} \tag{2}$$

$$V_{en} = -\frac{q_n q_e}{\|\mathbf{x}_e - \mathbf{X}_n\|} \tag{3}$$

The electron charge is represented by $q_e$ and the nuclei charge by $q_n = |q_e| \times Z$ (where $Z$ is the atomic number).

The time independent Schrödinger equation (from which one could determine the ground state, perform quantum static computations or accomplish separated representations of the time-dependent solution) is written:

$$-\frac{\hbar^2}{2m_e}\sum_{e=1}^{e=N}\nabla_e^2\Psi + \sum_{e=1}^{e=N-1}\sum_{e'=e+1}^{e'=N}V_{ee'}\Psi + \sum_{e=1}^{e=N}\sum_{n=1}^{n=M}V_{en}\Psi = E\,\Psi \tag{4}$$

where the ground state corresponds to the eigenfunction $\Psi_0$ associated with the most negative eigenvalue $E_0$.

### 2.1.2 Advanced models: The Dirac equation

The Schrödinger equation does not agree with the relativity principle (it does not verify the Lorentz transformation). Considering the relativistic expression of the Hamiltonian

$$H^2 = p^2c^2 + m^2c^4 \tag{5}$$

one could derive the so-called Klein-Gordon equation:

$$-\hbar^2\frac{\partial^2}{\partial t^2}\Psi = \left(-\hbar^2c^2\nabla^2 + m^2c^4\right)\Psi \tag{6}$$

that has been widely used to model bosonic systems. However, even if it agrees with the relativity constraints, it cannot ensure a positive definite electronic distribution.

To circumvent these difficulties, in 1928 Dirac proposed an alternative approach that agrees with the relativity constraints and leads to a positive definite electronic distribution. The Dirac equation for a system composed of a single particle states:

$$i\hbar\frac{\partial}{\partial t}\Psi = -i\,\hbar c\left(\hat{\alpha}_x\frac{\partial\Psi}{\partial x}+\hat{\alpha}_y\frac{\partial\Psi}{\partial y}+\hat{\alpha}_z\frac{\partial\Psi}{\partial z}\right)+\beta mc^2\Psi = H_e\,\Psi \tag{7}$$

where the quadrivector $\Psi$ represents the so-called spinnor, that contains for the particle and the associated anti-particle the distribution for both spin values, and $H_e$ denotes the one particle Hamiltonian.

When this equation is written in the Hartree atomic unit system, in which $\hbar = 1$, $m = 1$, $c = 137$, and applied for a one particle system within a potential $V$, the ground-state steady solution results from the solution of the eigenproblem:

$$-i\,c\left(\hat{\alpha}_x\frac{\partial\Psi}{\partial x}+\hat{\alpha}_y\frac{\partial\Psi}{\partial y}+\hat{\alpha}_z\frac{\partial\Psi}{\partial z}\right)+\beta c^2\Psi+V(x,y,z)\Psi = E\Psi \tag{8}$$

where the different matrices are defined by:

$$\beta = \begin{pmatrix} 1 & 0 & 0 & 0 \\ 0 & 1 & 0 & 0 \\ 0 & 0 & -1 & 0 \\ 0 & 0 & 0 & -1 \end{pmatrix} \tag{9}$$

$$\hat{\alpha}_x = \begin{pmatrix} 0 & 0 & 0 & 1 \\ 0 & 0 & 1 & 0 \\ 0 & 1 & 0 & 0 \\ 1 & 0 & 0 & 0 \end{pmatrix} \tag{10}$$

$$\hat{\alpha}_y = \begin{pmatrix} 0 & 0 & 0 & -i \\ 0 & 0 & -i & 0 \\ 0 & i & 0 & 0 \\ i & 0 & 0 & 0 \end{pmatrix} \tag{11}$$

$$\hat{\alpha}_z = \begin{pmatrix} 0 & 0 & 1 & 0 \\ 0 & 0 & 0 & -1 \\ 1 & 0 & 0 & 0 \\ 0 & -1 & 0 & 0 \end{pmatrix} \tag{12}$$

For a system composed of $N_e$ electrons and $N_n$ nuclei the atomic potential states:

$$V = \sum_{e=1}^{e=N_e}\sum_{n=1}^{n=N_n} V_{en} + \sum_{e=1}^{e=N_e-1}\sum_{e'=e+1}^{e'=N_e} V_{ee'} \tag{13}$$

where

$$V_{ee'} = \frac{1}{\|\mathbf{x}_e - \mathbf{x}_{e'}\|} - \frac{\hat{\alpha}_{xe}\hat{\alpha}_{xe'}}{\|\mathbf{x}_e - \mathbf{x}_{e'}\|} - \frac{\hat{\alpha}_{ye}\hat{\alpha}_{ye'}}{\|\mathbf{x}_e - \mathbf{x}_{e'}\|} - \frac{\hat{\alpha}_{ze}\hat{\alpha}_{ze'}}{\|\mathbf{x}_e - \mathbf{x}_{e'}\|} \tag{14}$$

and

$$V_{en} = -\frac{Z}{\|\mathbf{x}_e - \mathbf{X}_n\|} \tag{15}$$

where $Z$ is the atomic number.

Thus, the eigenproblem to be solved reads:

$$\underbrace{\left( \sum_{e=1}^{e=N_e} H_e + \sum_{e=1}^{e=N_e} \sum_{n=1}^{n=N_n} V_{en} + \sum_{e=1}^{e=N_e-1} \sum_{e'=e+1}^{e'=N_e} V_{ee'} \right)}_{\mathcal{H}} \Psi = E\Psi \tag{16}$$

where $H_e$ and $\mathcal{H}$ denote the single particle and the whole system Hamiltonians respectively. The ground-state is given by the spinnor $\Psi$ related to the lowest positive eigenvalue $E_{gs}$

Despite the fact that the Dirac formulation can address systems composed of heavy nuclei, whose treatment needs the consideration of relativistic effects, the Pauli's exclusion principle must be enforced explicitly as was the case within the Schrödinger framework. Thus, the antisymmetry of $\Psi$ must be enforced by using, for example, the Slater determinant.

Thus, in what follows, we are focussing on the numerical solution of the Schrödinger equation, because the vast majority of techniques that were applied for solving the Schrödinger equation were then extended for solving the Dirac one.

### 2.1.3 On the numerical solution of the Schrödinger equation

Several techniques have been proposed for solving the Schrödinger equation. Some of them lie in the direct solution of the (time-independent or time-dependent) Schrödinger equation. Due to the curse of dimensionality its solution was only possible for very reduced populations of electrons.

Another solution strategy is based on the Hartree-Fock (HF) approach and its derived approaches (post-Hartree-Fock methods). The main assumption of this approach lies in the approximation of the joint electronic wavefunction (related to the $N$ electrons) as a product of $N$ $3D$-functions (the molecular orbitals) verifying the antisymmetry restriction derived from the Pauli's principle. Thus, the original HF approach consists of writing the joint wavefunction from a single Slater's determinant. The Schödinger equation allows computing the $N$ molecular orbitals after solving the resulting strongly non-linear problem. This technique has been extensively used in quantum chemistry to analyze the structure and behavior of molecules involving a moderate number of electrons. Of course, the HF assumption sometimes represents too crude an approximation which invalidates the derived results.

To circumvent this crude approximation different multi-determinant approaches have been proposed. Interested readers can refer to the excellent overview of Cancès *et al.* [2] as well as the different chapters of the handbook on computational chemistry [3]. The simplest possibility consists in writing the solution as a linear combination of some Slater determinants built by combining $n$ molecular orbitals, with $n > N$. These molecular orbitals are assumed to be known (*e.g.* the orbitals related to the hydrogen atom) and the weights are searched to minimize the electronic energy. When the molecular orbitals are built from the Hartree-Fock solution (by employing the ground state and some excited eigenfunctions) the technique is known as Configuration Interaction method (CI). A more sophisticated technique consists in writing this many-determinants approximation of the solution by using a number of molecular orbitals $n$ (with $n > N$) assumed unknown. Thus, the minimization of the electronic energy leads the molecular orbitals as well as the associated coefficients of this many-determinants expansion to compute simultaneously. Obviously, each one of these unknown molecular orbitals are expressed in an appropriate functional basis (*e.g.* gaussian functions). This strategy is known as Multi-Configuration Self-Consistent Field (MCSCF).

All the strategies just mentioned (and others like the coupled cluster or the Moller-Plesset perturbation methods) belong to the family of the wavefunction based methods. In any case all these methods can be used only to solve quantum systems composed of a moderate number of electrons. As we confirm later the main difficulty is not in the dimensionality of the space, but in the use of the Slater determinants (needed to account for the Pauli's principle) whose complexity scales on the factorial of the number of electrons, *i.e.* in $N!$.

The second family of approximation methods, widely used in quantum systems composed of hundreds, thousands and even millions of electrons, are based on the density functional theory (DFT). These models, more than looking for the expression of the wavefunction (with the associated multi-dimensional issue) look for the electronic distribution $\rho(\mathbf{x})$ itself. The main difficulties of this approach are related to the expressions of both the kinetic energy of electrons and the inter-electronic repulsion energy. The second term is usually modelled from the electrostatic self-interaction energy of a charge distribution $\rho(\mathbf{x})$. On the other hand the kinetic energy term is also evaluated in an approximate manner (from the electronic distribution itself in the Thomas-Fermi and related orbital-free DFT models or from a system of $N$ non-interacting electrons - Kohn-Sham models). Obviously, due to the just mentioned approximations introduced in the kinetic and inter-electronic interaction energies, a correction term is needed, the so-called exchange-correlation-residual-kinetic energy. However, no exact expression of this correction term exists and then different approximate expressions have been proposed and used. Thus, the validity and accuracy of the computed results will depend on the accuracy of the the exchange-correlation term that must be fitted for each system.

The models related to the Thomas-Fermi, less accurate in practice because of the too phenomenological expression of the kinetic energy coming from the reference system of a uniform non-interacting electron gas, allows one to consider large multi-electronic systems. In a recent work, Gavini *et al.* [4] performed multi-million atom simulations by employing the Thomas-Fermi-Weizsacker family of orbital-free kinetic energy functionals. On the other hand, the Kohn-Sham based models are *a priori* more accurate, but they need the computation of the $N$ eigenfunctions related to the $N$ lowest eigenvalues of a non-physical atom composed of $N$ non-interacting electrons.

Transient solutions are very common in the context of quantum gas dynamics (physics of plasma) but are more infrequent in material science when the structure and properties of molecules or crystals are concerned. For this reason, in what follows, we are focusing on the solution of the time-independent Schrödinger equation which leads to the solution of the associated multidimensional eigenproblem, whose eigenfunction related to the most negative eigenvalue constitutes the ground state of the system.

Quantum chemistry calculations performed in the Born-Oppenheimer setting consist either (i) in solving the geometry optimization problem, that is, to compute the equilibrium molecular configuration (nuclei distribution) that minimizes the energy of the system, finding the most stable molecular configuration that determines numerous properties like for instance infrared spectrum or elastic constants; or (ii) in performing an *ab initio* molecular dynamics simulation, that is, to simulate the time evolution of the molecular structure according to the Newton law of classical mechanics. Molecular dynamics simulations allow computation of various transport properties (thermal conductivity, viscosity, *etc.*) as well as some other non-equilibrium properties.

## 2.2 From "ab initio" to molecular dynamics

Depending on the choice of the method, on the accuracy required, and on the computer facility available, the *ab initio* methods allow today for the simulation of systems up to ten, one hundred or some million atoms. In time dependent simulations, they are only convenient for small-time simulations, say not more than a picosecond. However, sometimes larger systems are concerned, and for this purpose one must focus on faster approaches, obviously less accurate. Two possibilities exist: the semi-empirical and the empirical approaches. The semi-empirical approaches speed up the *ab initio* methods by profiting from the information coming from experiments or previous simulations. Empirical methods go on by considering explicitly only the nuclei, by introducing "empirical" potentials leading to the forces acting on the nuclei. Thus, in the stationary setting only the stable configuration is searched, and for this a geometrical optimization (to computed the nuclei equilibrium distribution) is addressed leading to the so-called molecular mechanics. The transient setting results in the classical molecular dynamics but now the computation is speeded up by many orders of magnitude with respect to the molecular dynamics where the potentials are computed at the *ab initio* level.

Thus, if we assume a population of $M$ nuclei (of mass $m_n$) and a two-body potential (many-body potentials are also available), now the Newton's law writes for a generic nuclei $n$:

$$m_n \frac{d^2\mathbf{X}_n}{dt^2} = \sum_{k=1,k\neq n} \mathbf{F}_k^n, \quad \forall n \in [1, \cdots, M] \tag{17}$$

where $F_k^n$ denotes the force acting on nucleus $n$ originated by the presence of nucleus $k$. Obviously these forces can be computed from the gradient of the assumed inter-particle potentials.

Accurate algorithms for integrating these equations exist. The simplectic Verlet's scheme is one of the most used. Molecular dynamics simulations are confronted, despite their conceptual simplicity, with diverse difficulties of different nature:

- The first and most important comes, as previously indicated, from the impossibility of using an "exact" interaction potential derived from quantum mechanics. This situation is particularly delicate when we are dealing with some irregular nuclei distributions such as the ones encountered in the neighborhood of defaults in crystals (dislocations, crack tips, *etc.*), interfaces between different materials or in zones where different kinds of nuclei coexist.

- The second one comes from the units involved in this kind of simulations: the nuclei displacements are in the nanometric scale, the energies are of the order of the electron-volts, the time steps are of the order of $10^{-15}$s. Thus, because of the limits in the computer's precision, a change of units is required, which can be easily performed.

- In molecular dynamics the behavior of atoms and molecules is described in the framework of classical mechanics. Thus, the particles energy variations are continuous. The applicability of MD depends on the validity of this fundamental hypothesis. When we consider crystals at low temperature the quantum effects (implying discontinuous energy variations) are preponderant, and in consequence the matter properties at these temperatures cannot be determined by MD simulations. The use of MD is restricted to temperatures higher than the Debye's temperature (for example the Debye's temperature for the Fe $\alpha \approx 460K$). This analysis is in contrast to the vast majority of MD simulations carried out nowadays. In fact, higher is the temperature (kinetic energy) and higher results the velocity of particles, requiring shorter time steps in order to ensure the stability of the integration scheme. For this reason, nowadays most of the MD simulations in solid mechanics are carried out a zero degrees Kelvin or at very low temperatures but, as just pointed out, at these temperatures the validity of the computed MD solutions are polluted by the non negligible quantum effects.

- The prescription of boundary conditions is another delicate task. If the analysis is restricted to systems with free boundary conditions, then the MD simulation can be carried out without any particular treatment. On the other hand we must consider a system large enough to ensure that in the analyzed region the impact of the free surfaces can be neglected. Another possibility lies in the prescription of periodic boundary conditions, where an atom leaving the system for example through the right boundary is re-injected in the domain through the left boundary. Moreover, the particles located in the neighborhood of a boundary are influenced by the ones located in the neighborhood of the opposite boundary. The imposition of other boundary conditions is more delicate from both the numerical and the conceptual points of view. For example, what is the meaning of prescribing a displacement on a boundary? Each situation requires a deep analysis in order to define the best (the most physical) way to prescribe the boundary conditions.

- There are other difficulties related to the transient analysis, including the prescription of the initial conditions. Let a thermal system be in equilibrium (where the velocities distribution is in agreement with the Maxwell-Boltzmann distribution). Now, we proceed to heat the system. One possibility lies in increasing suddenly the kinetic energy of each particle. Obviously, even if the resulting velocities define a Maxwell-Boltzmann's distribution, the system remains off equilibrium because the partition between kinetic and potential energies is not the appropriate one. For this reason we must proceed to relax the system that evolves from this initial (non-physical) state to the equilibrium one. Other (more physical) possibility lies in the incorporation of a large enough ambient region around the analyzed system, whose particles are initially in equilibrium at the highest temperature. Now, both regions (the system and the ambient) interact, and the system initiates its heating process that reaches its equilibrium some time later. The final state of both evolutions is the same, but the time required to reach it depends on the technique used to induce the heating. The first transient is purely artificial whereas the second one is more physical allowing the identification of some transport coefficients (for example the thermal conductivity).

- Finally the CPU time continues to be the main limitation of MD simulations. The strongest handicap is related to the necessity of considering at each time step and for each particle the influence of all the others particles. Thus, the integration method seems to scale with the square of the number of particles. Even if some computing time savings can be introduced in the neighbors search, the extremely small time steps and the extremely large number of particles required to describe real scenarios, limit considerably the range of applicability of this kind of simulation, that has been accepted to be nowadays, in 2007, of the order of a cubic micrometer, even when the systems are considered at very low, and then non-physical, temperatures (close to zero degrees Kelvin). We can notice that, despite the impressive advances in the computational availabilities, the high performance computing and the use of massive parallel computing platforms, the state of the art does not allow the treatment of macroscopic systems encountered in practical applications of physics, chemistry and engineering.

The above mentioned difficulties in performing fully molecular dynamics simulations led to hybrid techniques that apply MD in the regions where the unknown field varies in a non-uniform way (molecular dynamics model) and a standard finite element approximation in those regions where the unknown field variation can be considered as uniform (continuous model). The main questions raised by these bridging strategies concern: (i) the kinematics representations in both models; (ii) the transfer conditions on the MD and continuous models interface and (iii) the macroscopic constitutive equation employed in the continuous model.

Different alternatives exist, and the construction of such bridges is nowadays one of the most active topics in computational mechanics. The spurious reflection of the high frequency parts of the waves is one of the main issues. We would like only to mention three "families" of bridging techniques, giving some key references in which the interested reader could find other extremely valuable references: (i) the quasi-continuum method proposed by Tadmor and Ortiz [5]; (ii) the multi-scale method proposed by Wagner and Liu [6] and (iii) the methods based on the "Arlequin" approach [7] like the one proposed by Belytschko in [8].

## 2.3 Coarse grained modelling: Brownian dynamics

Sometimes one is interested in analyzing the behavior of a system composed by a series of microscopic entities (particles assumed with a null extension) dispersed into another fluid (the solvent). The kinematic of such particles depends, of course, on their interactions with the solvent particles. A real molecular dynamics simulation is definitively forbidden (the MD simulation feasibilities nowadays rarely exceeds the number of particles contained within a cube of one micron of side).

One possibility to reduce the size of the discrete models lies in considering only the particles of interest. The other particles (the ones that constitute the solvent) are not considered explicitly and only their averaged effects are retained in the modelling.

Thus, the motion equation of a particle whose position is described by $\mathbf{x}_i$, is governed by the Langevin's equation:

$$m\frac{d^2\mathbf{x}_i}{dt^2} = \xi\left(\frac{d\mathbf{x}_i}{dt} - \mathbf{v}_f(\mathbf{x}_i)\right) + \mathbf{F}_i^{ext}(t); \quad \forall i \tag{18}$$

where $m$ denotes the particle mass, $\mathbf{x}_i$ the position of particle $i$, $\xi$ the friction coefficient, $\mathbf{v}_f(\mathbf{x}_i)$ the fluid velocity at position $\mathbf{x}_i$ and $\mathbf{F}_i^{ext}(t)$ all the other forces acting on the particle $i$ (coming from a external potential or from the solvent particles bombardment). We can see that even if this model doesn't incorporate explicitly the solvent particles population, their effects are taken into account from the drift term $\xi\left(\frac{d\mathbf{x}_i}{dt} - \mathbf{v}_f(\mathbf{x}_i)\right)$ as well as by the impact forces.

In the last expression the drift term is quite standard, however the external forces acting on each particle deserve some additional comment. In what follows and without any detriment of generality we assume that there is no other force than the one coming from the solvent particles bombardment and that the solvent is macroscopically at rest, *i.e.* $\mathbf{v}_f = 0$. The random nature of the interaction force is modelled from a statistical distribution function that becomes fully defined as soon as its mean value and its standard deviation are fixed. In respct of the mean value, one expects a null value if the microscopic dynamics is isotropic. Concerning the standard deviation one must proceed within the statistical mechanics framework. In what follows we summarize the main ideas for the derivation of the standard deviation expression.

If we define $B_{\Delta t}$

$$B_{\Delta t} = \int_0^{\Delta t} \frac{F^{ext}(t)}{m} dt = \sum_{k=1}^{k=p} \frac{F^{ext}(t_k)}{m} \delta t \tag{19}$$

where we assume that in $\Delta t$ the particle is subjected to $p$ impacts from the solvent particles, with $p >> 1$. These impacts are modelled by a constant force $F^{ext}(t_k)$ that applies for a time $\delta t$ ($\Delta t = p\delta t$). By invoking the central limit theorem we conclude that $B_{\Delta t}$ follows a gaussian distribution $\mathcal{N}(0, q\Delta t)$ because the number of impacts scales linearly with $\Delta t$.

Now, to compute $q$, one could integrate the Langevin's equation to obtain the equilibrium velocity distribution $W$:

$$W\left(\frac{dx}{dt}, t \to \infty\right) = \sqrt{\frac{\xi}{m\pi q}}\, e^{\frac{\xi\left(\frac{dx}{dt}\right)^2}{mq}} \tag{20}$$

that must coincide with the Maxwell-Boltzmann one (canonical ensemble), from which we deduce the expression of $q$:

$$q = \frac{2\xi k_b T}{m^2} \tag{21}$$

where $k_b$ is the Boltzmann constant and $T$ the absolute temperature.

Thus, the Langevin's equation is fully defined, by writing:

$$B_{\Delta t} = \int_0^{\Delta t} \frac{F^{ext}(t)}{m} dt = \mathcal{N}\left(0, 2\frac{\xi k_b T}{m^2}\Delta t\right) \tag{22}$$

expression that runs also in the presence of other forces like the ones coming from a gradient of a potential or when $\mathbf{v}_f \neq \mathbf{0}$.

Thus, one could track the movement of each particle $\mathbf{x}_i$ by considering a standard drift term and a random force whose distribution is perfectly known. This stochastic approach has been traditionally used also for solving deterministic models such as the advection-diffusion one, because one could compute some moments of the resulting distribution by tracking a moderate population of particles instead, to discretize the deterministic counterpart of the advection-diffusion model by using one of the standard mesh-based discretization techniques that could involve an excessive number of degrees of freedom in the case of 3D models.

To illustrate this procedure we consider the simplest form of the advection-diffusion equation:

$$\frac{\partial C}{\partial t} + \mathbf{v} \cdot \nabla C = D \Delta C \tag{23}$$

where $C = C(\mathbf{x}, t)$ is the concentration field and $D$ is the so-called diffusion coefficient.

Obviously, this simple parabolic equation could be solved by using any standard technique (finite differences, finite elements, spectral methods, finite volumes, the method of particles, ...), but in what follows we are solving it using a stochastic approach. For this purpose we assume the initial condition represented by $N$-Dirac masses:

$$C(\mathbf{x}, t = 0) \approx C^0(\mathbf{x}) = \sum_{i=1}^{i=N} c_i \, \delta(\mathbf{x} - \mathbf{x}_i(t = 0)) \tag{24}$$

Here, we are not discussing the choice of the coefficients $c_i$ and the locations $\mathbf{x}_i$. Several possibilities exist to perform a choice trying to represent, as precisely as possible, the initial concentration distribution. Most of them proceed by regularizing the Dirac distribution and then enforcing the minimization of $\|C(\mathbf{x}, t = 0) - C^0(\mathbf{x})\|$. From now on, $c_i$ and $\mathbf{x}_i(t = 0)$ are assumed known. Moreover, as the considered advection-diffusion equation does not contain source terms, the weights $c_i$ remain unchanged during the evolution of the pseudo-particles positions $\mathbf{x}_i(t)$.

Now, the simplest explicit integration algorithm proceeds by updating the particle position considering both the deterministic and the random contributions:

$$\mathbf{x}_i^{n+1} = \mathbf{x}_i^n + \mathbf{v}(\mathbf{x}_i^n)\Delta t + \mathcal{N}(0, 2D\Delta t)\mathbf{u}, \quad \forall i \tag{25}$$

where $\mathbf{x}_i^{n+1} \equiv \mathbf{x}_i(t = (n+1) \times \Delta t)$ and $\mathbf{u}$ is a unit random vector.

Now the distribution moments can be easily computed, and the concentration field could be reconstructed by employing some appropriate smoothing.

It is usual to find this kind of advection-diffusion equation in many branches of science and engineering. In particular they are encountered when one models macromolecular materials within the statistical mechanics framework, as we describe later. Despite its intrinsic simplicity, sometimes they arise defined in highly dimensional spaces including the physical and the conformation coordinates. To avoid the curse of dimensionality drawback characteristic of mesh-based techniques, different authors proposed the use of stochastic techniques exploiting the equivalence between the so-called Fokker-Planck equation and the Ito's stochastic equation [9], similar to the just described equivalence between models (23) and (25). Thus, if one is only interested in computing some moments of the resulting distribution function a moderate population of particles is enough to describe accurately the evolution of such moments. The size of the pseudo-particles population that must be considered for computing accurately the different moments of the solution scales linearly with the dimension of the space, however, if one wants to reconstruct the distribution itself, an impressive number of particles is required which do not scale any more linearly with the dimension of the space.

In general the technique just presented, is conceptually very simple and then easy to implement in a computer or in a parallel computing platform. Explicit integration schemes are usually employed, needing a careful choice and control of the time step.

It is nowadays widely recognized that the main drawbacks of stochastic techniques are: (i) the control of the statistical noise that makes the use of the stochastic approach to perform inverse parameter identification or optimization difficult , because the poor accuracy in the sensitivity analysis; (ii) the difficulty in reconstructing the model solution itself even in moderate multidimensional models; and (iii) the necessity of always solving the transient model, even if one is only interested in the steady state.

Moreover, in complex flow simulation using for example a finite element solver for the flow kinematics computation, one must ensure that all the elements contain a number of particles, at least enough to allow computation of the virial stress (the usual micro-macro bridge). Different possibilities exist, but all of them have a non negligible impact on the solution. Thus, if new particles are added (and probably others removed) the size of the model is changing, tracking procedures are time consuming, and the initialization of the just introduced particles induces a noticeable numerical diffusion.

One could think that the aforementioned difficulties could be circumvented by employing a Lagrangian description of the flow combined with a Lagrangian description of the microstructure evolution (stochastic approach), however usual mesh-based strategies fails because of the high distortion of the meshes during the flow. A first tentative attempt at coupling a meshless Lagrangian description of the flow kinematics (using the natural element method widely described in [10, 11]) and a Lagrangian microstructure description have been recently performed in [12]. This technique could be extended for coupling a Lagrangian flow description with a stochastic description of the microstructure evolution. In any case the difficulties just mentioned will persist.

To reduce the computational cost of numerical simulations different model reduction techniques have been proposed recently [13]. However, the coupling of such techniques (based on the proper orthogonal decomposition - also known as Karhunen-Loève decomposition) with a Lagrangian description of the microstructure evolution remains nowadays an open problem. However, sometimes the stochastic model can be written in an Eulerian form (Brownian Configuration Fields) and in that case, as illustrated in [14], the model reduction runs are opening some interesting perspectives.

## 2.4 Coming back to continuous descriptions: kinetic theory models

The next level of description concerns the statistical mechanics that we are going to illustrate by considering the dynamics of $N$ electrically charged particles of masse $m$. When $N$ becomes too large, direct molecular dynamics simulations are prohibitive from the computing time viewpoint. Thus, more than describing the system from the position and velocities of all the particles, one could introduce the function $f(t, \mathbf{x}, \mathbf{v})$ given the number of particles that at time $t$, are located within an elementary volume $d\mathbf{x} = (dx, dy, dz)^T$ placed at position $\mathbf{x}$ and having velocities within the volume defined by $d\mathbf{v} = (du, dv, dw)^T$ around $\mathbf{v}$. Now, the density balance is written:

$$\frac{\partial f}{\partial t} + \mathbf{v} \cdot \nabla_x f + \mathbf{a}(\mathbf{x}, t) \cdot \nabla_v f = S(t, \mathbf{x}, \mathbf{v}) \tag{26}$$

where $\nabla_x$ and $\nabla_v$ represent the gradient operator in the physical and velocity spaces respectively. We assumed that the acceleration $\mathbf{a} = \frac{d\mathbf{v}}{dt}$ does not depend on the velocity (by this reason it is is not affected by the velocity-gradient $\nabla_v$). The source term $S(t, \mathbf{x}, \mathbf{v})$ represents the so-called collision term and can be derived from an appropriate physical analysis.

We do not need any physics to model the velocity field $\mathbf{v}$ because now the velocity field is a real coordinate, like the spatial ones. On the contrary, we need to define the acceleration field $\mathbf{a}(\mathbf{x}, t)$. For this purpose we consider the Newton's law $\mathbf{a} = \frac{\mathbf{F}}{m}$, and compute the force acting on the particles by taking into account the nature of the system, that in the case addressed here consists of a population of charged particles interacting by means of the Coulomb's potential. The electrostatic potential $U(\mathbf{x}, t)$ can be computed by solving the associated Poisson's problem:

$$\Delta U(\mathbf{x}, t) = -4\pi k Q(\mathbf{x}, t) \tag{27}$$

where $k$ is the Coulomb's law constant and $Q(\mathbf{x}, t)$ is the electrical charge inside the volume $d\mathbf{x}$ around $\mathbf{x}$ that can be computed from:

$$Q(\mathbf{x}, t) = q \int_{\mathcal{R}^3} f(\mathbf{x}, \mathbf{v}, t) d\mathbf{v} \tag{28}$$

with $q$ the particle charge. The force acting on a particle can be finally computed by using:

$$\mathbf{F}(\mathbf{x}, t) = -q \, \nabla_x U(\mathbf{x}, t) \tag{29}$$

**Remarks:**

1. When the particles are not charged the steady state solution of Equation (26) leads to the Maxwell-Boltzmann distribution when the appropriate choice of the collision term representing the particles collisions is made.

2. When the interaction potential does not comply with Coulomb's law, model (27)-(29) cannot be employed. In this case we must compute the force by using:

$$\mathbf{F}(\mathbf{x}, t) = \int_{\mathcal{R}^3} \mathbf{F}(\mathbf{x}, t; \mathbf{x}') \, \tilde{f}(\mathbf{x}', t) d\mathbf{x}' \tag{30}$$

where

$$\tilde{f}(\mathbf{x},t) = \int_{\mathcal{R}^3} f(\mathbf{x},\mathbf{v},t)d\mathbf{v} \tag{31}$$

and $\mathbf{F}(\mathbf{x},t;\mathbf{x}')$ represents the force at position $\mathbf{x}$ originated by a particle located at position $\mathbf{x}'$. In any case this analysis fails when (i) the inter-particle potential leads to a non-definite integral (30); and (ii) in the case of dense systems where the movements of particles is correlated.

3. When the acceleration does not depend on the velocity (as was assumed in Equation (26)) and the collision terms vanish, an equivalence between the conservation equation and the Liouville's theorem in the phase space can be established.

4. In general the establishment of collision terms is quite difficult. To circumvent this difficulty and assuming that the equilibrium distribution is known $f_{eq}(\mathbf{x},\mathbf{v},t)$, one could approximate the collision term by

$$S(t,\mathbf{x},\mathbf{v}) = \frac{f_{eq}(\mathbf{x},\mathbf{v},t) - f(\mathbf{x},\mathbf{v},t)}{\tau} \tag{32}$$

   where $\tau$ represents a relaxation time. This approximation leads to the so-called BFK models.

5. Equation (26), also known as the Vlasov-Poisson-Boltzmann equation, is widely used to model quantum plasmas. One could imagine the application of this formalism to a variety of physical models: colloids, ferrofluids, coarse grained molecular dynamics, crystallization, *etc.* Some examples will be presented later and others are in progress. The associated kinetic theory descriptions constitute the Vlasov-Fokker-Planck formalism.

6. The kinetic theory formalism allows for transforming a discrete model into its continuous counterpart. However, in general, the continuous descriptions involve highly multidimensional spaces and their evolutions are governed by hyperbolic non-linear partial differential equations. To solve these kinds of models, appropriate stabilized solvers, able to proceed in highly multidimensional spaces, are needed. We come back to this point later.

In the previous paragraphs we introduced some ideas related to kinetic theory models of systems composed of particles. We are introducing in the next paragraphs some models describing the microscopic modelling of polymeric liquids. For the sake of simplicity we are focusing on polymer solutions (the entangled systems related to the polymer melts can be also described in the kinetic theory framework [15, 16]). We are addressing the Bead-Spring-Chain (BSC) model of polymer solutions. The BSC chain consists of $S+1$ beads connected by $S$ springs. The bead serves as an interaction point with the solvent and the spring contains the local stiffness information depending on local stretching (see [15] for more details). From now on we are also assuming a fully developed homogeneous flow. Thus, the microstructural state does not depend on the space coordinates.

The dynamics of the chain are governed by viscous drag, Brownian and connector forces. If we denote by $\dot{\mathbf{r}}_k$ the velocity of bead $k$ and by $\dot{\mathbf{q}}_k$ the velocity of the spring connector $\mathbf{q}_k$, we have

$$\dot{\mathbf{q}}_k = \dot{\mathbf{r}}_{k+1} - \dot{\mathbf{r}}_k \quad \forall k = 1, ..., S \tag{33}$$

The dynamics of each bead can be written as:

$$\underbrace{-\zeta(\dot{\mathbf{r}}_k - \mathbf{v}_0 - \nabla\mathbf{v}\cdot\mathbf{r}_k)}_{Viscous\ drag} - \underbrace{k_bT\frac{\partial ln(\psi)}{\mathbf{r}_k}}_{Brownian\ effects} + \underbrace{\mathbf{F}_k^c - \mathbf{F}_{k-1}^c}_{Interaction\ Forces} = 0, \tag{34}$$

where $\zeta$ is the drag coefficient, $\mathbf{v}$ is the velocity field, $\mathbf{v}_0$ is an average velocity, $k_b$ is the Boltzmann constant, $T$ is the absolute temperature and $\psi$ is the distribution function $\psi(\mathbf{r}_1, \cdots, \mathbf{r}_{S+1}, t)$. From equations (33) and (34) we obtain:

$$\dot{\mathbf{q}}_k = \nabla\mathbf{v}\cdot\mathbf{q}_j - \frac{1}{\zeta}\sum_{l=1}^{S}\mathbf{A}_{kl}\cdot\left(k_bT\frac{\partial ln(\psi)}{\partial\mathbf{q}_l} + \mathbf{F}_l^c\right), \tag{35}$$

where $\mathbf{A}_{kl}$ are the components of the Rouse matrix (see [15] for more details).

In the Rouse model the connector force $\mathbf{F}^c$ is a linear function of the connector vector, but we could use Finitely Extensible Non Linear springs, with a dimensionless connector force given by:

$$\mathbf{F}^c(\mathbf{q}_k) = \frac{1}{1 - \mathbf{q}_k^2/b}\mathbf{q}_k, \tag{36}$$

where $\sqrt{b}$ is the maximum dimensionless length of each spring connector of the chain.

Introducing Equation (35) in the equation governing the evolution of the distribution function, and considering homogeneous flows,

$$\frac{\partial\psi(\mathbf{q}_1, ..., \mathbf{q}_S, t)}{\partial t} = -\sum_{k=1}^{S}\left(\frac{\partial}{\partial\mathbf{q}_k}\left(\dot{\mathbf{q}}_k\psi(\mathbf{q}_1, ..., \mathbf{q}_S, t)\right)\right), \tag{37}$$

we obtain

$$\frac{\partial\psi}{\partial t} = -\sum_{k=1}^{S}\left(\frac{\partial}{\partial\mathbf{q}_k}\left(\left(\nabla\mathbf{v}\cdot\mathbf{q}_k - \frac{1}{\zeta}\sum_{l=1}^{S}\mathbf{A}_{kl}\cdot\mathbf{F}_l^c\right)\psi\right)\right) + \\ + \frac{k_bT}{\zeta}\sum_{k=1}^{S}\sum_{l=1}^{S}\mathbf{A}_{kl}\frac{\partial^2\psi}{\partial\mathbf{q}_k\partial\mathbf{q}_l}. \tag{38}$$

The micro-macro bridging is performed by computing the virial stress, that within the rheology community is known as the Kramer's rule. The main difficulty in using this description is the highly multidimensional problem defined by Equation (38) that needs for specific advanced solvers such as the ones that which we proposed in some of our former works and we revisit in the next section.

# 3 Advanced solvers for multi-dimensional models

The solution of models like the one just addressed in Equation (38) needs new strategies, because the standard ones suffer from the curse of dimensionality. Some strategies have been recently proposed for solving models defined in multi-dimensional spaces. The sparse grid techniques are one of the most popular [17], but as we described in the first section, separate representations were also used for solving the models encountered in quantum mechanics (Hartree-Fock and post-Hartree-Fock techniques).

The sparse grids [17] are restricted (as argued in [18]) for treating models involving up to twenty dimensions. On the other hand, separated representations like the one considered in the multi-configuration-self-consistent-fields (MCSCF) [2, 3], fix the number of products (anti-symmetrized) and also assume that those products are constructed by combining a certain number of unknown one-dimensional functions.

In the next section we are describing another alternative and efficient technique recently proposed and based on the use of separated representations within a variational framework.

## 3.1 A new strategy for circumventing curse of dimensionality by using separated representations

Recently we proposed a new strategy able to solve highly multi-dimensional models circumventing the curse of dimensionality. This technique was successfully applied for treating some multi-dimensional models encountered in the kinetic theory description of complex fluids [19, 20]. It allows for defining the optimal number of products containing the optimal one-dimensional functions. In what follows we are summarizing the main ideas that the aforementioned technique involves. For the sake of simplicity, we are illustrating the solution procedure by solving the Poisson problem defined in a space of dimension $D$:

$$\triangle T = -f(x_1, x_2, ..., x_D), \tag{39}$$

where $T$ is a scalar function of $x_1, x_2, ..., x_D$ . Problem (39) is defined in the domain $\Omega = ]-L, +L[^D$ with vanishing boundary conditions.
The problem solution can be written in the form:

$$T(x_1, x_2, ..., x_D) = \sum_{j=1}^{\infty} \alpha_j \prod_{k=1}^{D} F_{kj}(x_k), \tag{40}$$

where $F_{kj}$ is the $j^{th}$ approximation function, with unit norm and that depends only on the $k^{th}$ coordinate.

It is well known that the solution of numerous problems can be accurately approximated using a finite (sometimes very reduced) number of approximation functions, *i.e.*:

$$T(x_1, x_2, ..., x_D) \approx \sum_{j=1}^{Q} \alpha_j \prod_{k=1}^{D} F_{kj}(x_k). \quad (41)$$

The previous expression implies the same number of approximation functions in each dimension, but each one of these functions could be expressed in a discrete form using a different number of parameters (nodes of the one-dimensional grids).

Now, an appropriate numerical procedure is needed for computing the coefficients $\alpha_j$ as well as the $Q \times D$ one-dimensional approximation functions. The proposed numerical scheme consists of an iteration procedure that solves at each iteration $n$ the following three steps:

**Step 1:** *Projection of the solution in a discrete basis*

If we assume the functions $F_{kj}$, $\forall j \in [1, ..., n]$; $\forall k \in [1, ..., D]$ known (verifying the boundary conditions), the coefficients $\alpha_j$ can be computed by introducing the approximation of $T$ into the Galerkin variational formulation associated with Equation (39):

$$\int_\Omega \nabla T^* \cdot \nabla T d\Omega = \int_\Omega T^* f \, d\Omega. \quad (42)$$

Introducing the approximation of $T$ and $T^*$:

$$T(x_1, x_2, ..., x_D) = \sum_{j=1}^{n} \alpha_j \prod_{k=1}^{D} F_{kj}(x_k), \quad (43)$$

$$T^*(x_1, x_2, ..., x_D) = \sum_{j=1}^{n} \alpha_j^* \prod_{k=1}^{D} F_{kj}(x_k), \quad (44)$$

into Equation (42), we have

$$\int_\Omega \nabla \left( \sum_{j=1}^{n} \alpha_j^* \prod_{k=1}^{D} F_{kj}(x_k) \right) \cdot \nabla \left( \sum_{j=1}^{n} \alpha_j \prod_{k=1}^{D} F_{kj}(x_k) \right) d\Omega =$$

$$= \int_\Omega \left( \sum_{j=1}^{n} \alpha_j^* \prod_{k=1}^{D} F_{kj}(x_k) \right) f \, d\Omega. \quad (45)$$

Now, we assume that $f(x_1, \cdots, x_D)$ can be written in the form

$$f(x_1, \cdots, x_D) = \sum_{h=1}^{m} \prod_{k=1}^{D} f_{kh}(x_k). \tag{46}$$

Equation (45) involves integrals of products of $D$ functions each one defined in a different dimension. Let $\prod_{k=1}^{D} g_k(x_k)$ be one of these functions to be integrated. The integral over $\Omega$ can be performed by integrating each function in its definition interval and then multiplying the $D$ computed integrals according to:

$$\int_{\Omega} \prod_{k=1}^{D} g_k(x_k)\, d\Omega = \prod_{k=1}^{D} \int_{-L}^{L} g_k(x_k) dx_k, \tag{47}$$

which makes possible the numerical integration in highly dimensional spaces.

Now, due to the arbitrariness of the coefficients $\alpha_j^*$, Equation (45) allows computation of the $n$-approximation coefficients $\alpha_j$, solving the resulting linear system of size $n \times n$. This problem is linear and moreover rarely exceeds the order of tens of degrees of freedom in standard elliptic models. Thus, even if the resulting coefficient matrix is densely populated, the time required for its solution is negligible with respect to the one required for performing the approximation basis enrichment (step 3).

**Step 2:** *Checking convergence*

From the solution of $T$ at iteration $n$ given by Equation (43) we compute the residual $Re$ related to Equation (39):

$$Re = \frac{\sqrt{\int_{\Omega} \left(\Delta T + f(x_1, \cdots, x_D)\right)^2}}{\|T\|}. \tag{48}$$

The integral in Equation (48) can be written as the product of one-dimensional integrals by performing a separated representation of the square of the residual.

If $Re < \epsilon$ (epsilon is a small enough parameter) the iteration process stops, yielding the solution $T(x_1, \cdots, x_D)$ given by Equation (43). Otherwise, the iteration procedure continues.

**Step 3:** *Enrichment of the approximation basis*

From the coefficients $\alpha_j$ just computed the approximation basis can be enriched by adding the new function $\prod_{k=1}^{D} F_{k(n+1)}(x_k)$. For this purpose we solve the non-linear Galerkin variational formulation related to Equation (39):

$$\int_{\Omega} \nabla T^* \cdot \nabla T d\Omega = \int_{\Omega} T^* f \, d\Omega, \tag{49}$$

using the approximation of $T$ given by:

$$T(x_1, x_2, ..., x_D) = \sum_{j=1}^{n} \alpha_j \prod_{k=1}^{D} F_{kj}(x_k) + \prod_{k=1}^{D} R_k(x_k). \tag{50}$$

The solution of Equation (49) can be expressed from the stationarity of functional $J(T)$:

$$J(T) = \int_{\Omega} \frac{1}{2}(\nabla T)^2 d\Omega - \int_{\Omega} T\ f\ d\Omega, \tag{51}$$

which results in:

$$\delta J = \int_{\Omega} \nabla \delta T \cdot \nabla T d\Omega - \int_{\Omega} \delta T\ f\ d\Omega = 0, \tag{52}$$

that corresponds to Equation (49) by putting $\delta T = T^*$.

Now, taking the variation of $T$ according to its expression given by Equation (50), where the variation of the known functions $F_{kj}$ vanishes, we have:

$$\begin{aligned} \delta T &= \delta(R_1(x_1) \cdots R_D(x_D)) = \\ &= \delta R_1(x_1) R_2(x_2) \cdots R_D(x_D) + \cdots + R_1(x_1) R_2(x_2) \cdots \delta R_D(x_D), \end{aligned} \tag{53}$$

that can be written as

$$\begin{aligned} T^*(x_1, x_2, ..., x_D) &= \\ &= R_1^*(x_1) R_2(x_2) \cdots R_D(x_D) + \cdots + R_1(x_1) R_2(x_2) \cdots R_D^*(x_D). \end{aligned} \tag{54}$$

leading to a non-linear variational problem, whose solution is the $D$ functions $R_k(x_k)$, $k = 1, \cdots, D$. After convergence of the non-linear solver, functions $F_{k(n+1)}(x_k)$ are finally obtained by normalizing the functions $R_1, R_2, ..., R_D$.

The discretization needs an appropriate discrete approximation of functions $R_k(x_k)$ and $F_{kj}(x_k)$. Each one of these functions is approximated using a 1D finite element description. If we assume than $p_k$ nodes are used to construct the interpolation of functions $R_k(x_k)$ and $F_{kj}(x_k)$ in the interval $[-L, L]$, then the size of the resulting discrete non-linear problem is $\sum_{k=1}^{k=D} p_k$. The price to pay for avoiding a whole mesh in the multidimensional domain is the solution of a non-linear problem whose size scales linearly with the dimension of the space. However, even in high dimensions the size of the non-linear problems remain moderate and no particular difficulties have been found in its solution.

Concerning the computation time, even when the non-linear solver converges quickly, this step consumes the main part of the global computing time. Different non-linear solvers have been analyzed: fixed-point, Newton or one based on an alternating directions scheme [19–21].

Preliminary convergence analysis reveals that even when one uses piecewise linear one-dimensional finite elements to approximate all the one-dimensional functions involved by the separated representation, the convergence rate (in the $\mathcal{L}^2$-norm) seems to be fourth order. This result is a little bit strange because we expected the usual second order convergence. We are performing further analysis to check if there is some super-convergence phenomena. At the same time one could modify the one-dimensional approximations to enhance the approximation consistency and then the convergence rates by using spectral approximations or use wavelet approximation basis to take advantage of its multi-resolution properties, specially useful to define adaptive multi-scale strategies. Another possibility to enhance the convergence rate could be the use of moving least squares –MLS– based approximations (widely used in the context of meshless methods [22, 23]).

# 4 Conclusions

In this paper we presented different physical descriptions at different scales involved in material science. At the lowest scale the quantum mechanics involves high-dimensional models whose solution, when fermions are considered, must be anti-symmetric. We have pointed out that in fact more than the multidimensional character of quantum mechanics models, the real challenge is the anti-symmetry constraint. All the practical details concerning the discretization of the Schrödinger equation by using a wave-function separated representation can be found in [24]. However, when the solution of the Schrödinger equation is addressed a new difficulty appears, the one related to the Pauli's principle that requires the anti-symmetrization of each one of the functional products involved by the separated representation. This anti-symmetrization is performed by using the Slater's determinants, but we must recall that the complexity in the use of such determinants scales in the order of $N!$, being $N$ the number of electrons involved in the quantum system. In this way, we can affirm that the main difficulty in the solution of the Schrödinger equation is not its multi-dimensional character, but the necessity of enforcing the solution anti-symmetry when fermions are involved.

Thus, at present, the procedure just employed must be viewed as an enhanced variant of the MCSCF (multi-configuration self consistent field) technique that considers an optimal number of determinants and the optimal functions (in general all the functions involved by the different anti-symmetrized products –determinants– are different).

Coarser modelling involves molecular dynamics or Brownian dynamics but despite its conceptual simplicity its use is not exempted from computational and conceptual difficulties. The next level is the one that corresponds to statistical mechanics descriptions leading to continuous models described by non-linear and coupled multidimensional partial differential equations. Some of these models have been successfully solved in some of our former works [19, 20] by using the adaptive separated representation revisited in the present work. This technique seems a promising alternative for addressing highly-multidimensional models.

The main perspectives of this work are: (i) the possibility of performing a kinetic theory description of different models until now described in the context of molecular dynamics; (ii) the improving of separated representation by enhancing the convergence rate, by enriching the approximation basis and by stabilizing the advection terms that those models usually involve; and (iii) the exploration of new quantum mechanics descriptions as the ones derived from the Dirac formalism or by applying the Schrödinger one to systems composed of bosons where the Pauli's principle do not apply.

# References

[1] R.B. Laughlin, *"The theory of everything"*, Proceeding of the U.S.A. National Academy of Sciences, 2000.

[2] E. Cancès, M. Defranceschi, W. Kutzelnigg, C. Le Bris, Y. Maday, *"Computational Quantum Chemistry: a primer"*, Handbook of Numerical Analysis, Vol. X, Elsevier, 3-270, 2003.

[3] *"Handbook of numerical analysis, Vol. X: Computational chemistry"*, C. Le Bris editor, Elsevier, 2003.

[4] V. Gavini, K. Bhattacharya, M. Ortiz, *"Quasi-continuum orbital-free density-functional theory: A route to multi-million atom non-periodic DFT calculation"*, Journal of the Mechanics and Physics of Solids, 55/4, 697-718, 2007.

[5] V.B. Shenoy, R. Millera, E.B. Tadmor, D. Rodney, R. Phillips, M. Ortiz, *"An adaptive finite element approach to atomic-scale mechanics – the quasicontinuum method"*, Journal of the Mechanics and Physics of Solids, 36, 500-531, 1999.

[6] G.J. Wagner, W.K. Liu, *"Coupling of atomistic and continuum simulations using a bridging scale decomposition"*, Journal of Computational Physics, 190, 249-274, 2003.

[7] H. Ben Dhia, *"Multiscale mechanical problems: the Arlequin method"*, C. R. Acad. Sci., Paris 326, Ser-II b, 899-904, 1998.

[8] S.P. Xiao, T. Belytschko, *"A bridging domain method for coupling continua with molecular dynamics"*, Comput. Methods Appl. Mech. Engrg., 193, 1645-1669, 2004.

[9] H.C. Öttinger, M. Laso, *"Smart polymers in finite element calculation"*, In P. Moldanaers and R. Keunings (Editors) Theoretical and Applied Rheology, Vol. 1, Proceedings of the XIth International Congress on Rheology (Elsevier), 286-288, 1992.

[10] M.A. Martinez, E. Cueto, I. Alfaro, M. Doblare, F. Chinesta, *"Updated Lagrangian free surface flow simulations with the natural neighbour Galerkin methods"*, International Journal for Numerical Methods in Engineering, 60/13, 2105-2129, 2004.

[11] D. Gonzalez, E. Cueto, F. Chinesta, M. Doblare, *"An updated Lagrangian strategy for free-surface fluid dynamics"*, Journal of Computational Physics, 223, 127-150, 2007.

[12] M.A. Martinez, E. Cueto, M. Doblare, F. Chinesta, *"Natural element meshless*

*simulation of injection processes involving short fiber suspensions"*, Journal of Non-Newtonian Fluid Mechanics, 115, 51-78, 2003.

[13] A. Ammar, D. Ryckelynck, F. Chinesta, R. Keunings, *"On the reduction of kinetic theory models related to finitely extensible dumbbells"*, 134, 136-147, 2006.

[14] F. Chinesta, A. Ammar, A. Falco, M. Laso, *"On the reduction of stochastic kinetic theory models of complex fluids"*, Modelling and Simulation in Material Science and Engineering, 15, 639-652, 2007.

[15] B.B. Bird, C.F. Curtiss, R.C. Armstrong, O. Hassager, *"Dynamics of polymeric liquids"*, vol. 2, John Wiley & Sons, 1987.

[16] R.G. Larson, *"The structure and rheology of complex fluids"*, Oxford University Press, New York, 1999.

[17] H.J. Bungartz, M. Griebel, *"Sparse grids"*, Acta Numer., 13, 1-123, 2004.

[18] Y. Achdou, O. Pironneau, *"Computational methods for option pricing"*, SIAM Front. Appl. Math., 2005.

[19] A. Ammar, B. Mokdad, F. Chinesta, R. Keunings, *"A new family of solvers for some classes of multidimensional partial differential equations encountered in kinetic theory modeling of complex fluids"*, J. Non-Newtonian Fluid Mech., 139, 153-176, 2006.

[20] A. Ammar, B. Mokdad, F. Chinesta et R. Keunings, *"A new family of solvers for some classes of multidimensional partial differential equations encountered in kinetic theory modeling of complex fluids. Part II: transient simulation using space-time separated representations"*, J. Non-Newtonian Fluid Mech., 144, 98-121, 2007.

[21] G. Beylkin, M. Mohlenkamp, *"Algorithms for numerical analysis in high dimensions"*, SIAM J. Sci. Commun., 26/6, 2133-2159, 2005.

[22] B. Nayroles, G. Touzot, P. Villon, *"Generalizing the finite element method: diffuse approximation and diffuse elements"*, Computational Mechanics, 10, 307-318, 1992.

[23] T. Belytschko , Y.Y. Lu, L. Gu, *"Element-free Galerkin methods"*, International Journal for Numerical Methods in Engineering, 37, 229-256, 1994.

[24] A. Ammar, F. Chinesta, *"Circumventing curse of dimensionality in the solution of highly multidimensional models encountered in quantum mechanics using meshfree separated representations"*, Lecture Notes on Computational Science and Engineering, Springer, In press.

©Saxe-Coburg Publications, 2008.
Trends in Engineering Computational Technology
B.H.V. Topping and M. Papadrakakis, (Editors)
Saxe-Coburg Publications, Stirlingshire, Scotland, 271-294.

# Chapter 14

# Continuum and Atomic-Scale Modeling of Self-Positioning Microstructures and Nanostructures

**G.P. Nikishkov and Y. Nishidate**
**The University of Aizu**
**Aizu-Wakamatsu City, Fukushima, Japan**

## Abstract

This chapter presents investigations of self-positioning microstructures and nanostructures by analytical techniques, finite element analysis and atomic-scale modeling. Closed-form solutions for curvature radius of self-positioning hinge structures are obtained for plane strain and generalized plane strain deformation. The finite element method is used for predicting hinge curvature radius for self-positioning structures of variable width. Anisotropic finite element analysis of self-positioning structures with different orientation of material axes is performed to estimate the effect of material anisotropy on the self-positioning. An algorithm of the atomic-scale finite element method (AFEM) based on the Tersoff interatomic potential has been developed. The AFEM is applied to modeling of GaAs and InAs bi-layer self-positioning nanostructures. Nanohinge curvature radius dependence on the structure thickness and the material orientation angle is investigated. It was found that atomic-scale effects play a considerable role for nanostructures of small thickness less than 40 nm.

**Keywords:** nanostructure, self-positioning, finite element method, atomic-scale.

## 1 Introduction

Multilayer structures are used in microelectronics, optics and other engineering areas. An efficient method of three-dimensional micro- and nanostructure fabrication is based on a self-positioning phenomenon of multilayer epitaxial structures with lattice mismatch [1–3]. Figure 1 illustrates the self-positioning fabrication procedure. Three material layers are placed on a substrate using the molecular beam epitaxy method. Two upper material layers have different lattice periods. This leads to the existence of initial strains and initial stresses. The detachment of a strained bilayer from the substrate is done by etching out a sacrificial layer. Hinges, tubes and other three-dimensional micro- and nanostructures can be created by selecting appropriate physical properties and thickness for two or more lattice-mismatched epitaxial layers.

After etching away the sacrificial layer, the layered materials form hinges or tubes with diameter controllable by layer material properties and thickness. Strain-driven self-positioning can be used to create 3D nanoscale structures by folding 2D membran es as *origami* (Japanese paper craft work) [4, 5]. A possibility of creating rolled-up micro- and nanotubes from a single material layer is investigated in publication [6].

In order to avoid trials during nanostructure fabrication it is desirable to be able to predict the final shape of self-positioning structures. Analytical solutions or numerical modeling can be used. Closed-form solutions [7, 8] have been obtained for plane stress conditions and for biaxial free deformation of multilayer structures under thermal influence. However, these solutions have limited applicability for self-positioning hinged structures, which are characterized by a deformation constraint along the line of the hinge detachment from the substrate.

This article presents application of analytical techniques, finite element analysis and atomic-scale modeling to investigation of self-positioning microstructures and nanostructures.

Closed-form solutions for curvature radius estimation of self-positioning rolled-up structures for cases of generalized plane strain (wide strips with bending constraint in one direction) are derived. Plane strain deformation of multilayer self-positioning structures is also considered.

The finite element method is used for predicting hinge curvature radius of self-positioning structures of variable widths. Anisotropic finite element analysis of self-positioning structures with different orientation of material axes shows that the curvature radius can significantly vary with change of material orientation angle.

Algorithm of the atomic-scale finite element method (AFEM) based on the Tersoff interatomic potential has been developed. A Solution procedure for problems with large displacements is organized as the Newton-Raphson iteration procedure. The developed AFEM code is applied to modeling of GaAs and InAs bi-layer self-positioning nanostructures. A problem series includes investigation of nanohinge curvature radius dependence on the structure thickness and on the material orientation angle.

## 2 Closed-form solutions for multilayer structures

Closed-form solutions for elastic multilayer structures have been obtained for the case of infinite sheet, which experiences spherical bending due to initial strains, and for the case of two-dimensional plane stress geometry, which undergoes plane bending [7–9]. However, these solutions are not appropriate for estimation of curvature for hinges and tubes that have considerable width and a displacement constraint at least at one of the ends as shown in Figure 1. The displacement constraint does not allow the multilayer structure to bend in the transverse direction.

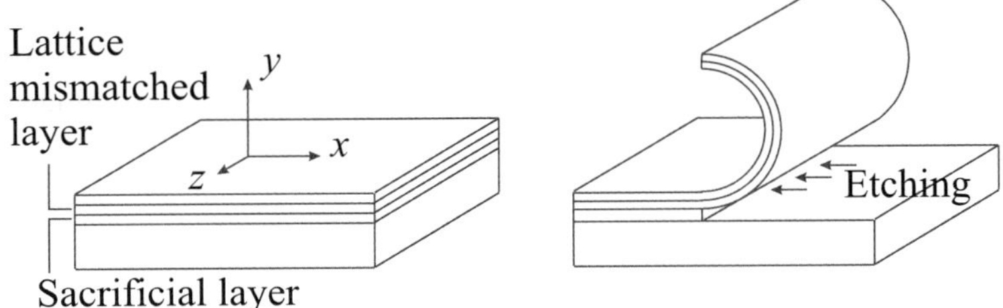


Figure 1: Self-positioning fabrication procedure

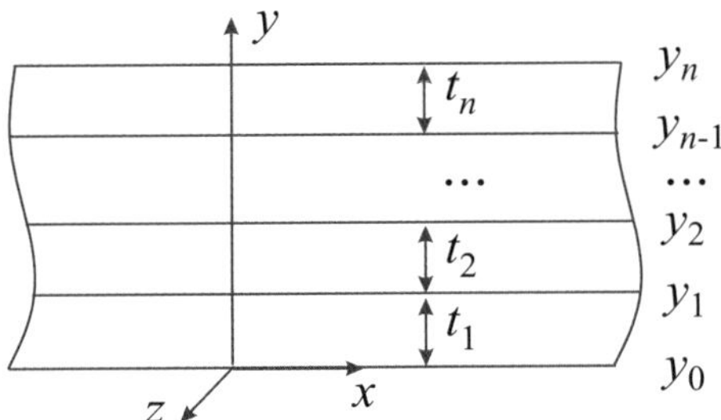


Figure 2: Multilayer structure

In order to obtain a curvature estimate suitable for multilayer hinges and tubes, let us consider an elastic multilayer structure shown in Figure 2 under conditions of generalized plane strain [10]. The structure consists of $n$ layers with thickness $t_i$, $i = 1, 2, ..., n$. Materials of layers have the elastic properties $E_i$, Young's modulus and $\nu_i$, Poisson's ratio. The layers are under the influence of initial strains $\varepsilon_i^0$, which are due to lattice mismatch or due to different thermal properties.

For generalized plane strain conditions, the total strain in $z$-direction is equal to some value $d$:

$$\varepsilon_z = \frac{1}{E}(\sigma_z - \nu\sigma_x) + \varepsilon^0 = d \tag{1}$$

which magnitude is determined by a condition of zero total force in $z$-direction:

$$\int_{y_0}^{y_n} \sigma_z dy = 0 \tag{2}$$

Under the plane section assumption, strain $\varepsilon_x$ is a linear function of the $y$-coordinate:

$$\varepsilon_x = c + \frac{y - y_b}{R} \tag{3}$$

where $c$ is the uniform strain component, $y_b$ is the $y$-level where the bending strain component is zero, and $R$ is the structure curvature radius. This expression is true for both generalized and ordinary plane strain conditions [11]. Using Eqs. (1) and (3) and

Hooke's law, it is possible to express normal stresses $\sigma_x$ and $\sigma_z$ through parameters $c$, $y_b$, $d$, and $R$:

$$\sigma_x = E'\left(c + \frac{y - y_b}{R} - \eta\varepsilon^0 + \nu d\right) \tag{4}$$

$$\sigma_z = \nu E'\left(c + \frac{y - y_b}{R} - \eta\varepsilon^0 + \nu d\right) + E(d - \varepsilon^0) \tag{5}$$

where $E' = E/(1 - \nu^2)$ and $\eta = 1 + \nu$.

For determination of the unknown parameters introduced in Eqs. (1–5), we follow the approach used in publications [7, 11]. Unknown parameters $c$, $y_b$, $d$, and $R$, are determined from the following equilibrium equations:

force due to bending fraction of stress $\sigma_x$:

$$\sum_{i=1}^{n} \int_{y_{i-1}}^{y_i} \frac{E_i'(y - y_b)}{R} dy = 0 \tag{6}$$

force due to uniform fraction of stress $\sigma_x$:

$$\sum_{i=1}^{n} E_i'\, t_i(c - \eta_i\varepsilon_i^0 + \nu_i d) = 0 \tag{7}$$

bending moment created by the normal stress $\sigma_x$ with respect to bending axis $z$:

$$\sum_{i=1}^{n} \int_{y_{i-1}}^{y_i} E_i'\left(c + \frac{y - y_b}{R} - \eta_i\varepsilon_i^0 + \nu_i d\right)(y - y_b)dy = 0 \tag{8}$$

total force in $z$-direction:

$$\sum_{i=1}^{n} \int_{y_{i-1}}^{y_i} \left[\nu_i E_i'\left(c + \frac{y - y_b}{R} - \eta_i\varepsilon_i^0 + \nu_i d\right) + E_i(d - \varepsilon_i^0)\right] dy = 0 \tag{9}$$

Parameter $y_b$ is directly obtained from Equation (6):

$$y_b = \frac{\sum_{i=1}^{n} E_i' t_i(y_i + y_{i-1})}{2\sum_{i=1}^{n} E_i' t_i} \tag{10}$$

where $t_i = y_i - y_{i-1}$ is a thickness of $i$th layer. The other three parameters are determined by the following system of linear algebraic equations:

$$\begin{bmatrix} a_{11} & a_{12} & 0 \\ a_{21} & a_{22} & a_{23} \\ a_{31} & a_{32} & a_{33} \end{bmatrix} \begin{Bmatrix} c \\ d \\ K \end{Bmatrix} = \begin{Bmatrix} b_1 \\ b_2 \\ b_3 \end{Bmatrix} \tag{11}$$

$$
\begin{aligned}
a_{11} &= a_{32} = \sum E_i' t_i \\
a_{12} &= a_{31} = \sum E_i' t_i \nu_i \\
a_{21} &= \sum E_i' t_i (y_i^m - y_b) \\
a_{22} &= a_{33} = \sum E_i' t_i \nu_i (y_i^m - y_b) \\
a_{23} &= \tfrac{1}{3} \sum E_i' t_i [4(y_i^m)^2 - y_i y_{i-1} - 3y_b(2y_i^m - y_b)] \\
b_1 &= b_3 = \sum E_i' t_i \eta_i \varepsilon_i^0 \\
b_2 &= \sum E_i' t_i \eta_i \varepsilon_i^0 (y_i^m - y_b)
\end{aligned}
\tag{12}
$$

where $K = 1/R$ is a structure curvature, and $y_i^m = (y_i + y_{i-1})/2$.

Solution of equation system (11) provides the following relation for curvature $K$:

$$
K = \frac{1}{R} = -\frac{3\sum_{i=1}^{n} E_i' t_i \left(y_i + y_{i-1} - 2y_b\right)\left(c - \eta_i \varepsilon_i^0 + \nu_i d\right)}{2\sum_{i=1}^{n} E_i' t_i \left[y_i^2 + y_i y_{i-1} + y_{i-1}^2 - 3y_b\left(y_i + y_{i-1} - y_b\right)\right]} \tag{13}
$$

$$
c = \frac{\sum_{i=1}^{n} E_i' t_i \left(\eta_i \varepsilon_i^0 - \nu_i d\right)}{\sum_{i=1}^{n} E_i' t_i} \tag{14}
$$

$$
d = \frac{(a_{12} - a_{11}) a_{23} b_1 - (a_{21} b_1 - a_{11} b_2) a_{22}}{(a_{11} a_{22} - a_{12} a_{21}) a_{22} + (a_{12}^2 - a_{11}^2) a_{23}} \tag{15}
$$

For a bilayer structure with $n = 2$, curvature $K$ is given by the following expression:

$$
K = -\frac{6E_1' E_2' t_1 t_2 (t_1 + t_2)[\eta_1 \varepsilon_1^0 - \eta_2 \varepsilon_2^0 + d(\nu_2 - \nu_1)]}{E_1'^2 t_1^4 + E_2'^2 t_2^4 + 2E_1' E_2' t_1 t_2 (2t_1^2 + 2t_2^2 + 3t_1 t_2)} \tag{16}
$$

Under generalized plane strain conditions, the curvature radius is affected by the difference of Poisson's ratio values $\nu_2$ and $\nu_1$.

The *plane strain* solution [11] characterized by zero strain $\varepsilon_z$ is a particular case of equations (13) and (16) with $d = 0$, $E_i' = E_i/(1 - \nu_i^2)$ and $\eta_i = 1 + \nu_i$.

The *plane stress* solution applicable to narrow strips is obtained from (13) and (16) when we set $d = 0$, $E_i' = E_i$, $\eta_i = 1$.

# 3 Finite element analysis

## 3.1 Finite element algorithm for anisotropic problems with large displacements

The following finite element procedure is used for three-dimensional modeling of anisotropic elastic geometrically nonlinear structures under the influence of initial

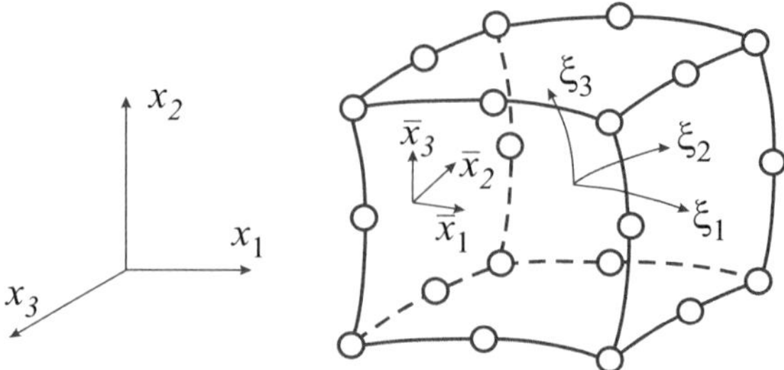


Figure 3: Three-dimensional finite element and coordinate systems: global coordinate system $x_1,\ x_2,\ x_3$, material coordinate system $\bar{x}_1,\ \bar{x}_2,\ \bar{x}_3$ and local element coordinate system $\xi_1,\ \xi_2,\ \xi_3$

strains. It is assumed that rotations and translations are large but strains are small. Our algorithm [12] includes both material anisotropy and large displacements. Usually, geometrical nonlinearity and anisotropy are treated separately [13, 14].

### 3.1.1 Coordinates

Three coordinate systems shown in Figure 3 are used in the finite element analysis:

1) $x_1,\ x_2,\ x_3$ is the global cartesian coordinate system (fixed in space, used for the whole structure);

2) $\xi_1,\ \xi_2,\ \xi_3$ is the local element coordinate system (nonorthogonal, movable, one for each finite element);

3) $\bar{x}_1,\ \bar{x}_2,\ \bar{x}_3$ is the material coordinate system (orthogonal, movable, defined at any point inside a finite element).

The global coordinate system is employed for formulating the global finite element equation system. The local coordinate systems are used for interpolation within three-dimensional isoparametric finite elements:

$$\begin{aligned} x_i(\xi_1,\xi_2,\xi_3) &= N_m(\xi_1,\xi_2,\xi_3)x_i^m \\ u_i(\xi_1,\xi_2,\xi_3) &= N_m(\xi_1,\xi_2,\xi_3)u_i^m \end{aligned} \tag{17}$$

Here $N_m$ are nodal shape functions; $x_i$ are the global coordinates of the point with local coordinates $\xi_1,\ \xi_2,\ \xi_3$; $x_i^m$ are the coordinates of the $m$th element node; $u_i$ are the displacements; $u_i^m$ are the displacements of the $m$th element node. The repeated indices imply summation.

Material coordinate axes are involved in anisotropic constitutive relations at any point of the structure. Material coordinates move and rotate with the material points.

Since the local coordinate system $\xi_i$ is not orthogonal, it cannot be used in material constitutive equations. However, the material coordinate system $\bar{x}_i$ at any point inside the finite element can be determined through the local coordinates $\xi_i$. Unit vectors $\vec{e}_{\xi_i}$ tangent to the local coordinates $\xi_i$ have the following components in the global coordinate system:

$$\begin{aligned} \vec{v}_{\xi_i} &= \left\{ \frac{\partial x_1}{\partial \xi_i} \;\; \frac{\partial x_2}{\partial \xi_i} \;\; \frac{\partial x_3}{\partial \xi_i} \right\} \\ \vec{e}_{\xi_i} &= \frac{\vec{v}_{\xi_i}}{|\vec{v}_{\xi_i}|} \end{aligned} \tag{18}$$

Derivatives of the global coordinates in respect to local coordinates are estimated using Equation (17):

$$\frac{\partial x_j}{\partial \xi_i} = \frac{\partial N_m}{\partial \xi_i} x_j^m \tag{19}$$

If we assume the unit vector $\vec{e}_{\bar{x}_1}$ of the material coordinate $\bar{x}_1$ coincides with the direction of the unit vector $\vec{e}_{\xi_1}$ (tangent to the local coordinate $\xi_1$) then two other unit vectors of the material coordinate system can be determined as vector products:

$$\begin{aligned} \vec{e}_{\bar{x}_1} &= \vec{e}_{\xi_1} \\ \vec{e}_{\bar{x}_3} &= \vec{e}_{\xi_1} \times \vec{e}_{\xi_2} \\ \vec{e}_{\bar{x}_2} &= \vec{e}_{\bar{x}_3} \times \vec{e}_{\bar{x}_1} \end{aligned} \tag{20}$$

Direction cosines $\alpha_{ij}$ for transformation from the global coordinate system $x_i$ to the material coordinate system $\bar{x}_i$ are expressed through components of the unit vectors $\vec{e}_{\bar{x}_i}$:

$$\alpha_{ij} = \cos(\bar{x}_i \; x_j) = e_{\bar{x}_i j} \tag{21}$$

Transformations of vectors from the global coordinate system to the material coordinate system and back are performed as follows:

$$\begin{aligned} \bar{x}_i &= \alpha_{ij} x_j \\ x_i &= \alpha_{ji} \bar{x}_j \end{aligned} \tag{22}$$

### 3.1.2 Anisotropic constitutive law

For an anisotropic elastic material, the stress tensor $\sigma_{ij}$ and the strain tensor $\varepsilon_{ij}$ are related through the Hooke's law

$$\sigma_{ij} = C_{ijkl}\varepsilon_{kl} \tag{23}$$

where $C_{ijkl}$ is the elasticity tensor. The symmetry properties allows to represent the Hooke's law in so-called contracted form using matrix-vector notations:

$$\begin{aligned} \sigma &= \mathbf{C}\varepsilon \\ \sigma &= \{\sigma_{11} \;\; \sigma_{22} \;\; \sigma_{33} \;\; \sigma_{12} \;\; \sigma_{23} \;\; \sigma_{31}\} \\ \varepsilon &= \{\varepsilon_{11} \;\; \varepsilon_{22} \;\; \varepsilon_{33} \;\; 2\varepsilon_{12} \;\; 2\varepsilon_{23} \;\; 2\varepsilon_{31}\} \end{aligned} \tag{24}$$

Here **C** is the contracted $6\times6$ elasticity matrix; $\sigma$, $\varepsilon$ are the contracted $1\times6$ stress and strain vectors. The contracted form of the Hooke's law is convenient for using in a finite element computation since it reduces the number of array dimensions.

Transformations from the global coordinate system to the material coordinate system for the stress and elasticity tensors are performed in the tensor form:

$$\begin{aligned}\bar{\sigma}_{ij} &= \alpha_{ip}\alpha_{jq}\sigma_{pq} \\ \bar{C}_{ijkl} &= \alpha_{ip}\alpha_{jq}\alpha_{kr}\alpha_{ls}C_{pqrs}\end{aligned} \tag{25}$$

During calculations of element matrices and vectors, it is more efficient to use matrix-vector notation. Contraction from the stress tensor to the stress vector and expansion from the stress vector to the stress tensor are done as follows:

$$\begin{aligned}\sigma_p &= \sigma_{m_p n_p}\,, \qquad p = 1...6 \\ \sigma_{m_p n_p} &= \sigma_p\,, \qquad p = 1...6\end{aligned} \tag{26}$$

Here two index vectors $\mathbf{m}$ and $\mathbf{n}$ are used:

$$\begin{aligned}\mathbf{m} &= \{1\ \ 2\ \ 3\ \ 1\ \ 2\ \ 3\} \\ \mathbf{n} &= \{1\ \ 2\ \ 3\ \ 2\ \ 3\ \ 1\}\end{aligned} \tag{27}$$

The expansion operation should be accompanied by filling symmetrical coefficients of the stress tensor $\sigma_{ij} = \sigma_{ji}$. For the elasticity tensor, contraction and expansion operations are performed in the following ways:

$$\begin{aligned}C_{pq} &= C_{m_p n_p m_q n_q}\,, \qquad p = 1...6, \qquad q = 1...6 \\ C_{m_p n_p m_q n_q} &= C_{pq}\,, \qquad p = 1...6, \qquad q = 1...6\end{aligned} \tag{28}$$

To finish expansion of the elasticity tensor it is necessary to fill symmetrical coefficients $C_{ijkl} = C_{jikl} = C_{ijlk}$.

### 3.1.3 Finite element equations

The incremental element equation relating load and displacement increments has the following form:

$$\mathbf{k}\Delta\mathbf{u} = \mathbf{f} - \mathbf{r} \tag{29}$$

Here $\mathbf{k}$ is the element tangent stiffness matrix at time $t$, $\Delta\mathbf{u}$ is the nodal displacement increment, $\mathbf{f}$ is the load vector at time $t + \Delta t$, and $\mathbf{r}$ is the vector of nodal internal forces corresponding to the current stress state at time $t$.

The element tangent stiffness matrix $\mathbf{k}$ is a sum of two matrices: the usual linear stiffness matrix $\mathbf{k}_e$ and the nonlinear addition $\mathbf{k}_\sigma$ depending on the stress state:

$$\mathbf{k} = \mathbf{k}_e + \mathbf{k}_\sigma \tag{30}$$

A $3\times3$ block of the linear element stiffness matrix $\mathbf{k}_e$ corresponding to a combination of nodes with local numbers $m$ and $n$ is estimated as follows:

$$(\mathbf{k}_e)^{mn} = \int_V \mathbf{B}_m^T\mathbf{C}\mathbf{B}_n dV \tag{31}$$

Here $\mathbf{B}_m$ is a 3×6 block of the displacement differentiation matrix containing derivatives of the shape function for node $m$; $\mathbf{C}$ is the contracted elasticity matrix in the global coordinate system. Integration is performed over the current element volume $V$ in the global coordinate system.

Coefficients of the nonlinear element stiffness matrix $\mathbf{k}_\sigma$ are given in index notation since it is simpler than matrix-vector form:

$$(k_\sigma)_{ij}^{mn} = \int_V \sigma_{kl} \frac{\partial N_m}{\partial x_k} \frac{\partial N_n}{\partial x_l} \delta_{ij} dV \tag{32}$$

Here $m, n$ are local node numbers; $i, j$ are indices related to the global coordinate axes $(x_1, x_2, x_3)$; $\sigma_{kl}$ are stresses in the global coordinate system; $N_m = N_m(\xi_1, \xi_2, \xi_3)$ are nodal shape functions and $\delta_{ij}$ is the Kronecker delta symbol. Stresses $\sigma_{kl}$ are originally calculated in the material coordinate system. Transformation from the material coordinate system to the global coordinate system at integration points should be done prior to integration.

Vector $\mathbf{f}$ in Equation (29) contains nodal equivalents of loads applied to the finite element model. A nodal block of the fictitious force vector $\mathbf{f}$ for modeling of lattice mismatch with the initial strain $\varepsilon_0$ is estimated as:

$$(\mathbf{f})^m = \int_V \mathbf{B}_m^T \mathbf{C} \boldsymbol{\varepsilon}_\mathbf{0} dV \tag{33}$$

where $\boldsymbol{\varepsilon}_\mathbf{0} = \{\varepsilon_0 \quad \varepsilon_0 \quad \varepsilon_0 \quad 0 \quad 0 \quad 0\}$ is the initial strain vector. Nodal internal forces $\mathbf{r}$ are obtained by integration of stresses over the current element volume in the global coordinate system:

$$(r)_i^m = \int_V \sigma_{ik} \frac{\partial N_m}{\partial x_k} dV \tag{34}$$

Element equations are assembled into a global equation system using element connectivity information:

$$\mathbf{K} \Delta \mathbf{U} = \mathbf{F} - \mathbf{R} \tag{35}$$

where $\mathbf{K}$ is the global stiffness matrix at time $t$, $\Delta\mathbf{U}$ is the nodal displacement increment, $\mathbf{F}$ is the nodal equivalent of the applied load at time $t + \Delta t$, and $\mathbf{R}$ is the vector of nodal internal forces corresponding to stresses at time $t$.

Predicted displacement increment $\Delta\mathbf{U}$ is not precise since the stiffness matrix $\mathbf{K}$ corresponds to the beginning of the time step. The following Newton-Raphson iteration procedure is used for improving solution at each time step:

$$\begin{aligned}
&\Psi^{(0)} = \mathbf{F} - \mathbf{R} \\
&\textbf{do} \\
&\qquad \mathbf{K} = \mathbf{K}(\mathbf{U}^{(i)}) \\
&\qquad \mathbf{R} = \mathbf{R}(\mathbf{U}^{(i)}) \\
&\qquad \Delta\mathbf{U}^{(i)} = \mathbf{K}^{-1}\Psi^{(i-1)} \\
&\qquad \mathbf{U}^{(i)} = \mathbf{U}^{(i-1)} + \Delta\mathbf{U}^{(i)} \\
&\qquad \Psi^{(i)} = \mathbf{F} - \mathbf{R}^{(i)} \\
&\textbf{while} \quad \left\| \Psi^{(i)} \right\| > \varepsilon
\end{aligned} \tag{36}$$

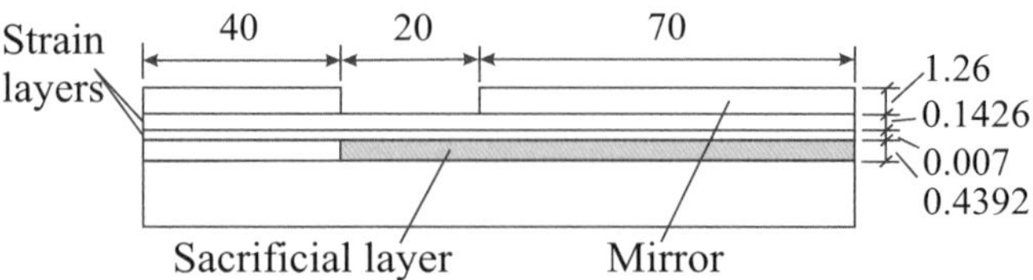


Figure 4: Schematic of a hinged mirror structure

Iterations are terminated when the norm of the nodal residual vector $\left\|\Psi^{(i)}\right\|$ becomes less than a specified tolerance $\varepsilon$. It was reported [16] that the Newton-Raphson iteration method is more efficient than other time integration schemes such as the Newmark method.

The finite element procedure for the solution of three-dimensional anisotropic geometrically nonlinear problems has been implemented as a C computer code. The 20-node hexahedral isoparametric element is used for discretization. The strains and the stresses are controlled at Gaussian integration points $2 \times 2 \times 2$ in the material coordinate frames. The initial strain $\varepsilon_0$ is divided into $n$ increments. At each step an increment of the initial strain $\varepsilon_0/n$ is applied, and the finite element equation (35) is solved.

## 3.2 Dependence of the curvature radius on the structure width

A closed-form solution for the curvature radius of a hinge structure under plane stress and plane strain conditions has been presented in Section 2. Plane stress solution corresponds to structures with small width and plane strain solution to structures with large width. Let us perform finite element analysis for structures of variable width and compare finite element results with closed-form predictions [11, 15].

An experimental procedure for creating a self-positioning hinged micromirror has been developed by Vaccaro *et al* [3]. A schematic of a hinged mirror structure is depicted in Figure 4 (sizes in $\mu$m). The structure consists of a mirror layer, two strain layers, a sacrificial layer and a substrate. The strain layers have different lattice periods that leads to self-positioning of the mirror after etching out the sacrificial layer.

The upper strain layer has a thickness $t_2$=142.6 nm and is composed of GaAs with a lattice constant $a_2$=0.56536 nm. The lower strain layer made of $In_{0.2}Ga_{0.8}As$ has the following parameters: $t_1$=7.0 nm, $a_1$=0.57347 nm. Ratio of elasticity modulus values for the upper and lower material layers are $E_2/E_1 = 1.076$. The difference in lattice constants creates the initial strain in the lower strain layer:

$$\varepsilon_0 = (a_1 - a_2)/a_2 = 0.014345.$$

Since we have estimates for curvature of a hinged structure for two limiting cases, it is of interest to investigate a dependence of curvature on the structure width using the

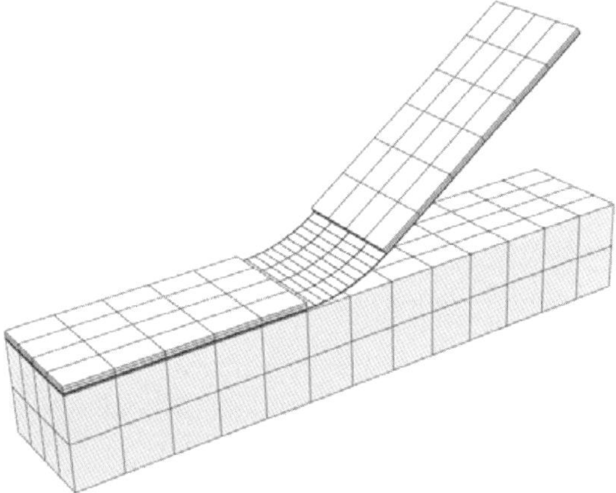

Figure 5: Final shape of the structure after self-positioning (the symmetrical half of the structure is shown)

finite element analysis. Finite element models for hinged structures of different widths are generated by sweeping a two-dimensional $xy$-mesh in the $z$-coordinate direction. A typical three-dimensional mesh for the symmetrical half of the structure consists of 484 hexahedral 20-node elements and 2868 nodes.

A series of problems for the self-positioning mirror of variable width is solved using 200 computational steps in each analysis. Typical appearance of the finite element model of the hinged mirror after its release from the substrate is shown in Figure 5. The curvature radius for hinges of different width is estimated as

$$R = \frac{(x_1^B - x_1^A)^2 + (x_3^B - x_3^A)^2}{2(x_3^B - x_3^B)} \tag{37}$$

Here $x$ and $y$ are horizontal and vertical coordinates of some point selected for the radius determination; $x_0$ and $y_0$ are coordinates of the hinge detachment from the substrate. The point located at the hinge center is used for the curvature radius estimation. The dependence of the hinge curvature radius on the structure width is presented in Figure 6. Good correspondence of the closed-form solution (16) and finite element results is observed for narrow strips (plane stress) and for wide strips (plane strain). So, analytical solution can be used for curvature radius estimation in the two limiting cases. Computational modeling can be used for hinges of intermediate width.

## 3.3 Effect of material anisotropy

Here we investigate the effect of material anisotropy on the curvature radius of self-positioning structures and compare results with experimental data [12]. The hinged micro-plate used in computational modeling and in the experiment has the hinge of the

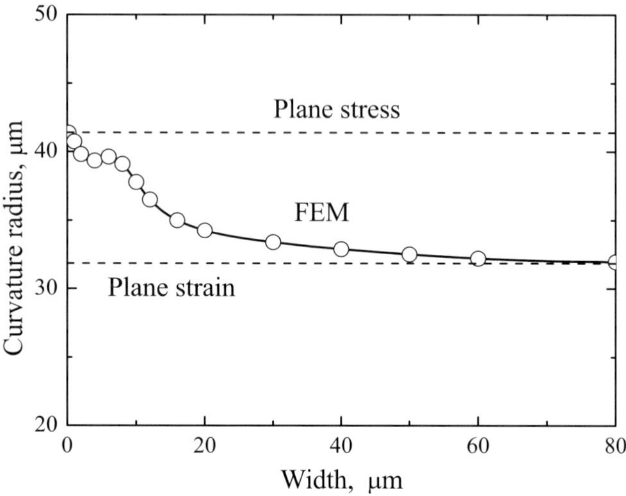


Figure 6: Dependence of the hinge curvature on the structure width

size 6 $\mu$m along $x_1$ by 100 $\mu$m along $x_3$ (see Figure 7). The upper strain layer made of GaAs has a thickness $t_2$=88 nm. The lower strain layer is composed of $In_{0.2}Ga_{0.8}As$ with thickness $t_1$=56 nm. Both materials are anisotropic with cubic crystal symmetry.

The elasticity matrix for materials with cubic crystal symmetry is composed of

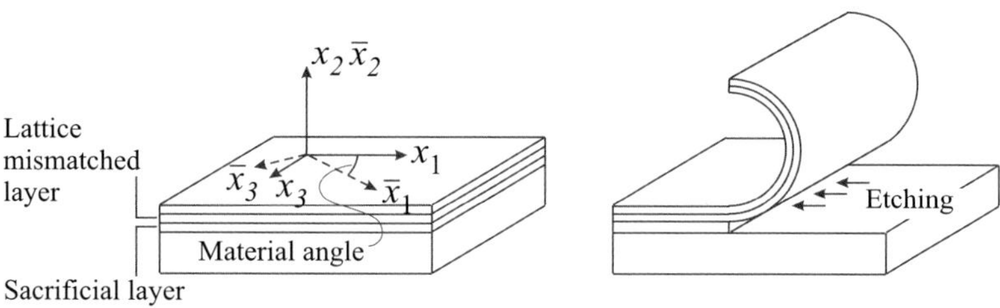


Figure 7: Multilayer structure before (left) and after (right) self-positioning. The hinge orientation angle is an angle between the material axis $\bar{x}_1$ at time $t = 0$ and the global axis $x_1$

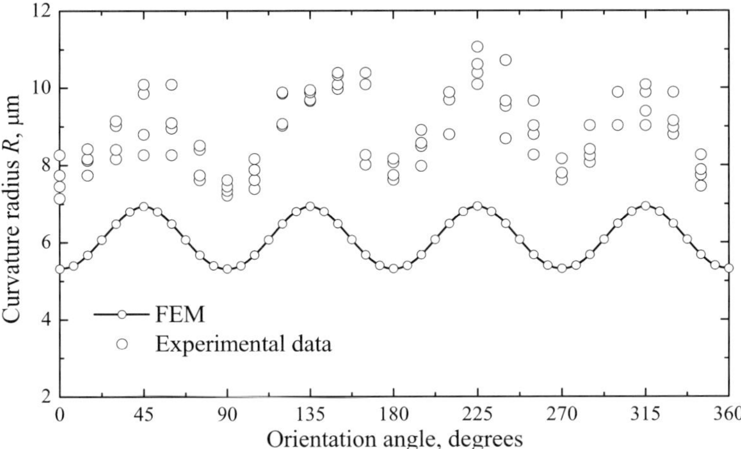


Figure 8: Dependence of the hinge curvature on the hinge orientation angle: comparison of finite element results with experimental data

three independent constants $C_{11}$, $C_{12}$ and $C_{44}$:

$$\mathbf{C} = \begin{bmatrix} C_{11} & C_{12} & C_{12} & & & \\ C_{12} & C_{11} & C_{12} & & & \\ C_{12} & C_{12} & C_{11} & & & \\ & & & C_{44} & & \\ & & & & C_{44} & \\ & & & & & C_{44} \end{bmatrix} \tag{38}$$

Elastic material constants (GPa) for two materials are as follows:

GaAs: $C_{11} = 119.00$, $C_{12} = 53.40$, $C_{44} = 59.60$

$In_{0.2}Ga_{0.8}As$: $C_{11} = 111.88$, $C_{12} = 51.80$, $C_{44} = 55.58$.

A three-dimensional mesh for the symmetrical half of the structure consists of 555 hexahedral 20-node elements and 3125 nodes. A series of problems for the self-positioning hinged nanostructure with different material orientations was considered. The hinge orientation is characterized by an angle between the material axis $\bar{x}_1$ (crystallographic axis <1 0 0>) at time $t = 0$ and the global axis $x_1$ (Figure 7). The problems with high nonlinearity are solved by dividing the total initial strain into increments.

The curvature radius $R$ of the hinge is estimated with the use of $x_1$- and $x_3$-coordinates of two points according to Equation (37). Experimental values of the hinge curvature radius are determined from the micro-plate elevation angle. Figure 8 presents the dependence of the hinge curvature radius $R$ on the hinge orientation angle. The values of $R$ determined by the finite element method are compared with experimental data. It is evident that the finite element results and experimental data have similar angular dependence. The ratio of maximum values to minimum values is about 1.3 for both numerical data and experimental data. However, absolute values obtained by computational modeling are noticeably lower than experimental values.

The difference between calculated and measured radius values may be explained by partial stress relaxation in the $In_{0.2}Ga_{0.8}As$ layer due to formation of threading dislocations. The critical thickness for strain relaxation by generation of dislocations in $In_{0.2}Ga_{0.8}As$ epitaxial layers deposited on GaAs according to the Matthews and Blakeslee [17] is 16 nm. The InGaAs thickness in experimental samples is much larger than the critical thickness. Experimental data with InGaAs thickness below the critical value would be desirable, but unfortunately was not available.

# 4 Atomic-scale modeling

## 4.1 Atomic-scale finite element method

The atomic-scale finite element method (AFEM) has been proposed for analysis of carbon nanotubes [18, 19]. The AFEM equation system is derived from the approximation of energy $E$ around current configuration $\mathbf{x}^{(i)}$:

$$\begin{aligned} E(\mathbf{x}) \approx\ & E(\mathbf{x}^{(i)}) + \frac{\partial E}{\partial \mathbf{x}}\Big|_{\mathbf{x}=\mathbf{x}^{(i)}} \cdot (\mathbf{x} - \mathbf{x}^{(i)}) \\ & + \frac{1}{2}(\mathbf{x} - \mathbf{x}^{(i)})^{\mathbf{T}} \cdot \frac{\partial^2 E}{\partial \mathbf{x}^2}\Big|_{\mathbf{x}=\mathbf{x}^{(i)}} \cdot (\mathbf{x} - \mathbf{x}^{(i)}) \end{aligned} \quad (39)$$

and its subsequent minimization:

$$\frac{\partial E}{\partial \mathbf{x}} = 0 \quad (40)$$

By substitution of equation (39) into (40), the AFEM global equation system can be expressed in a form similar to conventional finite element equation system:

$$\mathbf{KU} = \mathbf{F} \quad (41)$$

where $\mathbf{K}$ is a global stiffness matrix, $\mathbf{U}$ is a global displacement vector, and $\mathbf{F}$ is a global load (force) vector. The global stiffness matrix $\mathbf{K}$ and the load vector $\mathbf{F}$ are composed of second and first derivatives of the atomic structure energy with respect to atom positions:

$$\mathbf{K} = \begin{bmatrix} \left(\dfrac{\partial^2 E}{\partial \mathbf{x}_1 \partial \mathbf{x}_1}\right)_{3\times3} & \cdots & \left(\dfrac{\partial^2 E}{\partial \mathbf{x}_1 \partial \mathbf{x}_n}\right)_{3\times3} \\ \vdots & \ddots & \vdots \\ \left(\dfrac{\partial^2 E}{\partial \mathbf{x}_n \partial \mathbf{x}_1}\right)_{3\times3} & \cdots & \left(\dfrac{\partial^2 E}{\partial \mathbf{x}_n \partial \mathbf{x}_n}\right)_{3\times3} \end{bmatrix} \quad (42)$$

$$\mathbf{F} = \left\{ \begin{array}{c} \left(-\dfrac{\partial E}{\partial \mathbf{x}_1}\right)_{1\times 3} \\ \vdots \\ \left(-\dfrac{\partial E}{\partial \mathbf{x}_n}\right)_{1\times 3} \end{array} \right\} \tag{43}$$

Here $E$ is the total energy of the atomic system, $n$ is a number of atoms (AFEM nodes), and $\mathbf{x}_i$ is the coordinate vector of an $i$-th atom. Energy $E$ coincides with a potential energy for problems of finding the static equilibrium configuration of atomic structures without external loads.

Self-positioning is accompanied by large translational and rotational displacements, so it is necessary to divide loading into appropriate steps and to apply load gradually. At each load step, the Newton-Raphson iteration procedure is used to reach equilibrium. Gradual load application is performed with relaxation factor as shown below.

$$\begin{array}{l} \textbf{do} \\ \quad \mathbf{K} = \mathbf{K}(\mathbf{x}^{(i)}) \\ \quad \mathbf{F} = \mathbf{F}(\mathbf{x}^{(i)}) \\ \quad \alpha = g(\mathbf{K}, \mathbf{F}) \\ \quad \mathbf{F} = \alpha\, \mathbf{F}(\mathbf{x}^{(i)}) \\ \quad \Delta\mathbf{U} = \mathbf{K}^{-1}\mathbf{F} \\ \quad \mathbf{x}^{(i+1)} = \mathbf{x}^{(i)} + \Delta\mathbf{U} \\ \quad u^{(i+1)} = u^{(i)} + \|\Delta\mathbf{U}\| \\ \textbf{while} \quad \dfrac{\|\Delta\mathbf{U}\|}{u^{(i+1)}} > \varepsilon \end{array} \tag{44}$$

Here $\alpha$ is a load relaxation factor that is estimated by the function $g$; $\varepsilon$ is an error tolerance. Tangent stiffness matrix $\mathbf{K}$ and loading vector $\mathbf{F}$ are calculated at each iteration using equations (42) and (43).

Usually, derivative (slope) of the interaction potential energy becomes smaller as current configuration gets closer to equilibrium where loading becomes zero. The relaxation factor $\alpha$ is estimated using diagonal entries of the tangent stiffness matrix and components of the loading vector:

$$\begin{aligned} u_{\text{mean}} &= \frac{1}{n}\sum_{i=0}^{n} \left\| \left( \frac{F_1^i}{K_{11}^{ii}}, \frac{F_2^i}{K_{22}^{ii}}, \frac{F_3^i}{K_{33}^{ii}} \right) \right\| \\ \alpha &= \max\left(1, \frac{\delta a}{u_{\text{mean}}}\right) \end{aligned} \tag{45}$$

where $n$ is a number of atoms, $F_j^i$ is a $j$-th component of the full load vector acting on $i$-th atom, $K_{jj}^{ii}$ are corresponding diagonal entries of the tangent stiffness matrix, $a$ is a lattice period of the atomic system, and $\delta$ is a displacement suppression factor representing admissible mean displacement length. Stronger load relaxation is applied if smaller $\delta$ is selected, or calculated mean value of the solution guess $u_{\text{mean}}$ increases.

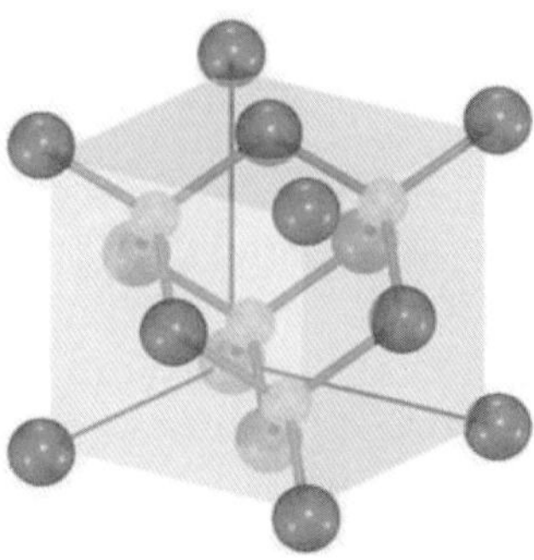

Figure 9: Atomic configuration and bonding in zincblende crystalline structures. There are eight corner and six face center atoms (bigger) and four inner atoms (smaller)

## 4.2 GaAs and InAs crystalline structures

Bi-layer self-positioning hinges consisting of GaAs upper and InAs lower layer (see Figure 1) are considered. Coordinate axes $x$, $y$, and $z$ are aligned to structure length, thickness, and width (bending axis) directions, respectively.

An AFEM mesh consisting of atoms is constructed in accordance with GaAs and InAs crystalline structures called zincblende crystal shown in Figure 9. Arsenide atoms occupy crystal corners and face centers of the unit crystal, and Gallium/Indium atoms are inside with positions (0.25, 0.25, 0.25), (0.75, 0.25, 0.75), (0.25, 0.75, 0.75), and (0.75, 0.75, 0.25) in the unit edge length crystal.

The GaAs and InAs possess material anisotropy depending on their crystal orientation. The effect of anisotropy is investigated depending on a material orientation angle, which characterizes crystal rotation around global $y$ axis. Figure 10 shows creation of AFEM meshes for modeling GaAs and InAs material anisotropy. First, an original crystalline structure with zero material orientation angle is prepared. Then the structure is rotated around $y$ axis and atoms outside the rectangular solution domain are removed.

The atomic-scale finite element method employs continuous empirical interatomic potential functions, which describe interatomic interactions. Several empirical interatomic potentials have been developed for modeling behavior of atomic systems [20–22]. The Brenner potential function is widely used and successfully applied for several types of atomic structures. However, its parameters for Indium, Gallium, and Arsenide systems are not available.

Another empirical potential energy model has been proposed by Tersoff [22]. In

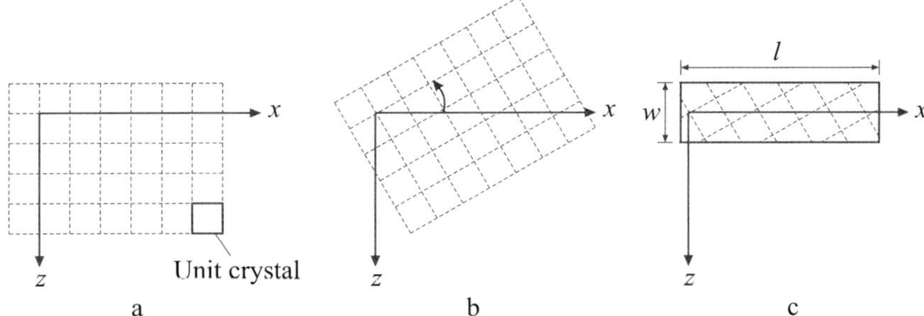


Figure 10: Modeling atomic structures with different material axes orientation. (a) Initial structure with unit crystals aligned to global axes. (b) Structure rotation around $y$ axis by specified orientation angle. (c) Removal of atoms outside a rectangular region

the Tersoff model, total potential energy $E$ is given by the following function:

$$E = \sum_i E_i = \frac{1}{2} \sum_i \sum_{j \neq i} V_{ij}$$
$$V_{ij} = f_C(r_{ij}) \left[ f_R(r_{ij}) + b_{ij} f_A(r_{ij}) \right] \tag{46}$$

where $E_i$ is the potential energy of atom $i$, $V_{ij}$ the potential energy of a bond $i$–$j$, $r_{ij}$ the distance from atom $i$ to atom $j$, $f_C$ the cut-off function to disregard effects from distant atoms, $f_R$ a reactive component, $f_A$ an attractive component, and $b_{ij}$ a bonding term to represent multi-atom interaction effects characterized by bonding angles. Reactive and attractive components are expressed through exponential functions:

$$f_R(r_{ij}) = A_{ij} \exp(-\lambda_{ij} r_{ij})$$
$$f_A(r_{ij}) = -B_{ij} \exp(-\mu_{ij} r_{ij}) \tag{47}$$

Parameterization of the potential function (46) for Indium, Gallium and Arsenide systems has been done by Nordlund *et al.* [23]. Numerical values of the parameters used in computations are given in publication [24].

The parameters obtained by Nordlund correspond to basic elastic and melting crystal properties. They were developed for investigation of damage at material interfaces. To confirm parameter suitability for modeling of self-positioning structures, we performed several computer experiments. The first test measured correspondence of elastic material properties obtained using the Nordlund parameters to macroelastic properties known from the literature. The second test involved calculation of lattice parameters for GaAs and InAs and their comparison to known values.

Using the AFEM with Tersoff potential and Nordlund parameters, elasticity modulus $E$ and Poisson's ratio $\nu$ of GaAs and InAs are estimated by applying tensile loads at the ends of a specimen shaped into a thin rod along its longitudinal direction. Strains and stresses are calculated at a position sufficiently far from the free end where external load is applied. Taking into account that the specimen is thin in transverse directions, Young's modulus $E$ and Poisson's ratio $\nu$ are determined by:

$$E = \frac{\sigma_x}{\varepsilon_x}, \qquad \nu = -\frac{\varepsilon_y}{\varepsilon_x} \tag{48}$$

For a cubic crystal with axes aligned with cube edges, estimation of Young's modulus and Poisson's ratio from constitutive tensor components $C_{11}$ and $C_{12}$ can be made as follows:

$$E = \frac{(C_{11} - C_{12})(C_{11} + 2C_{12})}{C_{11} + C_{12}}, \qquad \nu = \frac{C_{12}}{C_{11} + C_{12}} \tag{49}$$

The lattice period is estimated at the center of cube structures consisting of several crystals in all directions. Elastic properties and lattice periods provided by the AFEM modeling for GaAs and InAs are compared with experimental values.

A difference of 5% is observed for Young's modulus of GaAs, however the correspondence of other estimated elastic properties to their experimental values is within 1%. Estimated lattice periods are in very good agreement with known values. It was concluded that the Tersoff potential and parameters developed by Nordlund are suitable for AFEM simulation of nanostructures composed of GaAs and InAs.

## 4.3 Self-positioning GaAs–InAs hinges

A self-positioning structure consisting of GaAs upper and InAs lower layers is used for investigation of curvature radius dependence on structure thickness and crystal orientation angle (see Figure 11). The problem size parameter $c$ is used for characterizing the size of atomic systems.

GaAs and InAs bilayer structures with the problem size $c$ = 1, 2, 4, 8, 12, 16, 24, and 36 were modeled to find equilibrium configuration after self-positioning. Material orientation angles 0, 15, 30, 45, 60, 75, and 90 degrees were used for investigating the influence of material anisotropy on the self-positioning. The upper GaAs and lower InAs layers are composed of $3c$ and $c$ number of unit crystals in the thickness ($y$) direction. In the length ($x$) direction, the structure has length $16a_0c$ where $a_0$ is an initial lattice period (Figure 9).

We compare results obtained by the AFEM with the analytical continuum mechanics solution under plane strain conditions [11]. The plane strain conditions correspond to structures with infinite width in $z$ direction. In order to simulate atomic systems of infinite dimensions, periodic boundary conditions are usually employed in computational modeling. This helps to minimize the number of atoms in the models.

Periodic boundary conditions in the $z$ (width) direction are easy to apply to structures with orientation angles of 0, 45 and 90 degrees. Such structures consist of one complete and another incomplete crystal in the width direction, and connection across the periodic boundary is created when looking for neighboring atoms.

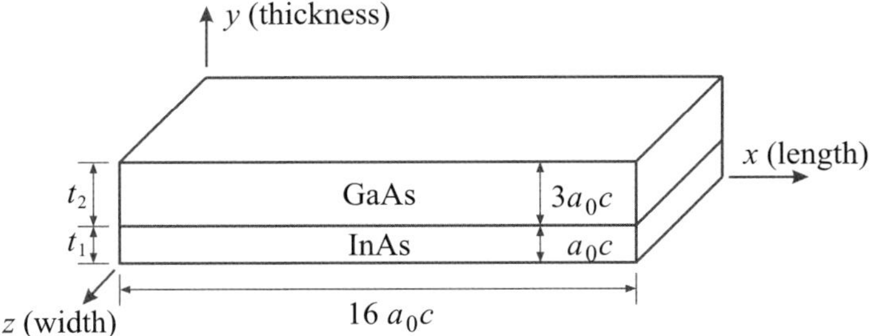


Figure 11: Schematic of a self-positioning problem. The number of atoms in the problem is determined by the parameter $c$

For hinges with the other orientation angles (15, 30, 60 and 75 degrees), a sufficient number of atomic layers corresponding to width $30a_0$ are prepared and displacement is constrained in the width direction to imitate plane strain conditions ($\varepsilon_z = 0$).

Boundary condition restricting displacements in $x$ direction at one end of the structure are also applied.

The analytical continuum mechanics solution [11] is used for estimating the initial configuration of atoms in AFEM meshes. Such configuration is created based on the curvature radius given by continuum mechanics solution under plane strain conditions. Using reasonable approximation for initial atom configuration may considerably reduce computing time.

Definition of thickness for structures consisting of just a few crystal layers in the thickness direction should be done with care when calculating curvature radius using an analytical technique. It is appropriate to add some offset equal to a 'radius' of an atom at each free surface. While adding such an offset is not significant for thick structures, it can be important for problems with a small number of unit crystals in the $y$ direction.

If we adopt half of the atom connectivity length as an offset, then for zincblende crystal structures the offset is equal to $\sqrt{3}a/8$, where $a$ is a lattice period. Corresponding offsets are $0.1224$ nm for GaAs and $0.1425$ nm for InAs.

Initial strains $\varepsilon_i^0$ in equation (16) are determined by initial ($a_0$) and material-specific lattice period ($a_i$) as:

$$\varepsilon_i^0 = \frac{a_i - a_0}{a_0} \tag{50}$$

We determine the initial lattice period for the bi-layer system using a weighted linear interpolation of GaAs and InAs specific lattice periods:

$$a_0 = \frac{a_1 n_1 + a_2 n_2}{n_1 + n_2} \tag{51}$$

where $n_1$ and $n_2$ are the number of unit crystals in InAs and GaAs layers. Lattice period $a_0$ is assumed to be $0.57546$ nm in our problems.

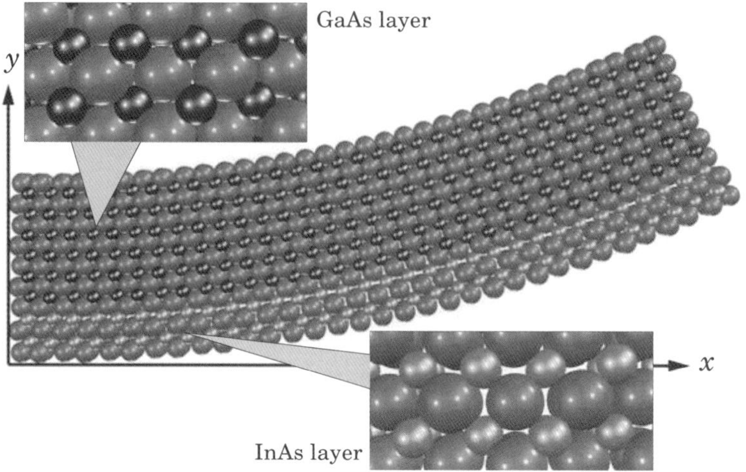


Figure 12: Final shape of atomic bi-layer structure with the size $c = 1$

The developed C++ AFEM computer program was used to solve two problem series for self-positioning nanostructures. The first one computes the curvature radius of bi-layer hinges with varying thickness. The second series includes problems with varying orientation of material axes for the same nanostructures.

### 4.3.1 Effect of the structure thickness

Bi-layer atomic structures of different thickness were modeled. In the problem series, size parameter $c$ is set to 1, 2, 4, 8, 12, 16, 24, and 36 that corresponds to thickness $2.56$, $4.86$, $9.46$, $18.65$, $27.84$, $37.03$, $55.41$ and $82.98$ nm. The largest AFEM model consists of 1,329,986 atoms that correspond to almost 4 million equations. The equilibrium configurations of bi-layer hinges are determined with the use of Newton-Raphson iteration procedure (44). The AFEM values of curvature radius are compared with the continuum mechanics solution for plane strain conditions [11]. In the continuum mechanics solution, elastic properties estimated by the AFEM on the tensile rod model are used.

Curvature radius values based on the AFEM modeling at the top and at the bottom of the atomic structures are calculated by using three neighbor nodes along the $x$ direction to fit a circle. These values are employed to calculate the curvature radius at the neutral layer by linear interpolation. The neutral layer is located at $0.54$ of the thickness from the bottom of the structure in our problems.

Figure 12 shows the final shape of an atomic model after self-positioning for problem size $c = 1$, which means totally four unit crystals in the thickness direction. Analysis reveals that spacing of atoms is smaller in GaAs and larger in InAs. It is also observed that the free end section is not flat due to expansion in the lower layer and

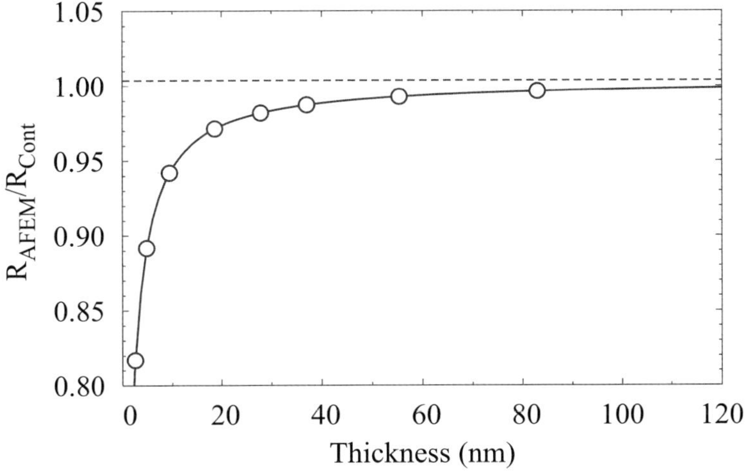


Figure 13: Ratio of the curvature radius determined by the AFEM and the continuum mechanics solution with varying thickness

compression in the top layer.

In order to estimate convergence of the curvature radius with thickness increase, the least square fit of the obtained eight numerical solutions is performed using power function $R(c) = \alpha_1 (\alpha_2 c + \alpha_3)^{-\beta} + \gamma$, where $c$ indicates the size of the atomic system, and $\alpha_1$, $\alpha_2$, $\alpha_3$, $\beta$, and $\gamma$ are parameters found by least square fit. Parameter $\gamma$ corresponds to converged value of the curvature radius for an infinite number of crystal layers.

Figure 13 depicts the ratio of the curvature radius determined by the AFEM and by the continuum mechanics solution with varying thickness. According to least square fit results, the computed curvature radius converges to $1.0037$ of the plane strain solution. For the biggest problem we investigated ($c = 36$), the difference between atomic-scale modeling and plane strain solution is $-0.36\%$. So, the AFEM and continuum mechanics solution are in agreement for large thickness. The difference between atomic-scale and continuum mechanics curvature radius increases with reduction of the structure thickness. This difference is $-18.4\%$ for $c = 1$, corresponding to four unit crystals in the thickness direction and thickness $2.56$ nm.

### 4.3.2 Effect of material anisotropy

To investigate the effect of material anisotropy on the self-positioning, a series of AFEM solutions for the specimen of Figure 11 with problem sizes $c = 1, 2, 4$ and $8$ and material orientation angles $0, 15, 30, 45, 60, 75$ and $90$ degrees is performed. Different orientation of crystallographic axes of GaAs and InAs with respect to hinge orientation is reached by creation of meshes with different atom arrangements as shown

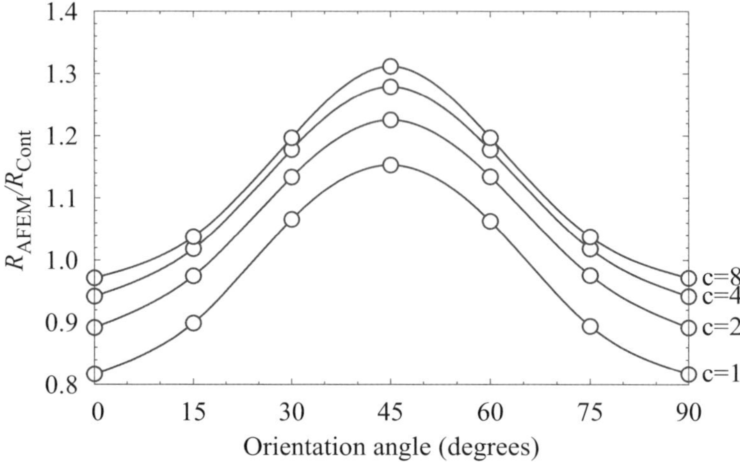


Figure 14: Dependency of $R_{\mathrm{AFEM}}/R_{\mathrm{Cont}(0°)}$ on orientation angle for problem sizes $c = 1$ to $8$ (4.86 nm to 27.8 nm)

in Figure 10.

Figure 14 presents dependency of the curvature radius ratio $R_{\mathrm{AFEM}}/R_{\mathrm{Cont}(0°)}$ on the material orientation angle. Here $R_{\mathrm{AFEM}}$ is the curvature radius determined by atomic-scale modeling for current material orientation and $R_{\mathrm{Cont}(0°)}$ is the continuum mechanics plane strain solution for zero material orientation angle. The curvature radius has its minimum at orientation angles $0$ and $90$ degrees and maximum at $45$ degrees. For varying orientation angles, the ratio of maximum and minimum values of the curvature radius is about $1.35$. This ratio is similar to experimental data and numerical finite element modeling of GaAs and $In_{0.2}Ga_{0.8}As$ bi-layer structures [12]. Dependency of the curvature radius on the material orientation angle is close to sinusoidal function with frequency $\pi$.

# 5 Conclusion

Analytical techniques, finite element analysis and atomic-scale modeling are used for the study of self-positioning structures with lattice mismatched material layers. Closed-form solutions are obtained for plane strain and generalized plane strain deformation of multi-layer structures with initial strains. An algorithm of the finite element method has been formulated for anisotropic problems with large displacements and rotations. Three-dimensional finite element analysis of bi-layer self-positioning structures with variable width confirmed applicability of closed-form solutions for estimation of curvature radius in plane stress and plane strain limiting cases. Anisotropic finite element modeling shows that curvature of self-positioning structures with different material orientation can differ by 35%.

Atomic-scale finite element method based on the Tersoff-Nordlund interatomic potential has been developed. It was applied to investigation of thickness effect on the curvature radius of self-positioning nanostructures. Atomic-scale modeling demonstrates that results obtained by continuum mechanics method coincide with those of atomic-scale approach for structures of large thickness. For nanostructures of small thickness (less than 40 nm) atomic-scale effects play considerable role.

# References


[1] S.V. Golod, V.Ya. Prinz, V.I. Mashnov, A.K. Gutakovsky, *"Fabrication of conducting GeSi/Si micro- and nanotubes and helical microcoils"*, Semicond. Sci. Technol., 16, 181–185, 2001.

[2] O.G. Schmidt, K. Eberl, *"Nanotechnology: Thin solid films roll up into nanotube"*, Nature, 410, 168, 2001.

[3] P.O. Vaccaro, K. Kubota, T. Aida, *"Strain-driven self-positioning of micromachined structures"*, Appl. Phys. Letters, 78, 2852–2854, 2001.

[4] H.J.In, S. Kumar, Y. Shao-Horn, G. Barbastathis,*"Origami fabrication of nanostructured, three-dimensional devices: Electrochemical capacitors with carbon electrodes"*, Appl. Phys. Lett., 88, 083104-1–3, 2006.

[5] W.J. Arora, A.J. Nichol, H.I. Smith, G. Barbastathis, *"Membrane folding to achieve three-dimensional nanostructures: Nanopatterned silicon nitride folded with stressed chromium hinges"*, Appl. Phys. Lett., 88, 53108-1–3, 2006.

[6] R. Songmuang, Ch. Deneke, O.G. Schmidt, *"Rolled-up micro- and nanotubes from single-material thin films"*, Appl. Phys. Letters, 89, 23109-1–3, 2006.

[7] C.-H. Hsueh, *"Modeling of elastic deformation of multilayers due to residual stresses and external bending"*, J. Appl. Phys., 91, 9652–9656, 2001.

[8] C.A. Klein, R.P. Miller, *"Strains and stresses in multilayered elastic structures: The case of chemically vapor-deposited ZnSZnSe laminates"*, J. Appl. Phys., 87, 2265–2272, 2000.

[9] P.H. Townsend, D.M. Barnett, T.A. Brunner, *"Elastic relationships in layered composite media with approximation for the case of thin films on a thick substrate "*, J. Appl. Phys., 62, 4438, 1987.

[10] Y. Nishidate, G.P.Nikishkov, *"Generalized plane strain deformation of multilayer structures with initial strains"*, J. Appl. Phys., 100, 113518-1–4, 2006.

[11] G.P. Nikishkov, *"Curvature estimation for multilayer hinged structures with initial strains"*, J. Appl. Phys., 94, 5333–5336, 2003.

[12] G.P. Nikishkov, Y. Nishidate, T. Ohnishi and P.O. Vaccaro, *"Effect of material anisotropy on the self-positioning of nanostructures"* Nanotechnology, 2006, 17, pp. 1128–1133.

[13] M.A. Criesfield, *"Non-Linear Finite Element Analysis of Solids and Structures. Volume 1: Essentials"*, Wiley, Chichester, 1996.

[14] T. Belitschko, W.K. Liu, B. Moran, *"Nonlinear Finite Elements for Continua and Structures"*, Wiley, Chichester, 2001.

[15] G.P. Nikishkov, I. Khmyrova, V. Ryzhii, *"Finite element analysis of self-positioning micro- and nanostructures"*, Nanotechnology, 14, 820–823, 2003.

[16] S. Okamoto and Y. Omura, *"Finite-element nonlinear dynamics of flexible structures in three dimensions"*, Computer Modeling in Engineering and Sciences, 4, 287–299, 2003.

[17] J.W. Matthews, A.E. Blakeslee, *"Defects in epitaxial multilayers: I. Misfit dislocations"*, J. Cryst. Growth, 27, 118, 1974.

[18] B. Liu, Y. Huang, H. Jiang, S. Qu, K.C. Hwang, *"The atomic-scale finite element method"*, Comput. Methods Appl. Mech. Engrg, 193, 1849–1864, 2004.

[19] B. Liu, H. Jiang, Y. Huang, S. Qu, M.-F. Yu, K.C. Hwang, *"Atomic-scale finite element method in multiscale computations with applications to carbon nanotubes"*, Phys. Rev. B, 72, 035435-1–8, 2005.

[20] D.W. Brenner, *"Empirical potential for hydrocarbons for use in simulating the chemical vapor deposition of diamond films"*, Phys. Rev. B, 42, 9458–9471, 1990.

[21] D.W. Brenner, O.A. Shenderova, J.A. Harrison, S.J. Stuart, B. Ni and S.B. Sinnott, *"A second-generation reactive empirical bond order (REBO) potential energy expression for hydrocarbons"*, J. Phys.: Condens. Matter, 14, 783–802, 2002.

[22] J. Tersoff, *"Modeling solid-state chemistry: Interatomic potentials for multicomponent systems"*, Phys. Rev. B, 39, 5566–5568, 2002.

[23] K. Nordlund, J. Nord, J. Frantz, J. Keinonen, *"Strain-induced Kirkendall mixing at semiconductor interfaces"*, Comput. Mater. Sci., 18, 283–294, 2000.

[24] Y. Nishidate, G.P. Nikishkov, *"Effect of thickness on the self-positioning of nanostructures"*, J. Appl. Phys., 102, 083501-1–5, 2007.

©Saxe-Coburg Publications, 2008.
Trends in Engineering Computational Technology
B.H.V. Topping and M. Papadrakakis, (Editors)
Saxe-Coburg Publications, Stirlingshire, Scotland, 295-319.

# Chapter 15

# Thermomechanical Deformation Processes of Rate Sensitive Solids: Material Issues and Stability

**I. Doltsinis**
**Faculty of Aerospace Engineering and Geodesy**
**University of Stuttgart, Germany**

## Abstract

Deformation of solids sensitive to the strain rate is considered under quasistatic conditions. The inelastic process implies the occurrence of thermal phenomena which interact with mechanical action such that they have to be considered at the same time, posing a coupled thermomechanical problem [1]. Material constitutive approaches are reviewed, which in addition to strain dependence are capable of accounting for rate effects and temperature changes; their impact on the response is detailed. In extension of the work in [2], the possibility of critical states to appear along the deformation path is examined under consideration of rate hardening and thermal softening of the material. In this connection stability of the evolving thermomechanical process is investigated from various points of view. A discussion of uniaxial tension under adiabatic conditions is followed by a study of complex configurations governed by finite element equations. Conditions that allow for fluctuations are stated and the significance for the numerical treatment is pointed out.

**Keywords:** inelastic solids, deformation processes, thermomechanical coupling, rate sensitivity, stability of evolution.

# 1 Introduction

Rate sensitivity becomes significant when metallic materials are processed at elevated temperature, and influences the deformation behaviour. Besides, rate effects may be activated under ambient conditions if deformation is fast. Accounting for rate sensitivity in the analysis of material deformation processes, implies considerations on the constitutive description of the material and the set up of a suitable algorithm for the numerical computer simulation. In addition, concurrent thermal phenomena that interact may be of importance. These are due to imposed boundary conditions and because of heat generation by the dissipation of mechanical work. The computer simulation of such thermomechanically coupled processes is a well developed subject [1]. An issue in the present account concerns the stability of the thermomechanical system along the deformation path. Thereby, conditions are explored which allow perturbations in the system to be augmented during the course of the process. For this purpose, the exposition in [2] is extended here to account for the thermal processes that interact with the deformation.

The pertinent equations for the thermomechanical deformation process are summarized in Section 2. A brief discussion on coupling effects refers to the continuum level, the equations governing a finite element representation of the system are listed. Section 3 reviews the constitutive modelling of the material from the mechanical point of view. The formalism for the viscous solid as a reference inelastic material is extended to incorporate a yield limit, thus modelling viscoplastic response. The dynamic flow stress appertaining to this model increases with material viscosity and deformation rate. The impact of an elastic constituent is demonstrated, but the material model is ultimately restricted to inelastic behaviour. Specification of the viscosity coefficient accounts for the actual flow characteristic of the material which may depend on the strain, strain rate and temperature. Section 4 deals with the stability issue. The considerations start with uniaxial tension, distinguishing between stability of loading and stability of deformation, the discussion bases on adiabatic processes. The obtained stability criterion reproduces those known for inviscid and viscous solids under isothermal conditions. Stability of more general systems regards the propagation of perturbations in the variables. Quantification for the finite element representation identifies the decisive system characteristics. The investigation is extended to the sensitivity of the solution of the rate problem and to the numerical stability of the temporal integration of the process.

The applications of thermomechanical analysis in Section 5 are selected such that rate effects become apparent. The analysis of thermal shock for elastic-viscoplastic material demonstrates the transition from elastic to plastic behaviour while a process of emergency cooling progresses. The computer simulation of a forging operation at different speeds focuses on the development of the temperature field in the rate sensitive material. The splashing of ceramic droplets on a cold surface is associated with the manufacturing of protective coatings by plasma spraying. The originally liquid material solidifies while deforming upon impact. The shape of the splat is influenced by the material viscosity.

# 2 Non-isothermal Deformation

## 2.1 Continuum Level

Isothermal deformation processes are characterized by a single field variable: the positions that material points occupy during the course of time. Non-isothermal deformation processes exhibit, in addition to the particle position $\mathbf{x}$, also the temperature

$T$ as a variable quantity. The following considerations refer to the model of a viscous material described by the constitutive law for the stress in the form

$$\boldsymbol{\sigma} = \boldsymbol{\mu}\left(\boldsymbol{\delta} - \boldsymbol{\beta}\dot{T}\right) \tag{1}$$

with the matrix $\boldsymbol{\mu}$ comprising the pertinent viscosity coefficients. The deformation rate $\boldsymbol{\delta}$ derives from the momentary velocity field $\mathbf{v}(\mathbf{x})$ by

$$\boldsymbol{\delta} = \boldsymbol{\partial}\mathbf{v} \tag{2}$$

where the matrix $\boldsymbol{\partial}$ is the conventional strain operator. The variation of temperature with time is denoted by $\dot{T}$. Part of the observed deformation rate is associated with the thermal expansion of the material. The vector array $\boldsymbol{\beta}$ comprises the differential coefficients of linear thermal expansion. With the stress from Equation (1), the condition of static equilibrium along with the static boundary conditions the mechanical equations governing the instantaneous velocity field in the deforming solid are

$$\boldsymbol{\partial}^{\mathrm{t}}\left[\boldsymbol{\mu}\left(\boldsymbol{\partial}\mathbf{v} - \boldsymbol{\beta}\dot{T}\right)\right] + \mathbf{f} = 0 \tag{3}$$

The temporal evolution of the thermomechanical process is governed by the equations

$$\begin{aligned} \frac{\mathrm{d}\mathbf{x}}{\mathrm{d}t} &= \mathbf{v} \\ \varrho c\frac{\mathrm{d}T}{\mathrm{d}t} &= \nabla\left(\lambda\nabla T\right) + \boldsymbol{\sigma}^{\mathrm{t}}\boldsymbol{\delta} \end{aligned} \tag{4}$$

The upper equation describes, in conjunction with the system for the velocity, the motion of material points, the lower equation describes the transient temperature accounting for unsteady heat conduction in the material. The time derivatives refer to positions fixed in the material and represent therefore *substantial derivatives*. The form of the heat conduction equation in Equation (4) allows for a variation of the heat conductivity $\lambda$ in the material. The last term on the right-hand side can be interpreted as an internal heat source from the mechanical power in the material.

The system of evolution equations in Equation (4) is coupled through the variables $\mathbf{x}$ and $T$. On the one hand, the thermal equation refers to the actual geometry that the material assumes during the course of the deformation. On the other hand, the material properties in the mechanical equations for the velocity are, as a rule, temperature dependent. In addition, the thermomechanical problem is coupled through the time rates $\mathbf{v} = \mathrm{d}\mathbf{x}/\mathrm{d}t$ and $\mathrm{d}T/\mathrm{d}t$ of the variables. We notice the presence of the mechanical work rate $\boldsymbol{\sigma}^{\mathrm{t}}\boldsymbol{\delta}$ in the thermal equation, while the temperature rate $\mathrm{d}T/\mathrm{d}t$ enters the mechanical part of the problem *via* the constitutive material law, Equation (1).

It is instructive to discuss Equation (4) for two limiting cases. When the deformation process is extremely slow ($\boldsymbol{\delta} \to \mathbf{0}$) the rate of mechanical work may become negligible. The instantaneous variation of the temperature with time can then be obtained independently from the mechanical solution for the velocity. A different situation is encountered when the deformation occurs very rapidly ($\mathrm{d}t \to 0$). In this case we discuss incremental changes of temperature as from

$$\varrho c \mathrm{d}T = \nabla\left(\lambda \nabla T\right)\mathrm{d}t + \boldsymbol{\sigma}^{\mathrm{t}}\boldsymbol{\delta}\mathrm{d}t \tag{5}$$

For a prescribed deformation $\boldsymbol{\delta}\mathrm{d}t$ in the increment, variation of the velocity of the process affects both the deformation rate $\boldsymbol{\delta}$ and the time increment $\mathrm{d}t$ in Equation (5), but not the product $\boldsymbol{\delta}\mathrm{d}t$. When $\mathrm{d}t \to 0$ the rate of work, the second term on the the right-hand side of the equation, remains finite while the first term, the spatial diffusion, vanishes. The process is adiabatic and the change in temperature is locally determined by the mechanical dissipation $\boldsymbol{\sigma}^{\mathrm{t}}\boldsymbol{\delta}\mathrm{d}t$ alone. Therefore only a mechanical field problem has to be solved in principle, the thermal one reducing to a local evaluation of Equation (5) for the temperature change.

## 2.2 Finite Element Representation

The finite element equation governing at instant $t$ the velocity field in quasistatic viscous deformation reads

$$\mathbf{S}(\mathbf{X}, \mathbf{V}, \mathbf{T}) = \mathbf{D}(\mathbf{X}, \mathbf{V}, \mathbf{T})\mathbf{V} = \mathbf{P}(t, \mathbf{X}, \mathbf{T}) \tag{6}$$

In the above, the vectors $\mathbf{X}, \mathbf{V}, \mathbf{S}$ and $\mathbf{P}$ comprise coordinates, velocities, stress resultants and applied forces at the mesh nodal points, respectively, the matrix $\mathbf{D}$ represents the viscosity matrix of the system. The dependence on the velocity $\mathbf{V}$ reflects nonlinear material behaviour. The pertinent equation for the instationary heat conduction in the finite element representation of the solid is

$$\mathbf{C}(\mathbf{X}, \mathbf{T})\dot{\mathbf{T}} + \mathbf{L}(\mathbf{X}, \mathbf{T})\mathbf{T} = \dot{\mathbf{Q}}(t, \mathbf{X}, \mathbf{V}, \mathbf{T}) \tag{7}$$

The vectors $\mathbf{T}$ and $\dot{\mathbf{Q}}$ comprise the nodal temperatures and thermal loading, respectively, $\mathbf{L}$ denotes the heat conductivity matrix, $\mathbf{C}$ the heat capacity matrix of the system. Apart from a mutual dependence on the actual geometry and temperature of the deforming solid, the individual Equations (6) and (7) are coupled through the time rates of the variables. In the vector $\mathbf{P}$ the mechanical loads are modified by the effect of thermal expansion, the thermal loads in the vector $\dot{\mathbf{Q}}$ account also for the effect of mechanical work. Solution algorithms for the coupled thermomechanical system are discussed in [1]. Beyond the rate problem, the interest lies on the evoluting deformation and temperature distribution that requires temporal integration

$$\mathbf{X} = {}^{o}\mathbf{X} + \int \mathbf{V}\mathrm{dt}, \qquad \mathbf{T} = {}^{o}\mathbf{T} + \int \dot{\mathbf{T}}\mathrm{dt}$$

which is effected numerically by incrementation. A simple approximation scheme advancing the variables from time $t = {}^{a}t$ to time $t = {}^{a}t + \tau = {}^{b}t$ is

$$\begin{aligned} {}^{b}\mathbf{X} &= {}^{a}\mathbf{X} + (1-\zeta)\,\tau\,{}^{a}\mathbf{V} + \zeta\tau\,{}^{b}\mathbf{V} \\ {}^{b}\mathbf{T} &= {}^{a}\mathbf{T} + (1-\zeta)\tau\,{}^{a}\dot{\mathbf{T}} + \zeta\tau\,{}^{b}\dot{\mathbf{T}} \end{aligned} \tag{8}$$

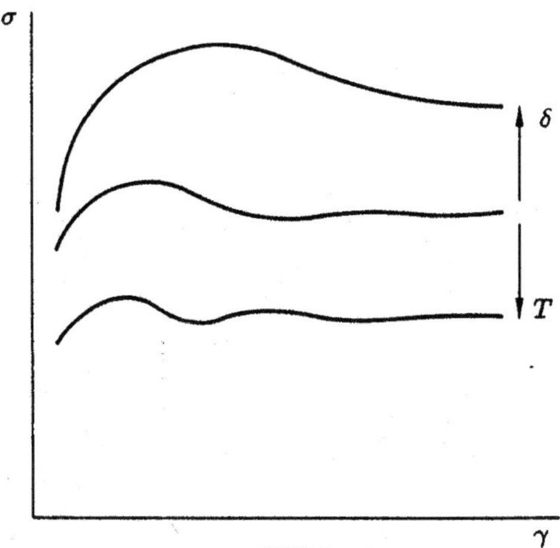


Figure 1: Dependence of flow stress $\sigma$ on strain $\gamma$ at different temperatures $T$ and deformation rates $\delta$ (steel material at elevated temperature, schematic).

The value of the collocation parameter $0 \leq \zeta \leq 1$ is chosen for best numerical properties, and Equations (6), (7) are treated in conjunction with Equations (8).

# 3 Rate Sensitive Material Response

## 3.1 Flow Stress

Under non-isothermal conditions the uniaxial flow stress $\sigma_\mathrm{f}$ of metallic materials recorded during inelastic deformation has to be considered a function of the axial strain $\gamma$ and temperature $T$. At elevated temperature as well as in fast deformation the viscous properties of the material become significant such that the flow stress depends on the deformation rate $\delta = \dot{\gamma}$. Thus

$$\sigma_\mathrm{f} = f(\gamma, \delta, T) \tag{9}$$

Figure 1 summarizes the dependence on deformation rate and temperature for steel materials. Figure 2 refers to a specific steel alloy. The curves in the uniaxial stress–strain diagram are for different values of the parameters $T$ = const. and $\delta$ = const. With increasing temperature the flow stress is lowered and its strain dependence becomes weaker. An increase of the strain rate implies higher stress. The waved shape of the flow curves reflects interactions between strain hardening and dynamic softening processes in the material which are thermally activated (recrystallization).

Stress–strain curves at constant temperature and deformation rate are obtained by

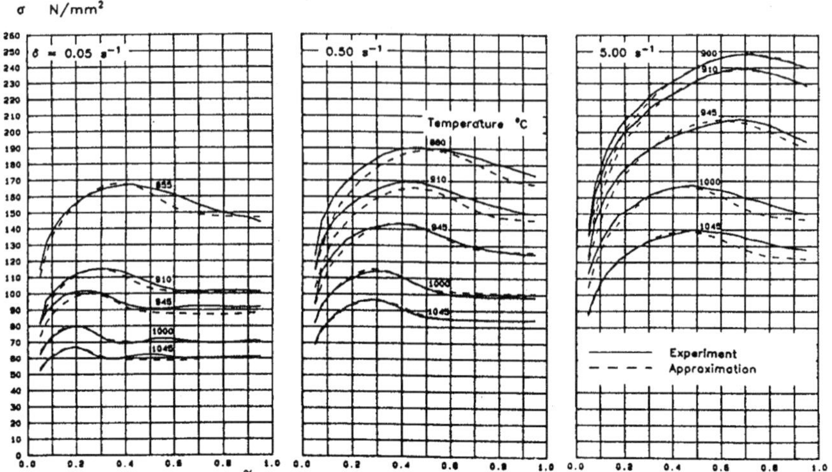


Figure 2: Flow stress of 50CrV4 steel material. Analytical approximation of laboratory data

laboratory measurements with specially prepared upsetting specimens that undergo homogeneous deformation. The results of such experimental tests form a basis for analytical approaches that describe the dependence of the flow stress on the local process parameters: strain, temperature and deformation rate. Analytical approximations intend to cover the description of the mechanical response completely over the domain under investigation. An example for the approximate description for the steel material in Figure 2 relying on experimental data and physical motivation is found in [1]. The development of a specific relationship for Equation (9) utilizes the apparent similarity of the flow curves. Frequently, an analytical approximation of the flow stress is attempted by the expression

$$\sigma_{\mathrm{f}} = k\mathrm{e}^{m_1 T}\delta^{m_2}\gamma^{m_3}\mathrm{e}^{m_4\gamma} = f(\gamma, \delta, T) \tag{10}$$

where the parameters $k, m_1, m_2, m_3, m_4$ are adjusted for the particular material.

The following discusses the extended viscoplastic model

$$\sigma_{\mathrm{f}} = \sigma_{\mathrm{f}0}(\gamma) + 3\mu(\gamma,\delta)\delta = \left(1 + \frac{3\mu}{\sigma_{\mathrm{f}0}}\delta\right)\sigma_{\mathrm{f}0} \tag{11}$$

The first part of the equation combines in parallel a plastic constituent with yield stress $\sigma_{\mathrm{f}0}(\gamma)$ given by the hardening characteristic of the material for vanishing viscosity, and a nonlinear viscous constituent with viscosity coefficient $\mu(\gamma,\delta)$. The viscous constituent is commonly presented in the power law form

$$\sigma = k\gamma^n\delta^m \tag{12}$$

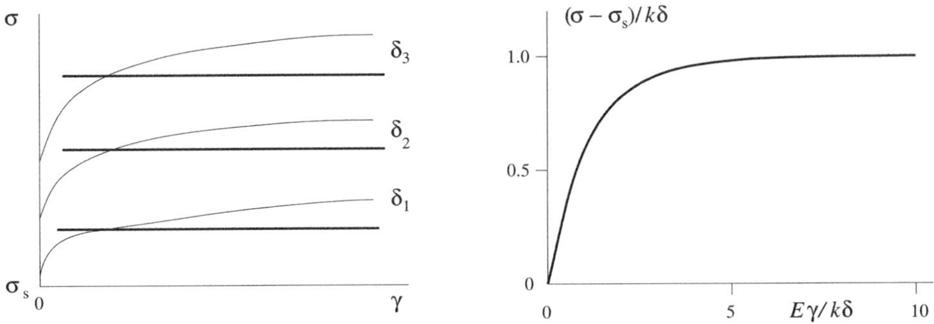


Figure 3: Behaviour of viscoplastic material model

which specifies the viscosity cofficient as a function of $\delta$ and $\gamma$. The second part of Equation (11) may be interpreted as the definition of a dynamic flow stress which depends on the viscosity coefficient and the deformation rate. For infinite viscosity ($\mu \rightarrow \infty$) or for sudden deformation ($\delta \rightarrow \infty$) the yield stress $\sigma_{\mathrm{f}}$ is not restricted. At the other end, in the absence of viscosity ($\mu \rightarrow 0$) or for very slow deformation ($\delta \rightarrow 0$), the flow stress becomes the inviscid yield stress $\sigma_{\mathrm{f0}}$.

If the viscous constituent in Equation (11) does not depend on the strain, deformation at different values of the strain rate effects a parallel shift of the flow curve (Figure 3). Increasing the magnitude of the stress does not induce a strain rate before reaching the static yield limit $\sigma_{\mathrm{f0}}(\gamma)$ apertaining to the actual strain. In fact, the deformation rate that follows as

$$\bar{\delta} = \frac{\bar{\sigma} - \sigma_{\mathrm{f}}}{3\mu} \geq 0 \tag{13}$$

is associated with an overstress $\bar{\sigma} - \sigma_{\mathrm{f}}$. Conversely, the stress difference $\bar{\sigma} - \sigma_{\mathrm{f}} = 3\mu\delta$ resp. the stress $\sigma$ is established instantaneously upon the imposition of a deformation rate $\delta$.

An elastic constituent as part of the deformation would delay the transition to the stationary stress. For a discussion of this effect assume the inelastic part of the deformation rate $\delta$ described by the Bingham material model

$$\sigma = \sigma_{\mathrm{s}} + k\delta \tag{14}$$

as a special case of Equation (11), with $\sigma_{\mathrm{s}}$ the conventional yield stress in perfectly plasticity and $k = 3\mu$ for the constant viscosity coefficient. The elastic part is related to the stress rate $\dot{\sigma}$ by means of the elastic modulus $E$. Thus, an extension rate $\delta$ imposed on the specimen consists of the sum

$$\delta = \frac{\dot{\sigma}}{E} + \frac{\sigma - \sigma_{\mathrm{s}}}{k} = \mathrm{const} \tag{15}$$

The solution of the differential equation (15) for the stress is

$$\sigma - \sigma_{\mathrm{s}} = \left[1 - \exp\left(-\frac{E}{k}\frac{\gamma}{\delta}\right)\right] k\delta \tag{16}$$

and describes the initial transient increase of the stress; the time is measured by the quotient $t = \gamma/\delta$. When the exponential in Equation (16) tends to zero the equation approaches Equation (14), which becomes the stationary limit of the time dependent solution. This transition is demonstrated in Figure 3, right, in a dimensionless manner. The stationary value of the stress is reached sooner the larger the quotient $E/k$. For an elastically rigid material ($E = \infty$) the stationary stress is achieved instantaneously upon imposition of the deformation rate as in Equation (14).

## 3.2 Multiaxial Approach

Under certain conditions, such as for metals at elevated temperature, the mechanical behaviour of solid materials can be considered purely inelastic and can be described as for a viscous fluid. Thereby, the Cauchy stress

$$\boldsymbol{\sigma} = \left\{ \sigma_{11}\ \sigma_{22}\ \sigma_{33}\ \sqrt{2}\sigma_{12}\ \sqrt{2}\sigma_{23}\ \sqrt{2}\sigma_{13} \right\} \tag{17}$$

is stated as a function of the deformation rate

$$\boldsymbol{\delta} = \left\{ \delta_{11}\ \delta_{22}\ \delta_{33}\ \sqrt{2}\delta_{12}\ \sqrt{2}\delta_{23}\ \sqrt{2}\delta_{13} \right\} \tag{18}$$

The constitutive law for a viscous isotropic material is given as a relationship between deviatoric stress and deformation rate, and a relationship between hydrostatic stress and volumetric deformation rate. Thus

$$\boldsymbol{\sigma} = 2\mu\boldsymbol{\delta}_{\mathrm{D}} + 3\kappa\boldsymbol{\delta}_{\mathrm{V}} = \boldsymbol{\mu}\boldsymbol{\delta} \tag{19}$$

The parameters $\mu$ and $\kappa$ are known as the viscosity coefficients which specify the deviatoric and the volumetric response of the material; the volumetric deformation rate is defined analogously to the hydrostatic stress. The viscosity matrix $\boldsymbol{\mu}$ of the material is

$$\boldsymbol{\mu} = 2\mu \left( \mathbf{I} + \frac{3\kappa - 2\mu}{6\mu}\mathbf{e}\mathbf{e}^{\mathrm{t}} \right) \qquad \text{and} \qquad \mathbf{e} = \{1\ 1\ 1\ 0\ 0\ 0\} \tag{20}$$

Usually, inelastic flow is an isochoric deformation process that preserves the volume. This is a material property; independent of the state of applied stress the material is incompressible. Hydrostatic stress and volumetric deformation rate can then not be related as in Equation (19). Instead, the isochoric condition

$$\mathbf{e}^{\mathrm{t}}\boldsymbol{\delta} = 3\delta_{\mathrm{V}} = 0 \tag{21}$$

must be imposed at each point of the velocity field. In order to utilize conventional, compressible solution techniques, however, a relaxed form of the strict isochoric condition is stated as

$$3\delta_{\mathrm{V}} = \frac{1}{\bar{\kappa}}\sigma_{\mathrm{H}} \rightarrow 0 \qquad (\bar{\kappa} \rightarrow \infty) \tag{22}$$

and approaches the isochoric condition as $\bar{\kappa}$ tends to infinity. In this connection, the parameter $\bar{\kappa}$ does not possess any physical meaning. In computer simulations it is selected under numerical aspects observing the available arithmetic precision.

The deviatoric response of the viscous material requires specification of the viscosity coefficient $\mu$ as the relevant material parameter. To this end, a relationship between the scalar equivalent deviatoric stress $\bar{\sigma}$ and rate of deformation $\bar{\delta}$ defined by

$$\bar{\sigma}^2 = \frac{3}{2}\boldsymbol{\sigma}_{\mathrm{D}}^{\mathrm{t}}\boldsymbol{\sigma}_{\mathrm{D}}, \qquad \bar{\delta}^2 = \frac{2}{3}\boldsymbol{\delta}_{\mathrm{D}}^{\mathrm{t}}\boldsymbol{\delta}_{\mathrm{D}} \tag{23}$$

is deduced as,

$$\bar{\sigma} = 3\mu\bar{\delta} = f\left(\bar{\delta}\right) \tag{24}$$

The relation $\bar{\sigma} = f\left(\bar{\delta}\right)$, obtained from laboratory tests, specifies the viscosity coefficient $\mu = f\left(\bar{\delta}\right)/3\bar{\delta}$ which accounts for any linear or nonlinear form of viscous behaviour described by Equation (24).

In general terms, isochoric inelastic deformation is considered co-axial with the deviatoric stress and *vice versa*. From this flow rule

$$\boldsymbol{\delta}_{\mathrm{D}} = \frac{3}{2}\frac{\bar{\delta}}{\bar{\sigma}}\boldsymbol{\sigma}_{\mathrm{D}}, \qquad \boldsymbol{\sigma}_{\mathrm{D}} = \frac{2}{3}\frac{\bar{\sigma}}{\bar{\delta}}\boldsymbol{\delta}_{\mathrm{D}} \tag{25}$$

The equations resemble a viscous material law with deviatoric viscosity coefficient

$$\mu \Leftarrow \frac{\bar{\sigma}}{3\bar{\delta}} = \frac{f(\bar{\gamma}, \bar{\delta}, T)}{3\bar{\delta}} \tag{26}$$

The transition to the second expression observes that $\bar{\sigma} = \sigma_{\mathrm{f}}$ during inelastic flow, the functional dependence of the flow stress $\sigma_{\mathrm{f}} = f(\bar{\gamma}, \bar{\delta}, T)$ on the local process variables characterising the actual material response. Implementation of plasticity, viscoplasticity or another constitutive approach via the flow characteristic is simple. Evaluation requires the deformation rate $\boldsymbol{\delta}$ as input. In the region of the solid where deformation does not progress, the stress cannot be determined by the model and causes numerical difficulties. For this reason, the magnitude of the deformation rate is not allowed to diminish below a lower limit selected with regard to the computer arithmetic.

Non-isothermal deformation introduces a volumetric rate as a consequence of the thermal expansion

$$\delta_{\mathrm{V}} = \beta\dot{T} \tag{27}$$

The hydrostatic stress in a compressible viscous material then becomes

$$\sigma_{\mathrm{H}} = 3\kappa\left(\delta_{\mathrm{V}} - \beta\dot{T}\right) \tag{28}$$

Use of $\kappa \leftarrow \bar{\kappa}$ $(\bar{\kappa} \rightarrow \infty)$ as the penalty factor in a relaxed incompressibility condition, requires caution because the thermal volumetric expansion is augmented beyond physical relevance. The non-isothermal material law in the form of the viscous Equation (1) can be detailed for isotropy as

$$\boldsymbol{\sigma} = \boldsymbol{\mu}\left(\boldsymbol{\delta} - \boldsymbol{\beta}\dot{T}\right) = 2\mu\boldsymbol{\delta}_{\mathrm{D}} + 3\kappa\left(\boldsymbol{\delta}_{\mathrm{V}} - \beta\dot{T}\mathbf{e}\right) \tag{29}$$

where the deviatoric viscosity coefficient may accommodate various inelastic models.

Accounting for elasticity requires a different constitutive description. In an hypoelastic approach, the Jaumann rate of Cauchy stress, $\breve{\boldsymbol{\sigma}}$, is related to the deformation rate $\boldsymbol{\delta}$ after subtraction of the inelastic part $\boldsymbol{\mu}^{-1}\boldsymbol{\sigma}$ and of the thermal contribution $\beta\dot{T}$. With $\boldsymbol{\kappa}$ as the elasticity matrix of the material,

$$\breve{\boldsymbol{\sigma}} = \boldsymbol{\kappa}\left(\boldsymbol{\delta} - \boldsymbol{\mu}^{-1}\boldsymbol{\sigma} - \beta\dot{T}\right) \tag{30}$$

For inelastic deformation in conjunction with hyperelasticity and thermomechanical coupling refer to the work of the author in [3], [4]. Consideration of elasticity is based on algorithmic concepts different from the flow approach [5].

# 4 Critical Conditions

## 4.1 Stability in Uniaxial Tension

The tensile force $P = \sigma A$ required for the extension of the specimen with length $l$ and cross-section $A$, attains a maximum when

$$\frac{l}{P}\frac{\mathrm{d}P}{\mathrm{d}l} = \frac{1}{\sigma}\frac{\mathrm{d}\sigma}{\mathrm{d}\gamma} + \frac{l}{A}\frac{\mathrm{d}A}{\mathrm{d}\gamma} = 0 \tag{31}$$

If the stress is sensitive to strain $\gamma$, strain rate $\delta = \dot{l}/l = \dot{\gamma}$ and temperature $T$,

$$\mathrm{d}\sigma = \sigma_\gamma \mathrm{d}\gamma + \sigma_\delta \mathrm{d}\delta + \sigma_\mathrm{T}\mathrm{d}T \tag{32}$$

Following [6] the sensitivity coefficients $\sigma_\gamma,\ \sigma_\delta$ and $\sigma_\mathrm{T}$ do not necessarily derive from a functional dependence $\sigma(\gamma, \delta, T)$.

From strain kinematics,

$$\frac{\mathrm{d}\delta}{\mathrm{d}\gamma} = \frac{\mathrm{d}\dot{l}}{\mathrm{d}l} - \delta, \qquad \frac{l}{A}\frac{\mathrm{d}A}{\mathrm{d}\gamma} = 3\beta\frac{\mathrm{d}T}{\mathrm{d}\gamma} - 1 \tag{33}$$

where thermal expansion is accounted for. Assuming adiabatic heating,

$$\frac{\mathrm{d}T}{\mathrm{d}\gamma} = \frac{\sigma}{\varrho c} \tag{34}$$

With the above relationships the condition of maximum force, Equation (31), gives for the critical stress

$$\sigma_\gamma + \sigma_\delta\left(\frac{\mathrm{d}\dot{l}}{\mathrm{d}l} - \delta\right) - \left(1 - \frac{\sigma_\mathrm{T}}{\varrho c}\right)\sigma + \frac{3\beta}{\varrho c}\sigma^2 = 0 \tag{35}$$

For constant extension velocity $\dot{l} = \mathrm{const.}$ this becomes

$$\sigma_\gamma - \sigma_\delta\delta - \left(1 - \frac{\sigma_\mathrm{T}}{\varrho c}\right)\sigma + \frac{3\beta}{\varrho c}\sigma^2 = 0 \tag{36}$$

while for constant strain rate $\delta = \dot{l}/l = \text{const.}$,

$$\sigma_\gamma - \left(1 - \frac{\sigma_\mathrm{T}}{\varrho c}\right)\sigma + \frac{3\beta}{\varrho c}\sigma^2 = 0 \tag{37}$$

Simple expressions for the critical stress are obtained if the thermal expansion is discarded.

To examine stability of the deformation process, a perturbation $\mathrm{d}A$ is introduced locally in the specimen and its evolution is investigated. In order that the perturbation does not increase, it is required that

$$\frac{\mathrm{d}\dot{A}}{\mathrm{d}A} \leq 0 \tag{38}$$

Cross-section perturbations must be counterbalanced by perturbations of stress such that the constancy of the force is maintained throughout the specimen. Instead of Equation (31),

$$\mathrm{d}P = A\mathrm{d}\sigma + \sigma\mathrm{d}A = 0 \tag{39}$$

and utilizing Equations (33), (34) in Equation (32) neglecting rate sensitivity, the adiabatic stress variation becomes

$$\mathrm{d}\sigma = \sigma_\gamma \mathrm{d}\gamma + \sigma_\mathrm{T}\mathrm{d}T = \left(\frac{3\beta}{\varrho c}\sigma - 1\right)^{-1}\left(\sigma_\gamma + \sigma_\mathrm{T}\frac{\sigma}{\varrho c}\right)\frac{\mathrm{d}A}{A} \tag{40}$$

Equation (39) requires

$$\left[\sigma_\gamma - \left(1 - \frac{\sigma_\mathrm{T}}{\varrho c}\right)\sigma + \frac{3\beta}{\varrho c}\sigma^2\right]\mathrm{d}A = 0 \tag{41}$$

and hence local perturbations $\mathrm{d}A \neq 0$ are tolerated when the condition of maximum force is met as stated by Equation (37), which is not affected by rate effects.

For the rate sensitive case, the deformation rate accounting for adiabatic thermal expansion is

$$\delta = \dot{\gamma} = \left(\frac{3\beta}{\varrho c}\sigma - 1\right)^{-1}\frac{\dot{A}}{A} \tag{42}$$

with variation

$$\mathrm{d}\delta = \left(\frac{3\beta}{\varrho c}\sigma - 1\right)^{-1}\left(\frac{\mathrm{d}\dot{A}}{\mathrm{d}A} - \frac{\dot{A}}{A}\right)\frac{\mathrm{d}A}{A} - \left(\frac{3\beta}{\varrho c}\sigma - 1\right)^{-2}\frac{\dot{A}}{A}\frac{3\beta}{\varrho c}\mathrm{d}\sigma \tag{43}$$

A simplification, rather conservative with regard to the progress of cross-section disturbances, neglects thermal expansion in which case $\delta = -\dot{A}/A$. Then, from Equation (39)

$$\frac{\mathrm{d}\dot{A}}{\mathrm{d}A} = \left[\left(1 - \frac{\sigma_\mathrm{T}}{\varrho c}\right)\frac{\sigma}{\sigma_\delta \delta} - \frac{\sigma_\gamma}{\sigma_\delta \delta} - 1\right]\delta \tag{44}$$

In order that the perturbation does not progress, Equation (38), the expression within the brackets must be not positive. Thus for stability the condition

$$\frac{\sigma_\gamma}{\sigma} + \frac{\sigma_\delta \delta}{\sigma} + \frac{\sigma_{\mathrm{T}}}{\varrho c} \geq 1 \tag{45}$$

must be fulfilled. For a specific form of the above stability condition, refer to Equation (10) as

$$\sigma = k\mathrm{e}^{m_1 T} \delta^{m_2} \gamma^{m_3} \tag{46}$$

from which $\sigma_{\mathrm{T}} = m_1 \sigma$, $\sigma_\delta \delta = m_2 \sigma$ and $\sigma_\gamma = m_3 \sigma / \gamma$. With this, stability of adiabatic deformation as by Equation (45) requires that

$$\frac{\sigma}{\varrho c} m_1 + m_2 + \frac{1}{\gamma} m_3 \geq 1 \tag{47}$$

Rate dependence may compensate thermal instability and the strain softening at higher strain (Figure 2).

## 4.2 The Thermomechanical Process of Deformation

In the following, stability of non-isothermal viscous deformation will be considered from a wider point of view considering thermal and mechanical coupling in complex systems.

### 4.2.1 Stability of Evolution

An examination of stability for viscous deformation under non-isothermal conditions is based on the evolution of perturbations in the system. In the representation by finite elements the position $\mathbf{X}$ and temperature $\mathbf{T}$ of the material nodal points are considered the process variables collected here in the vector $\mathbf{Y} = \{\mathbf{X}\ \mathbf{T}\}$. Their evolution is described by equations

$$\mathbf{Y} = \begin{bmatrix} \mathbf{X} \\ \mathbf{T} \end{bmatrix} = \begin{bmatrix} \boldsymbol{\phi}(t, \mathbf{X}_0, \mathbf{T}_0) \\ \boldsymbol{\psi}(t, \mathbf{T}_0, \mathbf{X}_0) \end{bmatrix} \tag{48}$$

The quantities $\mathbf{X}_0$, $\mathbf{T}_0$ define the geometrical configuration and the temperature in the solid at time $t = t_0$ when a disturbance $\boldsymbol{\alpha}_0 = \delta \mathbf{X}_0$, $\boldsymbol{\theta}_0 = \delta \mathbf{T}_0$ is introduced. Assuming the disturbed evolution be equally governed by Equation (48), the perturbation at $t > t_0$ becomes

$$\boldsymbol{\beta}(t) = \begin{bmatrix} \boldsymbol{\alpha}(t) \\ \boldsymbol{\theta}(t) \end{bmatrix} = \begin{bmatrix} \boldsymbol{\phi}(t, \mathbf{X}_0 + \boldsymbol{\alpha}_0, \mathbf{T}_0 + \boldsymbol{\theta}_0) - \boldsymbol{\phi}(t, \mathbf{X}_0, \mathbf{T}_0) \\ \boldsymbol{\psi}(t, \mathbf{T}_0 + \boldsymbol{\theta}_0, \mathbf{X}_0 + \boldsymbol{\alpha}_0) - \boldsymbol{\psi}(t, \mathbf{T}_0, \mathbf{X}_0) \end{bmatrix} \tag{49}$$

The Euclidean norm of the compound perturbation vector $\boldsymbol{\beta}(t) = \{\boldsymbol{\alpha}(t)\ \boldsymbol{\theta}(t)\}$ resp. its magnitude

$$\|\boldsymbol{\beta}\| = \sqrt{\boldsymbol{\beta}^{\mathrm{t}} \boldsymbol{\beta}} = \sqrt{\boldsymbol{\alpha}^{\mathrm{t}} \boldsymbol{\alpha} + \boldsymbol{\theta}^{\mathrm{t}} \boldsymbol{\theta}} \tag{50}$$

serves as a measure of the deviation from the undisturbed process. The evolution process by Equation (48) is termed stable if small perturbations remain small for all $t \geq t_0$. Asymptotic or complete stability is present if the disturbed motion converges to the original one as $t \to \infty$.

Sufficient for the above stability requirement is that the value of the positive definite expression $\boldsymbol{\beta}^{\mathrm{t}}\boldsymbol{\beta}$ does not increase. Therefore, if the time derivative satisfies for all $t \geq t_0$ the condition

$$\frac{\mathrm{d}}{\mathrm{d}t}(\boldsymbol{\beta}^{\mathrm{t}}\boldsymbol{\beta}) = 2\boldsymbol{\beta}^{\mathrm{t}}\dot{\boldsymbol{\beta}} \leq 0 \tag{51}$$

the evolution by Equation (48) is stable ($\leq 0$) or asymptotically stable ($< 0$).

### 4.2.2 Momentary State

It is convenient to consider the pertinent equations for the evolution of the thermomechanical process, Equations (6) and (7), in the residual form

$$\mathbf{R}(t, \dot{\mathbf{Y}}, \mathbf{Y}) = \begin{bmatrix} \mathbf{R}_{\mathrm{X}}(t, \mathbf{V}, \mathbf{X}; \dot{\mathbf{T}}, \mathbf{T}) \\ \mathbf{R}_{\mathrm{T}}(t, \dot{\mathbf{T}}, \mathbf{T}; \mathbf{V}, \mathbf{X}) \end{bmatrix} = \mathbf{0}, \tag{52}$$

where $\mathbf{R}_{\mathrm{X}} = \mathbf{P} - \mathbf{S} = \mathbf{P} - \mathbf{D}\mathbf{V}$ and $\mathbf{R}_{\mathrm{T}} = \dot{\mathbf{Q}} - \mathbf{C}\dot{\mathbf{T}} - \mathbf{L}\mathbf{T}$. A perturbation $\boldsymbol{\beta} = \delta\mathbf{Y}$ of geometry and temperature of the solid introduced at a certain instant during the course of the process, implies a perturbation $\dot{\boldsymbol{\beta}} = \delta\dot{\mathbf{Y}}$ in the rates. Both the actual state $\mathbf{Y}, \dot{\mathbf{Y}}$ and the perturbated one $\mathbf{Y} + \boldsymbol{\beta}, \dot{\mathbf{Y}} + \dot{\boldsymbol{\beta}}$, comply with Equation (52) so that for the excess quantities

$$\frac{\partial \mathbf{R}}{\partial \dot{\mathbf{Y}}}\dot{\boldsymbol{\beta}} + \frac{\partial \mathbf{R}}{\partial \mathbf{Y}}\boldsymbol{\beta} = \mathbf{0} \tag{53}$$

This relates the perturbations in the state variables and their rates by

$$\dot{\boldsymbol{\beta}} + \mathbf{N}\boldsymbol{\beta} = \mathbf{0} \tag{54}$$

where

$$\mathbf{N} = \left(\frac{\partial \mathbf{R}}{\partial \dot{\mathbf{Y}}}\right)^{-1} \frac{\partial \mathbf{R}}{\partial \mathbf{Y}} = -\frac{\mathrm{d}\dot{\mathbf{Y}}}{\mathrm{d}\mathbf{Y}} \tag{55}$$

With Equation (54) the stability condition of Equation (51) becomes

$$-\boldsymbol{\beta}^{\mathrm{t}}\mathbf{N}\boldsymbol{\beta} \leq 0 \tag{56}$$

Asymptotic stability ($< 0$) demands the symmetric part of the matrix $\mathbf{N}$ to be positive definite, with positive, non-zero eigenvalues. Given the variability of $\mathbf{N}$, conclusions on stability from the momentary conditions require caution.

For an alternative interpretation of stability the perturbation after a small interval in time $\tau$ is obtained with Equation (54) by linearization

$$\boldsymbol{\beta}(t + \tau) \cong \boldsymbol{\beta}(t) + \tau\dot{\boldsymbol{\beta}} = (\mathbf{I} - \tau\mathbf{N})\boldsymbol{\beta}(t) \tag{57}$$

The deformation process is stable at the considered instant if the perturbation decays. Taking the Euclidean norm $\|\boldsymbol{\beta}\|$ the requirement becomes

$$\|\boldsymbol{\beta}(t+\tau)\| = \|(\mathbf{I}-\tau\mathbf{N})\boldsymbol{\beta}(t)\| \leq \|(\mathbf{I}-\tau\mathbf{N})\|\|\boldsymbol{\beta}(t)\| < \|\boldsymbol{\beta}(t)\| \tag{58}$$

Hence, from the last inequality, for stability

$$\|(\mathbf{I}-\tau\mathbf{N})\| = \sigma(\mathbf{I}-\tau\mathbf{N}) < 1 \tag{59}$$

Here, $\sigma$ is the spectral norm

$$\sigma(\mathbf{A}) = \sqrt{\max \lambda_i(\mathbf{A}^{\mathrm{t}}\mathbf{A})} \tag{60}$$

of the amplification matrix $\mathbf{A} = (\mathbf{I}-\tau\mathbf{N})$ for the perturbation, and $\lambda_i$ denotes the eigenvalues of the symmetric form $\mathbf{A}^{\mathrm{t}}\mathbf{A}$. The amplification matrix and its properties depend on the extent of the time interval $\tau$ in the neighbourhood of the state in question.

In Equation (55) for $\mathbf{N}$, the matrices

$$\frac{\partial \mathbf{R}}{\partial \dot{\mathbf{Y}}} = \begin{bmatrix} \dfrac{\partial \mathbf{R}_{\mathrm{X}}}{\partial \mathbf{V}} & \dfrac{\partial \mathbf{R}_{\mathrm{X}}}{\partial \dot{\mathbf{T}}} \\ \dfrac{\partial \mathbf{R}_{\mathrm{T}}}{\partial \mathbf{V}} & \dfrac{\partial \mathbf{R}_{\mathrm{T}}}{\partial \dot{\mathbf{T}}} \end{bmatrix} \simeq \begin{bmatrix} -\mathbf{D} & \dfrac{\partial \mathbf{P}}{\partial \dot{\mathbf{T}}} \\ \dfrac{\partial \dot{\mathbf{Q}}}{\partial \mathbf{V}} & -\mathbf{C} \end{bmatrix} \tag{61}$$

and

$$\frac{\partial \mathbf{R}}{\partial \mathbf{Y}} = \begin{bmatrix} \dfrac{\partial \mathbf{R}_{\mathrm{X}}}{\partial \mathbf{X}} & \dfrac{\partial \mathbf{R}_{\mathrm{X}}}{\partial \mathbf{T}} \\ \dfrac{\partial \mathbf{R}_{\mathrm{T}}}{\partial \mathbf{X}} & \dfrac{\partial \mathbf{R}_{\mathrm{T}}}{\partial \mathbf{T}} \end{bmatrix} \simeq \begin{bmatrix} -\mathbf{K}_{\mathrm{G}} & -\dfrac{\partial \mathbf{S}}{\partial \mathbf{T}} \\ \dfrac{\partial \mathbf{R}_{\mathrm{T}}}{\partial \mathbf{X}} & -\mathbf{L} \end{bmatrix} \tag{62}$$

comprise the partial derivatives of the residual functions of the two sub-problems with respect to geometry and temperature and their time rates. The simplified expressions on the right-hand side of the equations point out main coupling terms disregarding nonlinear dependence on the variables. In Equation (61), the term $\partial\mathbf{P}/\partial\dot{\mathbf{T}}$ in the deformation problem is from thermal loading, the term $\partial\dot{\mathbf{Q}}/\partial\mathbf{V}$ refers to the effect of mechanical work on the thermal problem. In Equation (62), $\mathbf{K}_{\mathrm{G}} = \partial\mathbf{S}/\partial\mathbf{X}$ is the geometric matrix of the mechanical system accounting for the effect of geometry changes on the stress resultants while velocity is held constant. The term $\partial\mathbf{S}/\partial\mathbf{T}$ accounts for temperature dependent mechanical properties of the material.

### 4.2.3 Integration of the Process

The incrementation schemes for the temporal integration of temperature and deformation in Equation (8) lead for the compound variable $\mathbf{Y} = \{\mathbf{X}\ \mathbf{T}\}$ of the coupled evolution process to

$${}^{b}\mathbf{Y} = {}^{a}\mathbf{Y} + (1-\zeta)\tau\,{}^{a}\dot{\mathbf{Y}} + \zeta\tau\,{}^{b}\dot{\mathbf{Y}} \tag{63}$$

The time increment is $\tau = {}^bt - {}^at$, the value of the parameter $0 \le \zeta \le 1$ is chosen as for accuracy and numerical stability.

The issue of numerical stability refers to the integration of the equations of quasistatic motion and transient temperature. These equations are solved for the velocity at the beginning $t = {}^at$ and the end $t = {}^bt$ of the incremental step. Stability examines the sensitivity of the approximate integration to numerical disturbances of the coupled solution. A numerical disturbance ${}^a\boldsymbol{\beta}$ at the beginning of the increment is propagated by the integration scheme to ${}^b\boldsymbol{\beta}$ at the end, as

$$ {}^b\boldsymbol{\beta} = {}^a\boldsymbol{\beta} + (1-\zeta)\tau{}^a\dot{\boldsymbol{\beta}} + \zeta\tau{}^b\dot{\boldsymbol{\beta}} \tag{64} $$

At the same time the governing equations equally holding for the actual and the disturbed solution relate $\boldsymbol{\beta}$ and $\dot{\boldsymbol{\beta}}$ via the matrix $\mathbf{N}$ as by Equation (54). Substitution in Equation (64) gives

$$ {}^b\boldsymbol{\beta} = (\mathbf{I} + \zeta\tau{}^b\mathbf{N})^{-1}\left[\mathbf{I} - (1-\zeta)\tau{}^a\mathbf{N}\right]{}^a\boldsymbol{\beta} = \mathbf{A}\,{}^a\boldsymbol{\beta} \tag{65} $$

and the apertaining incremental amplification matrix of the disturbance is

$$ \mathbf{A} = (\mathbf{I} + \zeta\tau{}^b\mathbf{N})^{-1}\left[\mathbf{I} - (1-\zeta)\tau{}^a\mathbf{N}\right] \tag{66} $$

Stability demands the magnitude of $\boldsymbol{\beta}$ to decay, which restricts the spectral norm of the amplification matrix to $\sigma(\mathbf{A}) < 1$. This requirement establishes a relationship between the length of the time increment $\tau$ and the parameter $\zeta$ of the integration scheme. More specifically, assume the matrix $\mathbf{A} = \text{const.}$ and compose the initial disturbance $\boldsymbol{\beta}_0$ as a linear combination of the eigenvectors of the matrix. Then, after $n$ incremental steps

$$ \boldsymbol{\beta}_n = \mathbf{A}^n\boldsymbol{\beta}_0 = \sum_{i=1}^{m} \lambda_i^n c_i \mathbf{q}_i \tag{67} $$

The disturbance will decay with progressing $n$ if the magnitude of all eigenvalues $\lambda_i(\mathbf{A})$ is less than unity. This leads to a condition that involves the eigenvalues of the matrix $\mathbf{N}$:

$$ 0 < \tau\lambda_i(\mathbf{N}) < \frac{2}{1-2\zeta} \tag{68} $$

Accordingly, stability of the numerical integration requires that the eigenvalues $\lambda_i(\mathbf{N})$ are all positive for physical stability of the system, and restricts in addition the time increment in dependence of the selected value for the parameter $\zeta$.

The rate solution has not been questioned that far. It is of interest, however, to know whether the momentary state is receptive to fluctuations in the deformation velocity and the temperature rate. For this the perturbed Equation (52) at fixed time and state, gives

$$ \mathbf{R} + \frac{\partial \mathbf{R}}{\partial \dot{\mathbf{Y}}}\delta\dot{\mathbf{Y}} = \mathbf{0} \quad \text{and} \quad \frac{\partial \mathbf{R}}{\partial \dot{\mathbf{Y}}}\delta\dot{\mathbf{Y}} = \mathbf{0} \tag{69} $$

which requires the derivative matrix of the compound residuum $\mathbf{R}$ with respect to the rate quantities in the vector $\dot{\mathbf{Y}}$ to be singular. Apart from physical significance, this

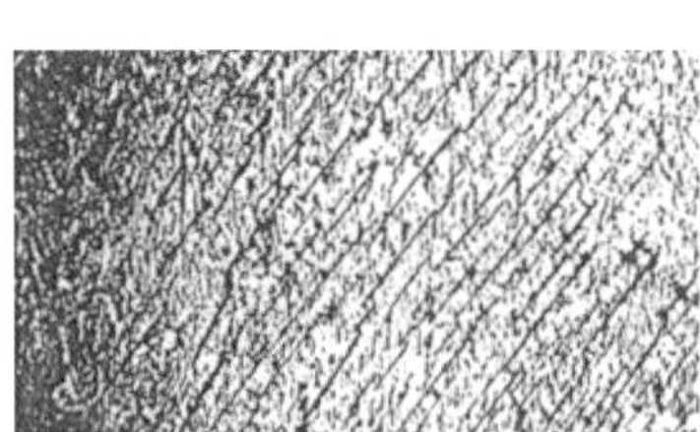

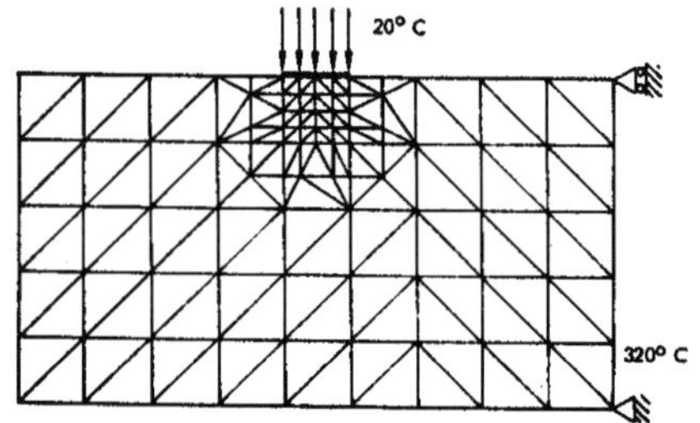


Figure 4: Crack pattern after 22,000 thermal shock cycles (left). Definition of plane thermal shock problem (right)

constellation is of importance to the numerical solution of the coupled rate problem. When the incremental integration is implicit the solution for the rates is at the end of the time step where the variables in $\mathbf{Y}$ depend on the time rate $\dot{\mathbf{Y}}$ by the integration scheme of Equation (63). The possibility of numerical fluctuations in the rates is then given if

$$\left(\frac{\partial \mathbf{R}}{\partial \dot{\mathbf{Y}}} + \zeta\tau\frac{\partial \mathbf{R}}{\partial \mathbf{Y}}\right)\delta^b\dot{\mathbf{Y}} = \mathbf{0} \tag{70}$$

which is associated with a singular matrix expression in the parentheses at $t = {}^b t$. For explicit integration $(\zeta = 0)$ the situation is as in Equation (69).

# 5 Applications

## 5.1 Stress Analysis of Thermal Shock

This application exposed in [5] refers to emergency cooling in nuclear reactor components inducing temporarily high temperature gradients responsible for local stress concentrations and irreversible deformation. The permanent straining accumulated during repeated cooling shock enbrittles the material and induces cracking, Figure 4. The study elucidates the performance of the viscoplastic constitutive approach. It deals with the thermomechanical processes associated with the local cooling of a metal block at a temperature of 320°C by a water jet of 20°C striking the surface of the metal over an area of 10 mm diameter. It is assumed that strains are small and the mechanical dissipation does not appreciably modify the development of the temperature in the solid thus allowing for an analysis of the thermal problem separately from the mechanical part where inertia effects are negligible, justifying quasistatic analysis.

The numerical model refers to a two-dimensional configuration, Figure 4; the effect of the remaining portions is substituted by appropriate boundary conditions for the temperature- and the stress analysis under the assumption of plane strain. The thermoelastic parameters of the material (elastic modulus $E$, Poisson ratio $\nu$, coeffi-

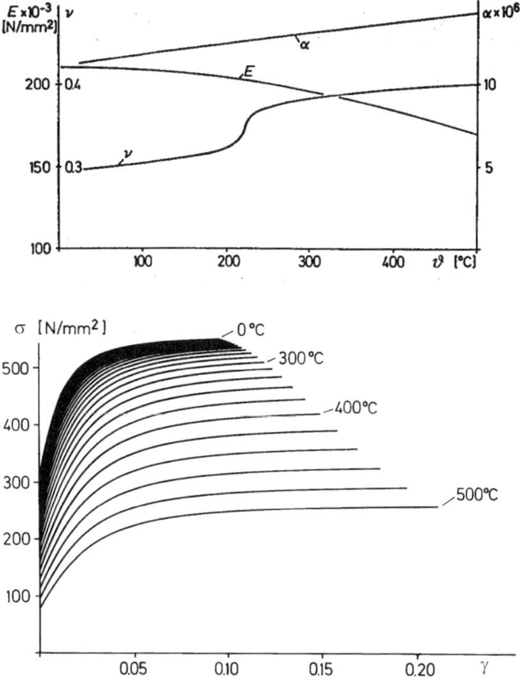


Figure 5: Thermoelastic parameters (top) and plastic hardening characteristic at different temperatures (bottom)

cient of linear thermal expansion $\alpha$) depend on the temperature as does the hardening characteristic in Figure 5. The material exhibits rate dependence modelled by the viscoplastic approach within the framework of an elastoplastic simulation algorithm. The viscosity parameter is determined by utilizing the laboratory information that in the tensile test a 25% increase of the static yield stress is observed at the strain rate $\delta = 1.08 \cdot 10^{-2}\,\mathrm{s}^{-1}$.

As a result of the computation Figure 6 illustrates the development of the stresses at the central point of the impingement of the water jet. The viscoplastic solution is compared with the elastic and the plastic ones. At the begining of the sudden cooling process, the viscoplastic stress state develops close to the elastic stress state. The two solutions diverge while inelastic deformation progresses. At the end of the transient process the plastic solution does not reproduce the stationary state of the viscoplastic response except approximately for the equivalent deviatoric stress $\bar{\sigma}$. As a matter of fact, the equivalent deviatoric stress is bounded by the flow stress of the material in both cases, but $\sigma_{\mathrm{f}}$ is not the same in the two solutions because the evolution of the inelastic strain differs. This is documented in Figure 7. As a consequence of the difference in permanent strain, the individual stress components deviate markedly for

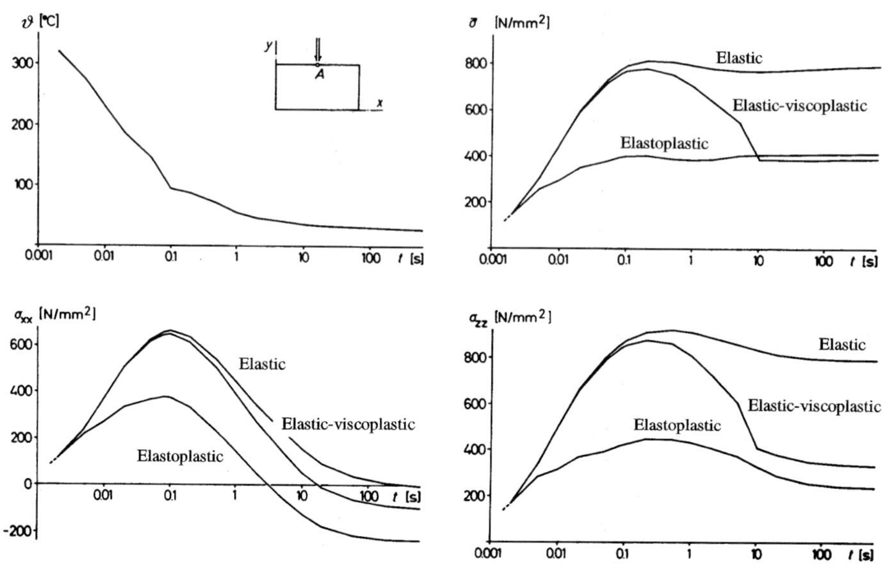


Figure 6: Temporal variation of temperature and stress at point A

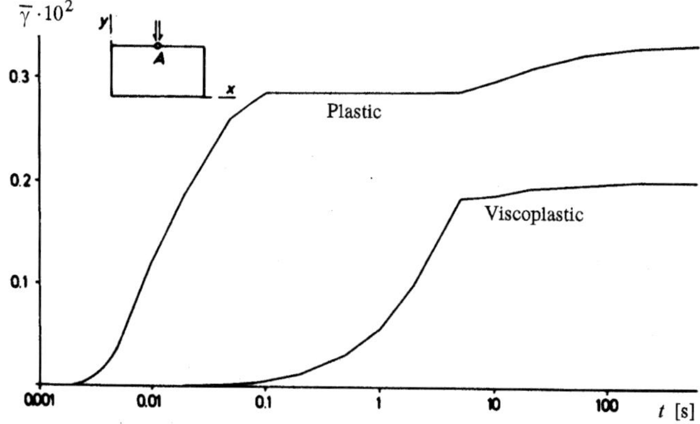


Figure 7: Development of the equivalent inelastic strain at location A

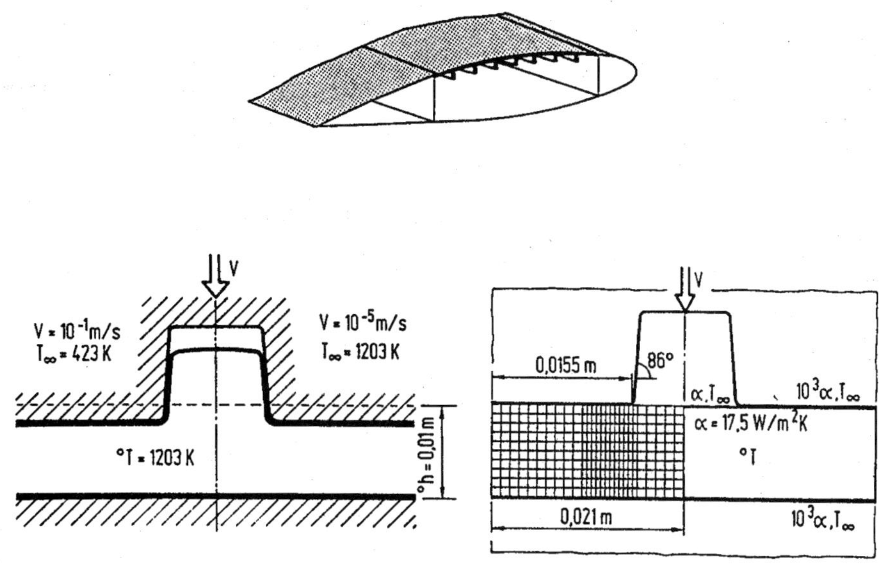


Figure 8: Hot forging of ribbed plate

plastic and viscoplastic response.

The study allows some conclusions to be drawn regarding the design for thermal shock. The comparison with the viscoplastic analysis shows that the elastic approach overestimates the stress and is therefore conservative for design relying on maximum stress criteria. Plasticity overestimates the inelastic deformation and is conservative for design based on maximum strain.

## 5.2 Hot Forging of Ribbed Plate

The case study taken from [1] is described in Figure 8. The work-piece material of Ti–6Al–4V alloy is originally flat with a thickness $^{\circ}h = 0.01$ m, the initial temperature is $^{\circ}T = 1203$ K. It is formed to a ribbed plate of thickness $h = 0.008$ m by the vertical motion of the tool. The numerical investigation of the forging deals with two process variants: the conventional and the isothermal forging. In the conventional process the tool is relatively cold ($T_\infty = 423$ K). In order to prevent extensive heat loss the forging time is kept short: the tool is moved rapidly with the velocity $V = 10^{-1}$ m/s. Due to the rate sensitivity of the flow stress of the material the high velocity of the process implies high tool forces. The other so-called isothermal forging process aims at lower forging forces by operating at the considerably lower tool velocity $V = 10^{-5}$ m/s. In this second variant the tool temperature is maintained at $T_\infty = 1203$ K, the initial

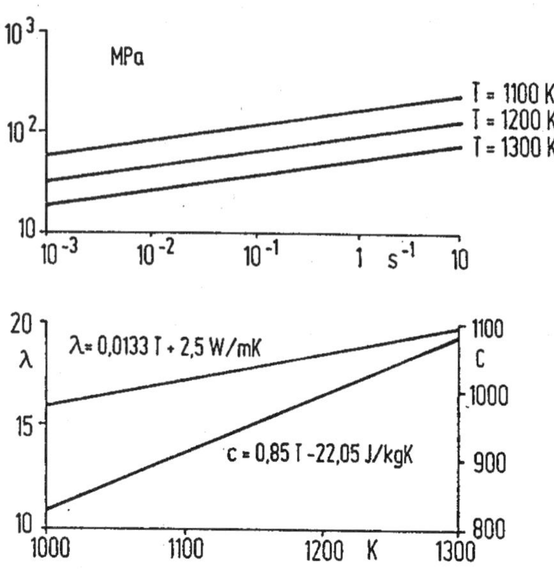


Figure 9: Thermomechanical properties of the Ti–6Al–4V alloy

temperature of the work-piece material. Isothermal forging is alternatively investigated also for the higher velocity $V = 10^{-1}$ m/s.

The simulation model is shown in Figure 8, right. The figure defines the geometry, initial and boundary conditions, and the finite element discretization. Plane conditions are assumed for deformation and temperature. The heat transfer coefficients to the ambient air and the tool are different. The respective heat transfer surfaces change following the evolution of the contact between work-piece material and tool. Sliding of the work-piece material along the tool surface is assumed frictionless. Figure 9 specifies the thermomechanical properties of the Ti–6Al–4V alloy in the domain of the process. The flow stress of the material (in MPa) is described by the power law of Equation (10) with the parameters $k = 8.26 \cdot 10^4, \;\; m_1 = -5.64 \cdot 10^{-3}, \;\; m_2 = 0.152$, while $m_3 = m_4 = 0$ (no strain dependence). The specific heat capacity $c$ and the thermal conductivity $\lambda$ of the material are functions of the absolute temperature.

The thermomechanically coupled simulation of the instationary process of hot forging uses 100 incremental time steps mainly required for tracing the evolution of contact between work-piece material and tool. Figure 10, left, shows results of the numerical simulation for the conventional forging process. The states shown refer to 20, 60 and 100% of the tool motion. The deformed finite element mesh is on the left-hand side, the temperature distribution on the right-hand side of the figure. The work-piece material locally cools down because of the contact with the cold tool, while the temperature is raised by the mechanical dissipation at locations of high deformation. The

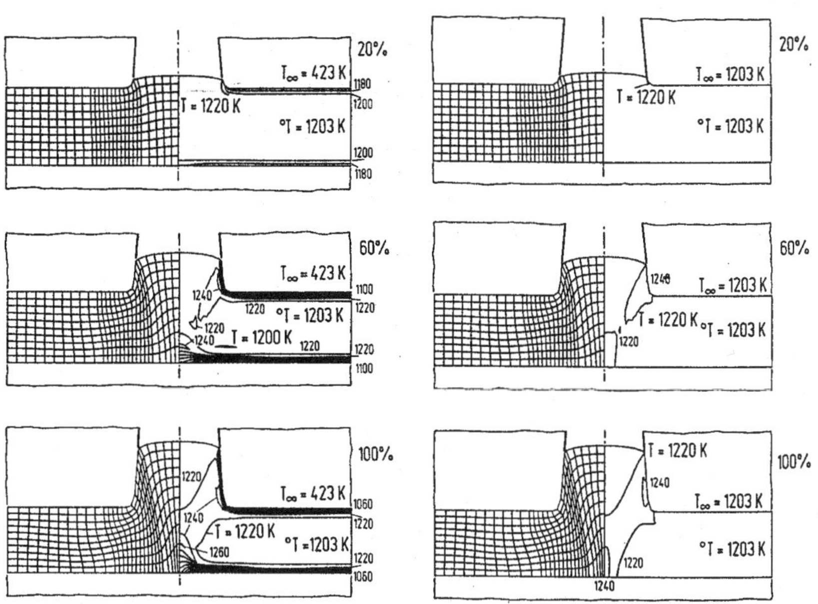


Figure 10: Deformation and temperature distribution. Conventional forging, left, rapid isothermal forging, right

numerical analysis predicts a local temperature increase beyond the phase transition of the material ($T_{\text{ph}} = 1253$ K), which is undesirable in forming practice.

A second investigation is concerned with the process of isothermal forging ($V = 10^{-5}$ m/s, $T_\infty = 1203$ K). The thermomechanically coupled analysis confirms that the higher tool temperature in fact maintains practically isothermal conditions in the work-piece material throughout the process. The deformations are seen not to be affected by this alternative forging process. As a third variant, we consider the temperature of the tool as for the isothermal case in conjunction with the velocity of the conventional process ($V = 10^{-1}$ m/s, $T_\infty = 1203$ K). Results of this study are given in Figure 10, right, and underline in particular the non-homogeneous temperature fields that occur in the present case.

## 5.3 Splashing of Ceramic Droplet

The process of plasma spraying is a convenient technique to coat structural parts with a material layer protecting from thermal shock or aggressive environment. This coating layer is formed by the impingement of small molten particles on the substrate to be coated. The material particles are injected and melt in the plasma torch at very

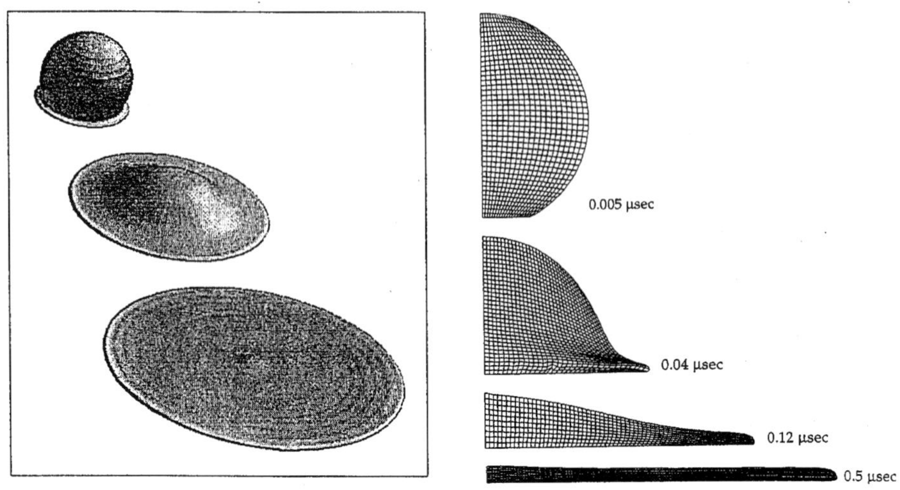


Figure 11: Simulation of splashing and mesh adapted to deformation

high temperature (approx. 20,000 K). One issue that influences the properties of ceramic coatings is related to the shape of the splats as a result of material flow and heat transfer when liquid droplets impinge on cool solid substrates. The modelling and computation of single splashing events for ceramic droplets (Figure 11) outlined in the following refer to [1].

The process parameters for normal impact of spherical ceramic droplets (yttria stabilized zirconia, 8%wt $Y_2O_3$): initial diameter, impact velocity and initial temperature, are taken as $d_0 = 20\ \mu\mathrm{m},\ V_0 = 180\ \mathrm{m/s},\ T_0 = 3400\ \mathrm{K}$. The mass density, specific heat and thermal conductivity are: $\varrho = 5.55 \cdot 10^3\ \mathrm{kg/m}^3,\ c = 500\ \mathrm{J/kgK},\ \lambda = 2\ \mathrm{W/mK}$. Latent heat of solidification, solidus temperature and liquidus temperature are: $q_\mathrm{L} = 6.1 \cdot 10^5\ \mathrm{J/K},\ T_\mathrm{S} = 2880\ \mathrm{K},\ T_\mathrm{L} = 2980\ \mathrm{K}$. The material viscosity follows the temperature dependent law

$$\mu = A \exp\left(\frac{4620}{T}\right)$$

as shown in Figure 12. The rigidity of solidified material is simulated by an augmented vicosity coefficient.

Thermal and mechanical boundary conditions during impact are indicated in Figure 13. They account for heat transfer from the droplet to the surrounding gas and heat exchange with the substrate by setting $T_\mathrm{Gas} = 1800\ \mathrm{K},\ \varepsilon_\mathrm{Rad} = 0.4;\quad T_\infty = 1200\ \mathrm{K},\ \alpha = 1.10^6\ \mathrm{W/m^2K}$. The mechanical boundary conditions consist of surface tension $\sigma = 0.5\ \mathrm{J/m}^2$ and Coulomb friction upon contact with the substrate with coefficient $C = 0.2$. Surface tension is seen to have negligible influence on the relaxation of the splashing process as compared to viscous dissipation. The friction conditions affect the material flow in the vicinity of the contact surface, but not the final shape

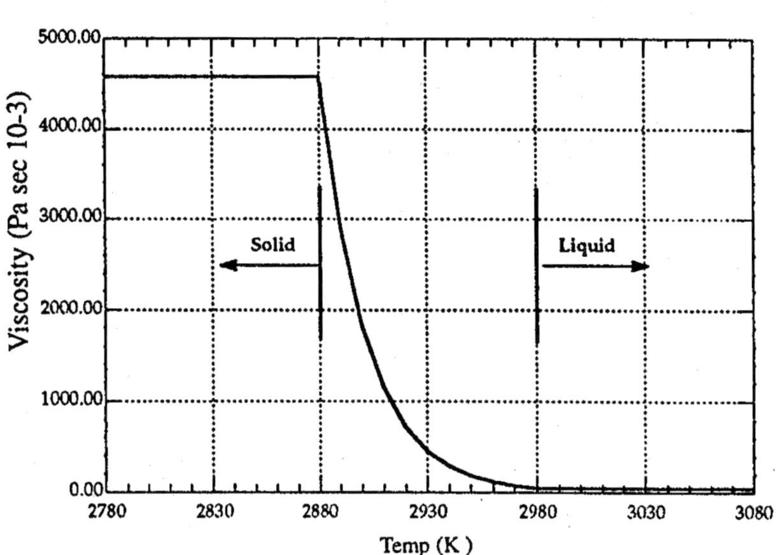


Figure 12: Variation of viscosity with temperature and transition to solid phase

of the splat. Effects of wetting have not been investigated. In the discretized representation of the droplet, momentary motion of the mesh nodal points and temperature variation are governed by the dynamic thermomechanical equations. Denoting the mass matrix of the discretized system by $\mathbf{M}$:

$$\mathbf{M}\dot{\mathbf{V}} + \mathbf{S} = \mathbf{P}, \qquad \mathbf{C}\dot{\mathbf{T}} + \mathbf{L}\mathbf{T} = \dot{\mathbf{Q}} \tag{71}$$

In the present problem the stress vector $\mathbf{S} = \mathbf{D}(\mathbf{X}, \mathbf{V}, \mathbf{T})\,\mathbf{V}$ in the deforming medium is a result of the temperature dependent material viscosity, and of forces from a kinematic approach to contact with the substrate and friction. The force vector $\mathbf{P}(\mathbf{X})$ emanates from the surface tension applied via superposed surface elements. In the thermal equation, the latent heat of solidification is accounted for by modifying the entries of the heat capacity matrix $\mathbf{C}(\mathbf{X}, \mathbf{T})$ within the interval between the liquid and the solid phase. Heat conductivity matrix $\mathbf{L}(\mathbf{X}, \mathbf{T})$ and thermal load vector $\dot{\mathbf{Q}}(\mathbf{X}, \mathbf{V}, \mathbf{T})$, comprise surface heat exchange terms depending on the temperature accounting for radiation. A heat source arises from viscous dissipation in the melt.

The dynamic thermomechanical simulation is performed until the splat material comes to rest. The motion is considered completed when the kinetic energy of the splat falls below $10^{-4}$ times the initial value. In the case study this occurs after $0.49\ \mu s$ upon collision. The thermal calculation is continued in order to investigate cooling down until solidification of the entire splat material, which starts at $0.5\ \mu s$

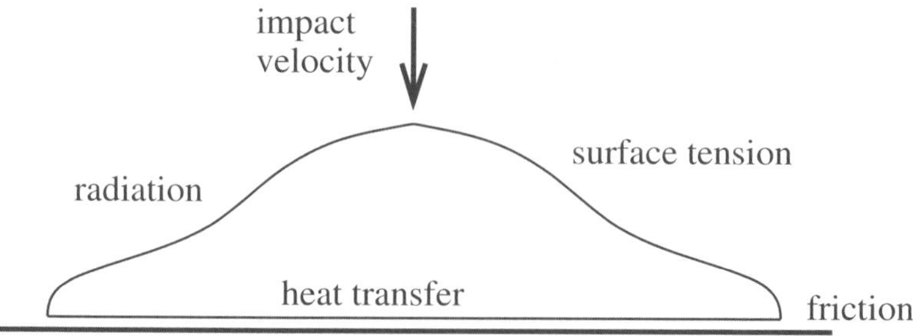


Figure 13: Boundary conditions in splashing

and ends at 5.9 $\mu$s.

The transient processes that rapidly evolve during impact of the droplet require a fine temporal resolution of the process involving tens of thousands of time steps. Furthermore, the severe deformations necessitate repeated modification of the computation mesh. Application of overall remeshing helps to comply with pronounced changes in shape of the droplet contour. More frequently local mesh modification is due because of the material flow close to the impact front (Figure 11). Decreasing viscosity enhances fluidity and can lead to deterioration of the material mesh by localized deformation, which questions the numerical model (Figure 14). Increasing fluidity may suggest continuous mesh modification rather than occasional if nodal points follow the material motion (Lagrange description). Alternatively, the Euler–Lagrange approach implies a computation mesh that deforms independently of the material such that the quality of the discretization is maintained.

# 6 Conclusion

Materials that exhibit pronounced sensitivity to the deformation rate suggest constitutive description by a viscous model. The approach is representative of inelastic behaviour, and may accommodate via the specification of the viscosity coefficient any scalar flow characteristic given as a function of strain, deformation rate and temperature as the local process variables. When the effects of elasticity can be discarded, a transparent algorithm solves for the deformation velocity, approximate time integration advances incrementally the evolution of the geometry of the solid. Coupling with thermal analysis provides a poweful tool for the computer simulation of a wide class of engineering problems of practical relevance.

Rate sensitivity contributes to the stability of the deformation process while temperature dependence has the opposite effect. In the absence of strain hardening, stability requires the rate effect to be stronger than the thermal degradation. At higher strain the stress-strain dependence may exhibit a waved shape as a result of hardening and softening processes in the material structure. In regions with strain softening

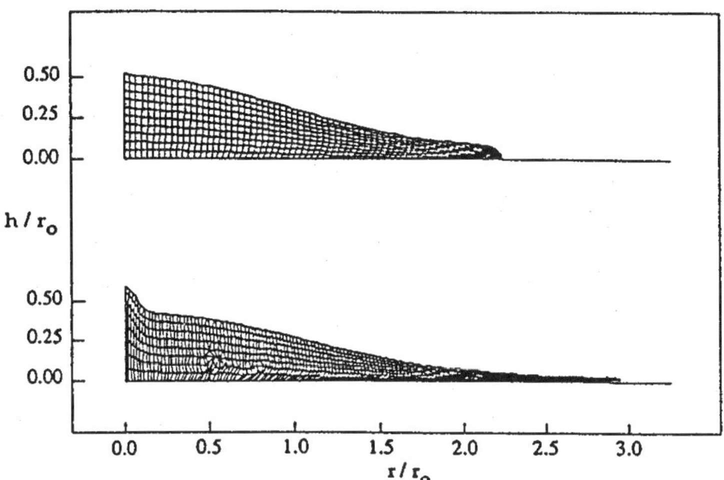


Figure 14: Effect of fluidity on deformation. Viscosity in lower frame ten times smaller

the deformation process will be stable if the rate effect compensates both, the thermal degradation and the strain softening. The stability conditions under more general situations involve the characteristics of the coupled thermomechanical system. Physical stability is seen to be of relevance for the sensitivity of the rate solution to fluctuations and for the numerical stability of the temporal integration.

# References

[1] I. Doltsinis, *"Large deformation processes of solids – From fundamentals to computer simulation and engineering applications"*, WIT Press, Southampton, 2003.

[2] I. Doltsinis, *"Critical states in deformation processes of inelastic solids"*, in: B.H.V. Topping and C.A. Mota Soares (eds), Progress in Computational Structures Technology, Saxe-Coburg Publications, Stirling, Scotland, 2004.

[3] J. Argyris, I. Doltsinis, *"On the large strain inelastic analysis in natural formulation – Part I. Quasistatic problems"*, Comput. Meths Appl. Mech. Engrg, **20**, 213–252, 1980.

[4] J. Argyris, I. Doltsinis, *"On the natural formulation and analysis of large deformation coupled thermomechanical problems"*, Comput. Meths Appl. Mech. Engng, **25**, 195–253, 1981.

[5] I. Doltsinis, *"Elements of plasticity – Theory and computation"*, WIT Press, Southampton, 1999.

[6] E.W. Hart, *"Theory of the tensile test"*, Acta Met. 15, 351-355, 1967.

©Saxe-Coburg Publications, 2008.
Trends in Engineering Computational Technology
B.H.V. Topping and M. Papadrakakis, (Editors)
Saxe-Coburg Publications, Stirlingshire, Scotland, 321-334.

# Chapter 16

# Computational Biomechanics of Sensory Organs in Spiders

**F.G. Rammerstorfer[1], B. Hößl[1,2], H.-E. Dechant[1,2]**
**H.J. Böhm[1] and F.G. Barth[2]**
**[1] Institute of Lightweight Design and Structural Biomechanics**
**Vienna University of Technology, Austria**
**[2] Department of Neurobiology and Cognition Research**
**Vienna University, Austria**

## Abstract

The chapter shows how computational mechanics can contribute to exploring and understanding the impressively sophisticated "design" and functionality of biological sensors. Two types of sensory systems are considered that are found on the legs of arachnids: tactile hairs and, in more detail, slit sensilla.

**Keywords:** biomechanics, sensors, spiders, arachnids, finite element analyses.

## 1 Introduction

"Sensors and sensing are essential for all forms of life. Correspondingly, there is a fascinating richness and diversity of sensory systems throughout the animal kingdom" [1]. The present paper deals with two types of sensory systems that are found on the legs of arachnids: tactile hairs and, in more detail, slit sensilla [2].

Spiders have a large number of mechanosensitive cuticular hairs. So-called "tactile hairs" are stimulated by contact forces. It was found by computational simulations that these hairs are "optimally designed" both from a structural mechanics point of view and with respect to their neurobiological sensitivity [3,4]. Mathematical models were developed and tested to study the force and directional sensitivity of the hair deflections [5].

Another type of sophisticated sensory system of spiders and other arachnids are strain sensitive slit sensilla. These sensilla are elongated openings in the cuticle with aspect ratios of up to 100, appearing as single slits or as complex arrangements of a number of mechanically interacting slits, which form so-called lyriform organs. The mechanical stimulus is encoded by the compression of the slits (*i.e.*, the reduction of the slits' gap width) close to the location of the sensory cell ending in the slit. In order to obtain basic information about the mechanical behavior of interacting slits and in view of their high aspect ratios a first approach used Kachanov's method [6] for interacting cracks. Because of the limitations of this method with respect to cracks positioned at small distances to each other and in order to get deeper insight into the deformation patterns leading to sensory cell stimulation, discretization methods were applied [7]. By using finite element models with realistic slit arrangements and taking the layered micro-structure of the cuticle into account it was possible to study the amplitude and directional sensitivity of the slits' deformation responses to mechanical loads for different slit configurations in the lyriform organs. Even minor morphological variations in the arrangements of the slits can substantially influence the stimulus transformation [8].

# 2 Sensory hairs of *Cupiennius salei*

## 2.1 General description

The exoskeleton of arthropods shows a wealth of hair sensilla differing in structure and function. Trichobothria (Figure 1, right) are extremely sensitive to fluid flow; they are stimulated by the slightest movement of the surrounding air (*e.g.*, caused by a fly or another prey). These hair sensilla have already been extensively investigated, see e.g., [2, 9].

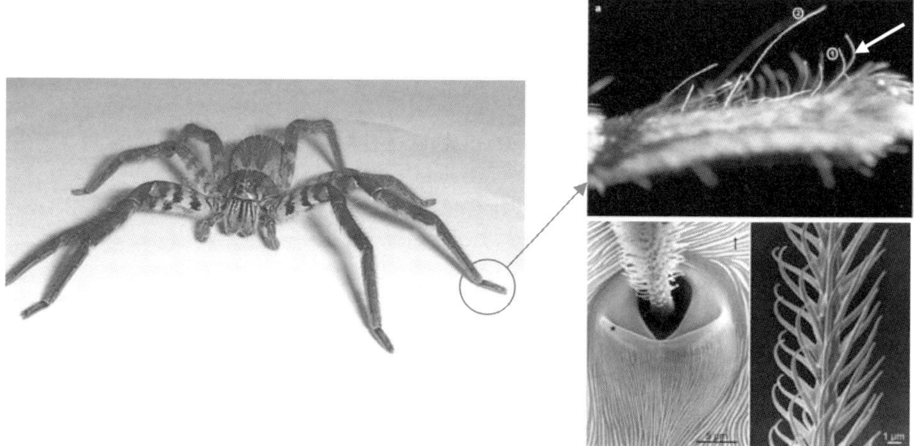

Figure 1: left: *Cupiennius salei*; right: Trichobothria (arrow); (reprinted from [9] with permission from Elsevier).

In contrast to trichobothria, tactile hairs (Figure 2) are stimulated by contact forces, leading to a deflection and bending of the hair shaft and deformations in the hair's anchoring system, where the dendrites of sensory cell are located and the mechanical stimulus is transduced into nervous signals. From the velocity and orientation of this deflection the required information regarding the stimulus is extracted.

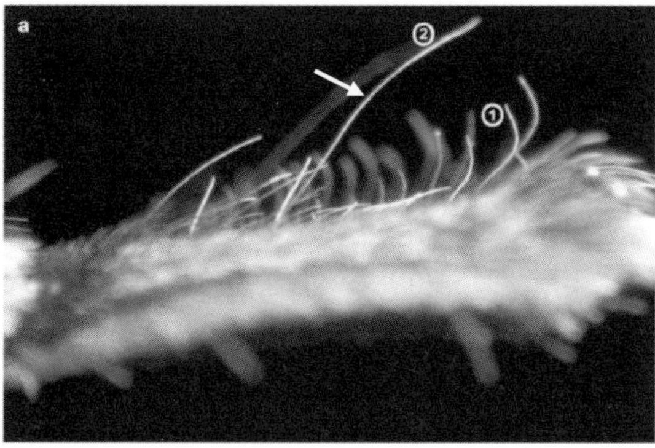

Figure 2: Left: Long tactile hair (arrow), right: its anchoring in the skeleton (left: reprinted from [9] with permission from Elsevier, right: from Figure 2b of [4] with permission from Springer-Verlag).

Because of the strong nonlinearities involved (frictional contact conditions, large deformations, nonlinear stiffness of the anchoring system) computational methods are required for analysing the behavior of such hairs.

## 2.2 Finite Element model and results

As described in detail in [3,10] the finite element model takes into account the sophisticated variation of the dimensions and shape of the cross section of the hair shaft along its length as well as the nonlinear character of the anchoring system, for which a special "joint-element" was developed in [10]. Figure 3 shows the variation of a hair shaft's longitudinal section along its length.

The finite element simulations were performed using the program CARINA [11]. A wide range of stimulus magnitudes (loads) was considered [3, 10]. Figure 4 shows comparison between measured and computed hair deflections and deformations due to contact forces from above.

It is worth noting that the hair represents an "optimized structure" in the following sense: It is quite clear, that during the highly nonlinear process (contact, large deformations) of increasing stimulus the position where the maximum longitudinal stress appears moves along the hair's length. However, it is not trivial that the value of the maximum stress that occurs anywhere along the length of the hair shaft under changing load conditions is constant over a wide range of stimuli (Figure 5).

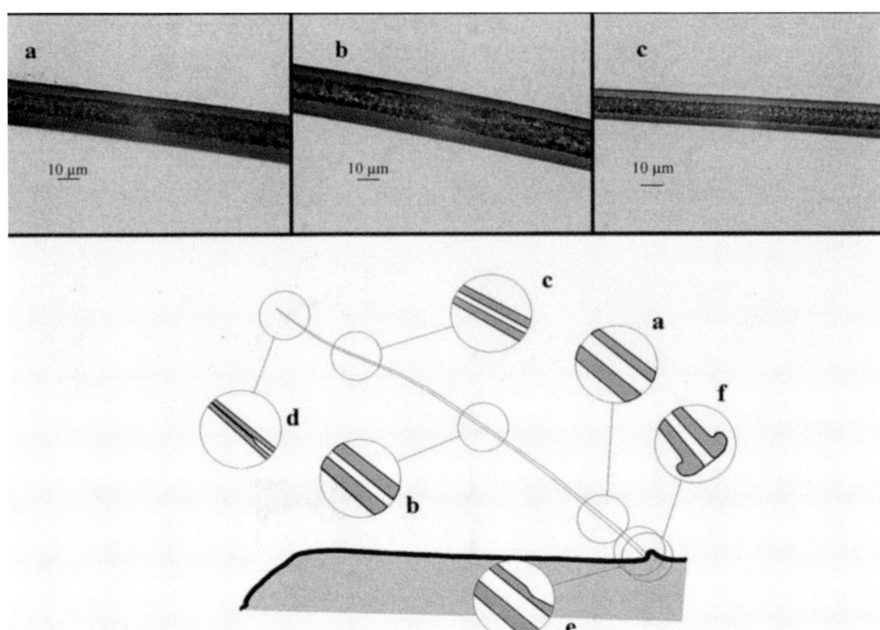


Figure 3: Variation of a hair's longitudinal section along its length (from Figure 3 of [3] with permission from Springer-Verlag).

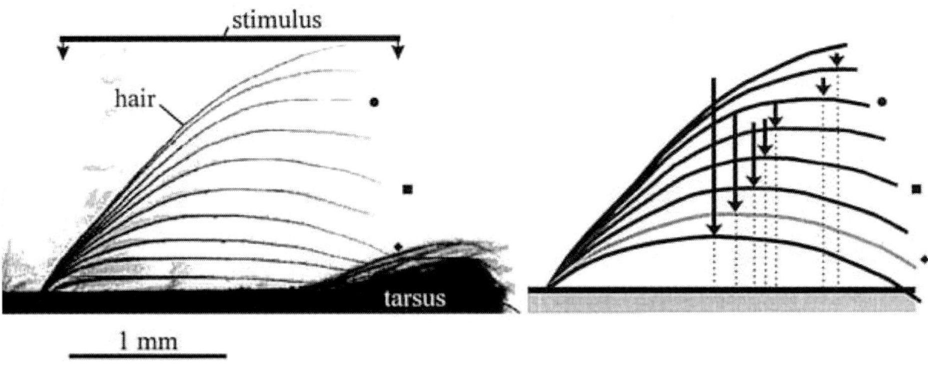


Figure 4: Deflections of the hair under stimulation by a flat surface from above; left: measured, right: calculated, arrows show how the "point of contact" moves along the hair with increasing load (from Figure 5 of [3] with permission from Springer-Verlag).

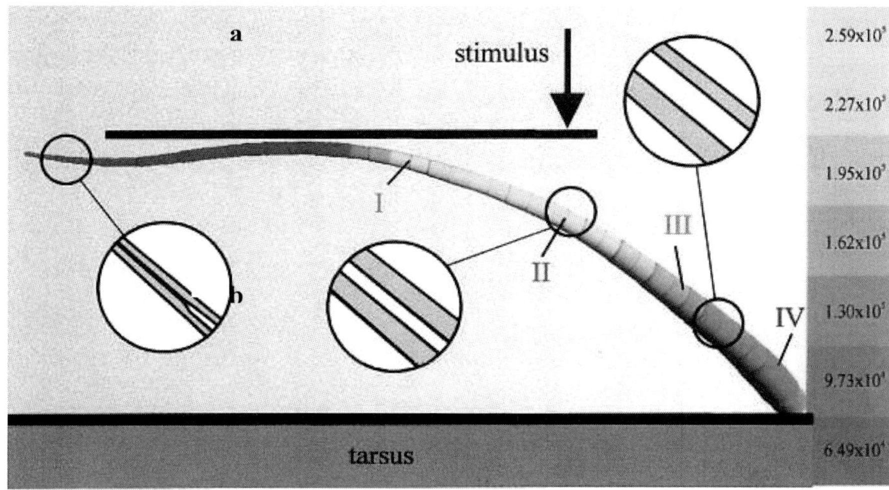


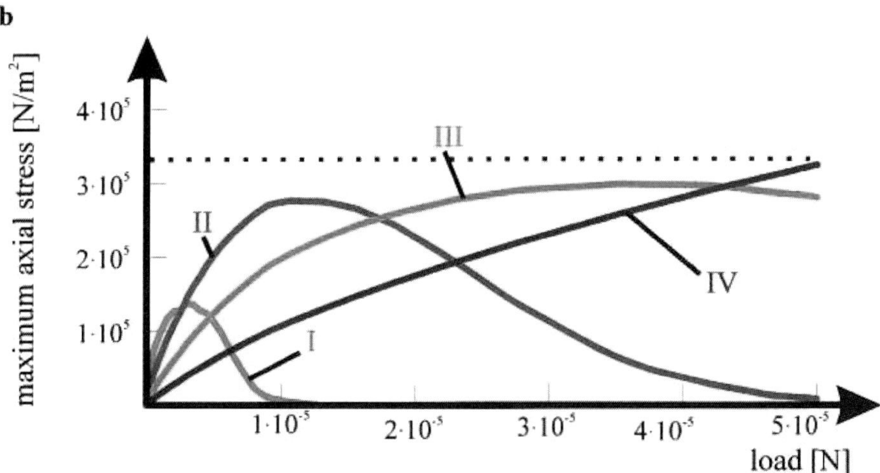


Figure 5: a) Deflected hair (thickness exaggerated for presentation reasons only) with positions of four longitudinal sections I to IV, for which in b) the histories of the maximum axial stress are shown for increasing stimulus (from Figure. 6 of [3] with permission from Springer-Verlag)

From these results one can conclude that the hair is "designed" in a way that ensures the avoidance of stresses beyond a critical value and at the same time saves material and minimises hair mass [3,10].

# 2 Slit sensilla of *Cupiennius salei*

## 2.1 General description

For monitoring the effects of various mechanical loads acting on their exoskeleton arachnids use mechanosensitive slit sense organs. These organs "measure" local strains from which the magnitude and the direction of mechanical forces can be deduced. Slit sensilla can be arranged in the form of isolated slits, as loose groups, and in the form of lyriform organs, i.e., close side-by-side arrangements of many slits (Figure 6).

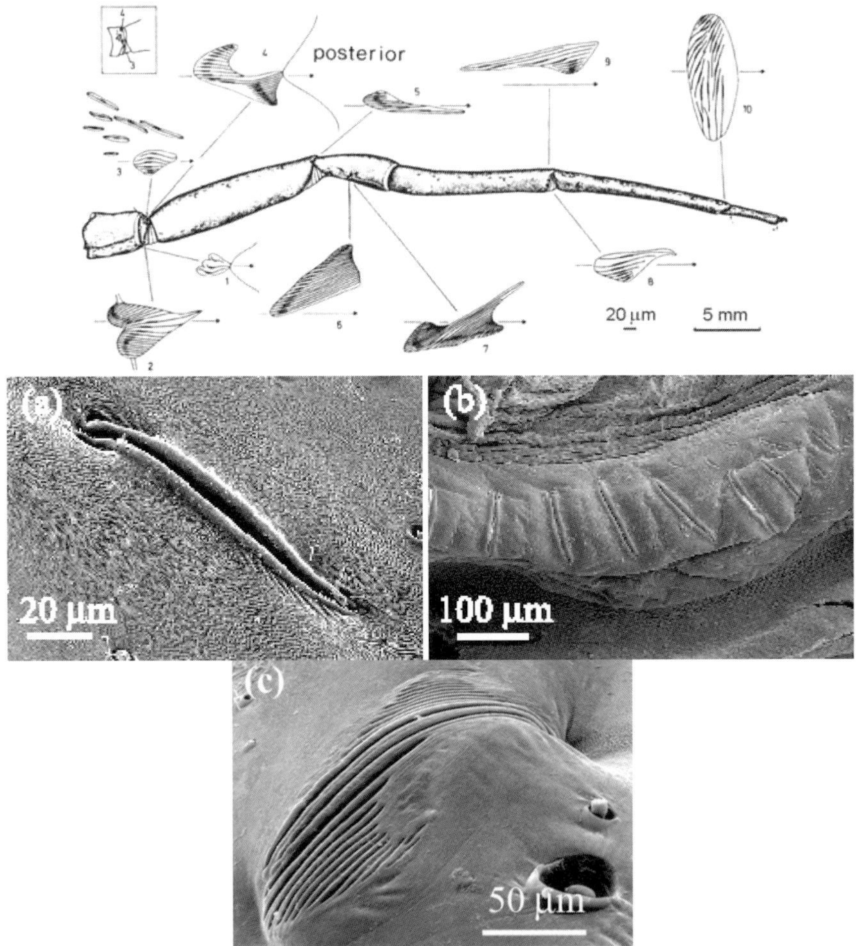


Figure 6: Slit sense organs on a leg of *Cupiennius salei* (from Figure 3b of [12] with permission from Springer-Verlag); a) isolated single slit, b) loose group, c) lyriform organ (micrographs at bottom: Courtesy R. Müllan).

Each of the slits is covered on the outside by an inwardly bulging, trough-shaped membrane. At a specific position (which is related to the mechanical sensitivity of the slit) a special structure, the coupling cylinder, is contained in the membrane for the attachment of the dendritic ending of a sensory cell (Figure 7). Only compressive in-plane deformations of the cuticle, which may result from different kinds of loadings, lead to a response of the dendrite.

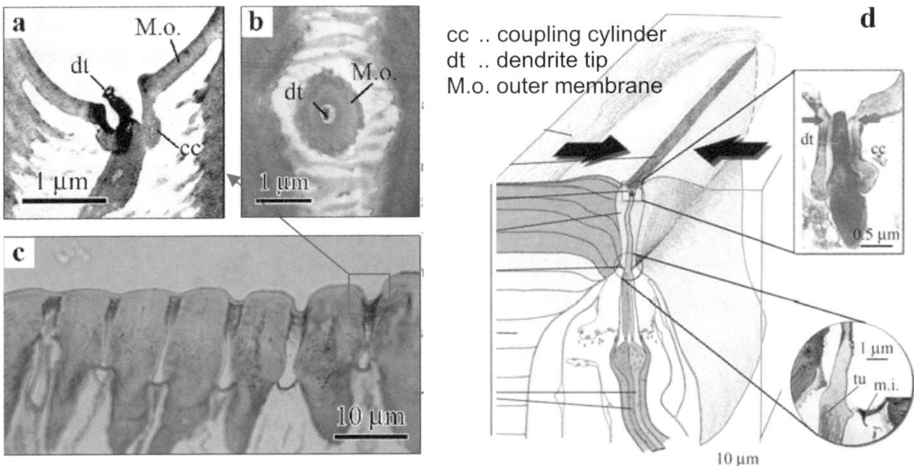


Figure 7: a) Cross section of a slit, b) view from above on to the membrane: coupling cylinder and dendrite, c) transverse section across the slits, d) sketch of a single slit;
(a) and d) from Figures 4, 7 and 8 of [13] with permission from Springer-Verlag, b) and c) from [14]).

In Figure 8 three typical arrangements of lyriform organs are presented. It will be shown that each of them leads to significantly different mechanical responses to one and the same stimulus.

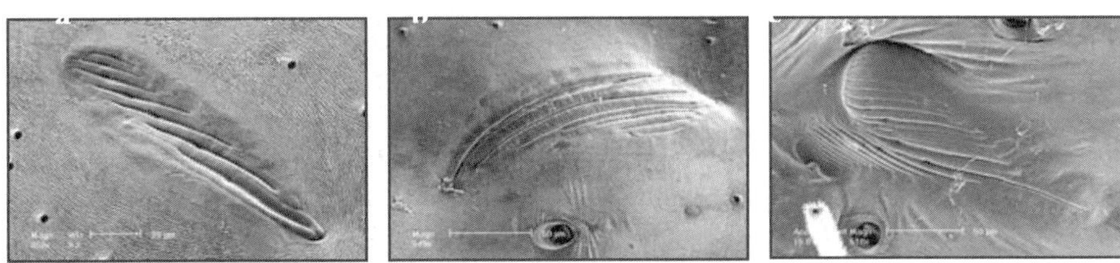

Figure 8: Typical geometries of lyriform organs: "oblique bar" arrangement, "triangular" arrangement, "heart-shaped" arrangement; (from [14]).

From the viewpoint of structural mechanics the exocuticle of arachnids is a layered composite shell with a large number of layers consisting of a protein matrix reinforced by embedded chitin fibres. The orientations of the fibres change helicoidally through the thickness of the shell which in *Cupiennius salei* is approximately 10 μm thick. The sequence of fibre-angles from layer to layer leads, according to lamination theory, to a quasi-isotropic laminate, i.e. the mechanical material behavior is isotropic for in-plane loadings; see [15].

Based on Huang's laminate analogy model [16], which uses a Mori-Tanaka approach [17], the effective material properties (in-plane and out-of-plane Young's moduli as well as corresponding Poisson's ratios) of the outer region of the cuticle were estimated in [18] (Figure 9) and applied in the mathematical models described in the following .

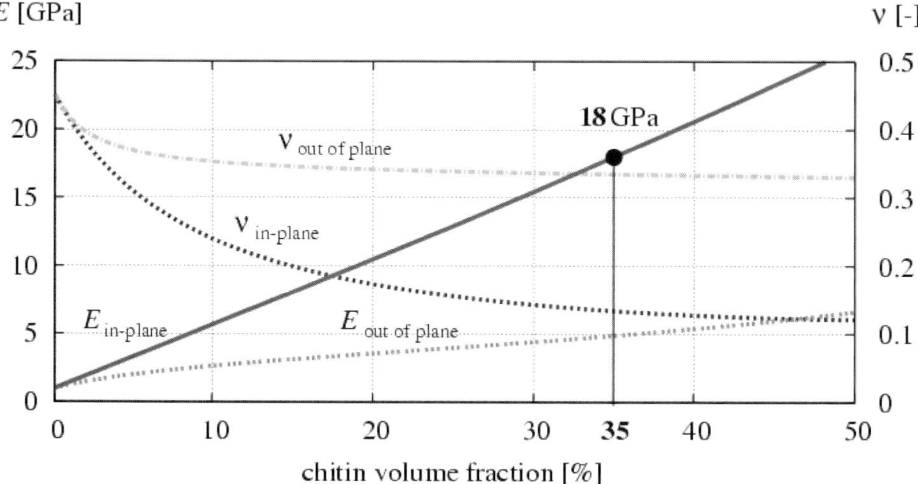


Figure 9: Estimates of the effective Young's moduli $E$ and Possion ratios $\nu$ of the outer region of the cuticle as a function of the chitin volume fraction; (from [18]).

## 2.2 Investigations using Kachanov's approach

Kachanov and Gorbatikh [6,19] developed an analytical approach to describing the stress intensity factors at the tips of interacting cracks situated in an infinitely extended disc. There are some critical assumptions in this approach: The interaction between the cracks is introduced via averaged tractions along the crack faces, and the displacements of the crack faces are approximated by ellipses that are enriched by quadratic polynomials (for details see [7,18]).

For basic studies of arrangement effects this analytical approach was applied in [7,18]. The fact that Kachanov's method assumes sharp crack tips and the tips of sensory slits are rounded can be expected to play a minor role as long as the deformation behavior in the vicinity of the coupling cylinder with the sensory dendrite is concerned. Furthermore, the stiffness of the membranes covering the slits is assumed to be very small compared to the stiffness of the cuticle shell. Hence the membranes were neglected in the models.

The results obtained in [7,8,18] lead to the following conclusions:

1. As exemplarily shown in Figure 10, the three typical arrangements of lyriform organs displayed in Figure 8 lead to different results regarding the ratios of compressions (*i.e.*, reduction of the width of the gap) $D/D_{sc}$ of the individual slits within the organs. $D$ is the reduction of the gap width due to displacements of the slit faces under compressive loading. $D_{sc}$ is a reference value, which is the reduction of the gap width at the mid length of a single reference slit of length $l_0$ under normal loading. This single slit has an aspect ratio $l_0/b = 100$ ($b$ is the gap width of the slits in the absence of loads) and is situated at the center of a square disc subjected to a uni-axial far-field stress equal to the far-field stress acting on the actual configuration.
2. Comparison with finite element calculations showed that the assumptions on which Kachanov's method is based lead to considerable errors when very closely neighboring slits (*i.e.*, small values of $S/l_0$; $S$ being the lateral distance of neighboring slits) are considered as found in lyriform organs (Figure 11a) .
3. Furthermore, the finite element computations showed that the distribution of the slit face displacement $D$ along the length of the slit is not resolved in sufficient detail when closely neighboring slits are considered (compare Figure 11 b).

Items 2 and 3 in the above list indicate that the described analytical method is inadequate for achieving sufficient accuracy in describing the behavior of lyriform organs, and, accordingly, finite element models were developed and solved, using ABAQUS.

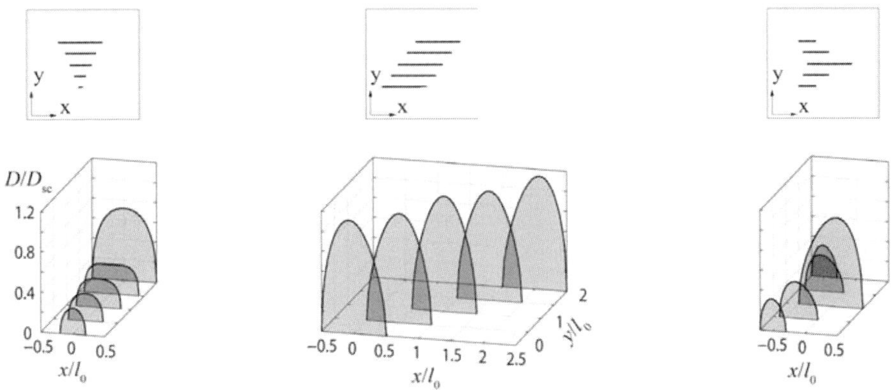


Figure 10: Relative compression (change in the gap width) under in-plane loading in *y*-direction, calculated by Kachanov's method (compare [7,18]).

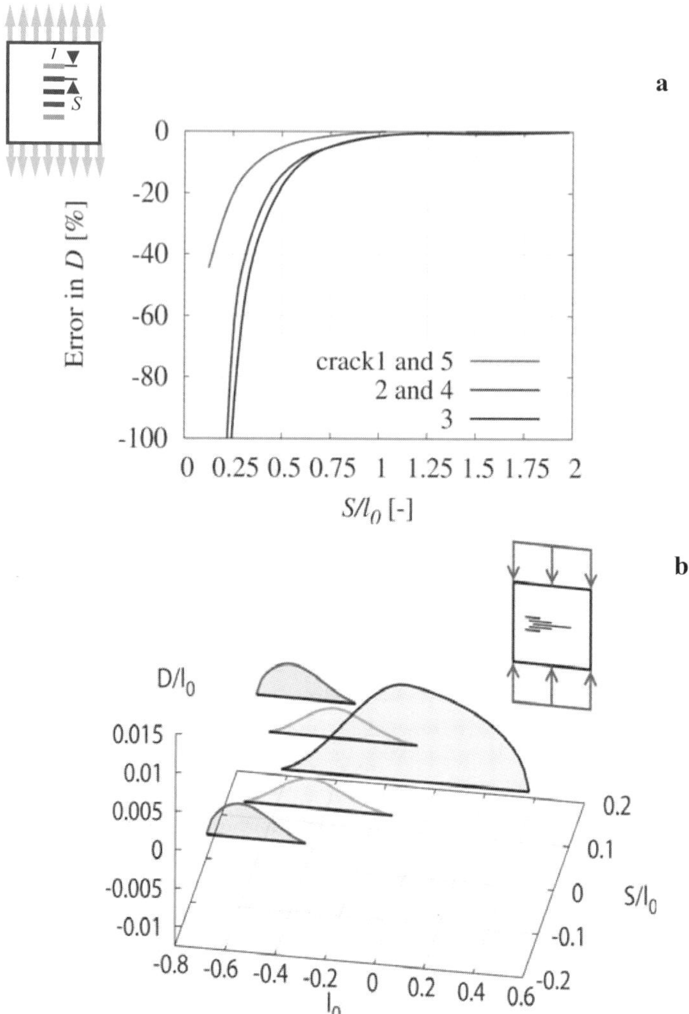


Figure 11: a) Error in maximum slit face displacement calculated by Kachanov's method, depending on the lateral distance between neighboring slits, b) Distribution of compression of the gap due to in-plane loading, calculated by finite elements – compare Figure 10 right. (Figures modified from [18]).

## 2.3 Finite element analyses

Studying the functionality of the different lyriform organs requires a careful representation of the particular shapes and geometrical arrangements of the slits. As shown above, analytical methods, such as Kachanov's method, are not adequate to accurately calculate the displacements of a slit's faces close to the position of the coupling cylinder in the slits' covering membrane which activate the dendrite of the sensory cell. Extremely fine meshes, as shown in Figure 12, are required to resolve with sufficient accuracy the interacting deformation behavior of the individual slits within the organ from which the spider obtains information used in determining the magnitude and direction of mechanical strains resulting from mechanical loads.

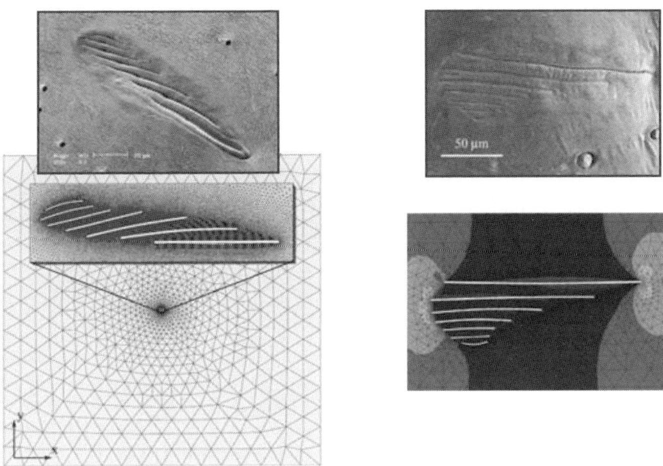


Figure 12: Finite element models for studying the behavior of "oblique bar" arrangements (left) and "triangular" arrangements of slits (right); (Figures modified from [18]).

For the activation of the individual sensory cells supplying the individual slits in a lyriform organ a certain threshold of slit compression at the coupling cylinder must be surpassed. Furthermore, from the sequence and the extent of activation of the individual neurons the information about the size and direction of the mechanical stimulus can be recovered. In [18] the results of finite element studies are presented in terms of polar diagrams of the compression of the individual slits of a number of models of lyriform organs (responses to far-field strains from different directions). Results from models with a rather generic geometry are compared to models of the same organ with realistic geometry representation.

These comparisons have shown that the geometry of the slit arrangements, *i.e.* the morphology of the organ, must be represented very accurately in the finite element models in order to resolve, on the one hand, the sequences in which the thresholds of the individual slits of the organ are exceeded and, on the other hand, the degrees of compressive displacements of the faces of each slit.

As reported in [18], good agreement between the predictions of simulations and the results of interferometric experiments (Schaber *et al.* in prep.) on an individual *Cupiennius salei* was obtained for a wide variety of lyriform organs if the real morphologies were carefully represented by the models.

Some of the single slits and of the lyriform organs are located at strongly curved regions of the cuticle, *i.e.*, on dimples or ridges of the spider's leg (Figure 13).

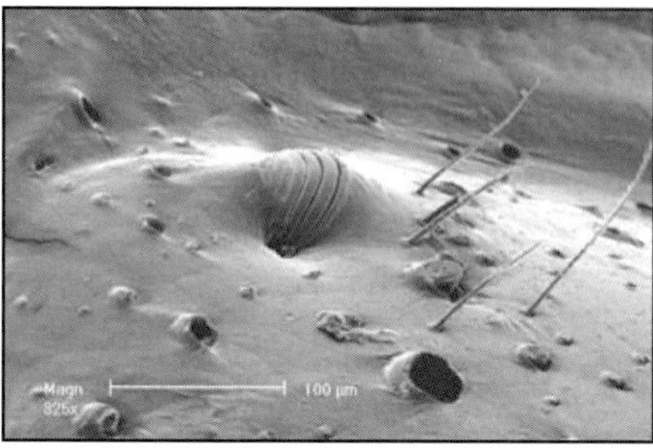

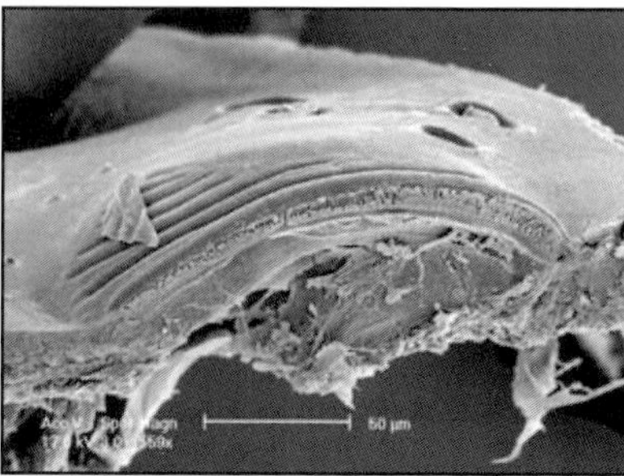


Figure 13: Lyriform organs located at strongly curved regions, (from [14]).

This kind of geometry, which in many cases increases the sensitivity of the organ, increases the complexity of the finite element models; for more details see [18].

# 3 Conclusions

The computational biomechanical investigations described in the present paper have substantially contributed to exploring and understanding the impressively sophisticated "design" and functionality of two types of sensory systems that are found on the legs of arachnids: tactile hairs and slit sensilla.

It was shown that the geometric and mechanical properties of these organs are not at all arbitrary but must rather be viewed as "optimized", with respect to both their sensory purposes and their mechanical behavior as structural elements. For instance, the variation of the shape of the cross-section of tactile hairs follows principles which in structural mechanics are related to "fully stressed design". The arrangements of the slits in lyriform organs are very specific, aiming at sensing mechanical force stimuli with respect to their magnitude and, at the same time, their orientation. Small deviations from the natural geometric arrangements can lead to significant deterioration of their mechanical behavior and, as a consequence, of their nervous responses.

# Acknowledgements

The financial support by grant FWF P12192-Bio to FGB and FWF P16348 to FGB and FGR of the Austrian Science Foundation (FWF) is gratefully acknowledged.

## References


[1] F.G. Barth, "Sensors and Sensing: A Biologist's View" in *Sensors and Sensing in Biology and Engineering* (F.G. Barth, J.A.C. Humphrey, T.W. Secomb, Eds.), pp. 3–15, Springer, Wien, 2003.

[2] F.G. Barth: "A Spider's World: Senses and Behavior", Springer, Wien, 2002.

[3] H.-E. Dechant, F.G. Rammerstorfer, F.G. Barth, "Arthropod Touch Reception: Stimulus Transformation and Finite Element Model of Spider Tactile Hairs"; J. Comp. Physiol. A, 187, 313 – 322, 2001.

[4] J.T. Albert, O.C. Friedrich, H.-E. Dechant, F.G. Barth, "Arthropod Touch Reception: Spider Hair Sensilla as Rapid Touch Detectors", J. Comp. Physiol. A, 187, 303-312, 2001.

[5] H.-E. Dechant, B. Hößl, F. G. Rammerstorfer, F.G. Barth: "Arthropod Mechanoreceptive Hairs: Modeling the Directionality of the Joint"; J. Comp. Physiol. A, 192, 1271–1278, 2006.

[6] M. Kachanov, "Elastic Solids with Interacting Cracks and Related Problems" in: *Advances in Applied Mechanics* (J. Hutchinson, T. Wu, Eds.), pp. 259–426, Academic Press, San Diego, 1994.

[7] B. Hößl, H. J. Böhm, F. G. Rammerstorfer, R. Müllan, F.G. Barth, "Studying Arachnid Slit Sensilla by a Fracture Mechanical Approach: Kachanov's Method and Slit Face Displacements"; J. Biomech., 39, 10; 1761 – 1768, 2006.

[8] B. Hößl, H.J. Böhm, F.G. Rammerstorfer, F.G. Barth, "Finite Element Modeling of Arachnid Slit Sensilla – I. The Mechanical Significance of Different Slit Arrays"; J. Comp. Physiol. A, 193, 4; 445–459, 2007.

[9] J.A.C. Humphrey, F.G. Barth, "Medium Flow-sensing Hairs: Biomechanics and Models", in: *Advances in Insect Physiology. Insect Mechanics and Control* (J. Casas , S.J. Simpson, Eds.), Vol. 34, pp. 1-80, Elsevier Ltd, 2008.

[10] H.-E. Dechant, "Mechanical Properties and Finite Element Simulation of Spider Tactile Hairs", Doctoral Thesis, TU Vienna, 2001.

[11] CARINA – A nonlinear finite element program, Institute for Lightweight Design and Structural Biomechanics, TU Vienna.

[12] F.G. Barth, W. Libera, "Ein Atlas der Spaltsinnesorgane von *Cupiennius salei* Keys., Chelizerata (Araneae)", Z. Morph. Tiere, 68, 343–369, 1970.

[13] F.G. Barth, "Der sensorische Apparat der Spaltsinnesorgane (*Cupiennius salei* Keys. Araneae)", Z. Zellforsch., 112, 212–246, 1971.

[14] R. Müllan, "Feinbau und Rekonstruktion der Cuticulastrukturen lyraförmiger Sinnesorgane der Spinne *Cupiennius salei*", Master's thesis, University of Vienna, 2005.

[15] F.G. Barth, "Laminated Composite Material in Biology. Microfiber Reinforcement of an Arthopod Cuticle", Z. Zellforschung, 144, 409-433, 1973.

[16] J.H. Huan, "Some Closed-form Solutions for Effective Moduli of Compos-ites Containing Randomly Orientated Short Fibers", Mater. Sci. Engng. A, 315, 11–20, 2001.

[17] T. Mori, K. Tanaka, "Average Stress in Matrix and Average Elastic Energy of Materials with Misfitting Inclusions", Acta Met., 21, 571–574, 1973.

[18] B. Hößl, "Mechanical Simulation of Slit Sensors of Arachnids", Doctoral Thesis, TU Vienna, 2007.
[19] L. Gorbatikh, M. Kachanov, "A Simple Technique for Constructing the Full Stress and Displacement Fields in Elastic Plates with Multiple Cracks", Eng. Fract. Mech., 66 , 51–63, 2000.

©Saxe-Coburg Publications, 2008.
Trends in Engineering Computational Technology
B.H.V. Topping and M. Papadrakakis, (Editors)
Saxe-Coburg Publications, Stirlingshire, Scotland, 335-353.

# Chapter 17

# Three-Dimensional Numerical Analysis of a Dynamic Structure, Saturated Soil and Pore Fluid Interaction Problem

**A.H.C. Chan[1] and J. Ou[2]**
**[1] Department of Civil Engineering
University of Birmingham, United Kingdom**
**[2] Department of Civil Engineering
University of Dundee, United Kingdom**

## Abstract

A numerical method for the calculation of the overall behaviour of a non-trivial three-dimensional saturated soil-structure system is given in this paper. The numerical modelling of saturated soil is discussed in detail together with comparisons with soil-structure physical experiments performed on the centrifuge facilities at the Rensselaer Polytechnic Institute for the VELACS project. The numerical procedure involves the Biot formulation which is the basic mathematical framework and the Finite Element method for discretisation. Displacement (u) and pore pressure (p) are the primary unknowns and the constitutive relationship used is the Pastor-Zienkiewicz mark III model. Very good comparisons with the results from the physical centrifuges test have been obtained.

**Keywords:** three-dimensional, soil-structure interaction, earthquake, Biot formulation, soil dynamics.

# 1 Introduction

Dynamic soil-structure interaction is one of the more complex problems in geotechnical engineering. For most dynamic soil-structure interaction problems, the structure can be considered essentially as a single phase material. With its relatively high stiffness, the structure could be modelled as a linear elastic perfectly plastic material. In some of the problems such as the concrete dyke on soil foundation, the structure can be modelled as a rigid block with no deformation to be considered. However, for problems where bending is important such as for a retaining wall, quadratic finite element is required to permit sufficient flexural flexibility in the structure. Under most practical circumstances, the hysteretic damping found in the soil should dominate the energy dissipation characteristic of the soil-structure system. However, if the motion in the structure is relatively large, substantial energy may be dissipated within the structure. If this dissipation is not accounted for within the analysis, the displacement and acceleration of the system would be overestimated.

It is common in most static soil-structure interaction problems to consider the soil as springs similar to that of Wrinker foundation. However, this approach is inadequate for some steady state problems and this is even more pronounced for dynamic problems. The main reason is that the stiffness and damping characteristic of the soil is highly non-linear and dependent on strain level. They are also inter-related and should not be simplified into a non-linear elastic stiffness and viscous damping coefficient. The modelling of soil is further complicated by the fact that soil is a two-phase material and if partially saturated, a three-phase material of solid, water and air. For simplicity, only fully saturated soils will be considered in this paper and the Terzaghi effective stress principle applies.

# 2 Analysis

## 2.1 Behaviour of saturated soils

Saturated soils and other two phase media have been the subject of much investigation both experimentally and numerically over many decades. Soils in particular have received considerable attention because of the vital role they play in foundations, dams, dykes and pavement construction. The interaction of soil and pore fluid can be strong enough, due to the build-up of the excess pore pressure, to lead to a catastrophic material softening, a phenomenon known as liquefaction, which occurs frequently in saturated loose granular material under earthquake and other dynamic loading such as blasting. The potential consequences of liquefaction can be illustrated by the near collapse of the Lower San Fernando dam near Los Angeles during the 1971 earthquake. This failure fortunately did not involve any loss of life as the level to which the dam "slumped" still contained the reservoir. Had this been but a few feet lower, the over-topping of the dam indeed would have caused a major catastrophe with the flood hitting a densely populated area of Los Angeles.

A further example to highlight the difference between the typical behaviour of soil and structure is the response of a series of residential buildings in Niigata, Japan during the 1964 earthquake. The buildings mostly survived the earthquake sustaining minimal structural damages and some have even been jacked back into vertical to be reused after the earthquake. However, their foundations completely liquefied leading to sinking and tilting of a number of them.

It is evident that the examples quoted above involved the interaction of pore water pressure and the soil skeleton. Perhaps the particular feature of this interaction however escapes immediate attention. This feature relates to the "weakening" of the soil-fluid composite during the cyclic motion such as that which is involved in an earthquake. However it is this rather than the overall acceleration forces which caused the slumping of the Lower San Fernando dam. What appears to have happened here is that during the motion the interstitial pore pressure increased thus reducing the inter-particle forces in the solid phase of the soil and causing a loss of strength. Therefore it is unsafe to separate the behaviour of structure and soil in a soil-structure interaction problem.

A qualitative, and if possible quantitative prediction of the phenomena leading to permanent deformation or unacceptably high build up of pore pressures, is therefore essential in order to guarantee the safe behaviour of such structures. In the analysis of such dynamic behaviour, the usual decoupled and factor of safety approach may not be most appropriate.

For very slow phenomena with adequate drainage, drained static behaviour can be assumed. The behaviour of the two phases, *i.e.* soil skeleton (the deformable porous solid) and water (the incompressible pore fluid), decouples and solutions can be found separately for the soil skeleton and pore fluid via usual mechanics and effective stress principles even for non-linear problems. On the other hand, if the loading is applied very rapidly and drainage is prevented, an assumption of undrained conditions can be made and the pore pressure can be calculated via the Bulk Modulus of the fluid and again a single set of field equations needs to be solved. However, under transient consolidation and dynamic conditions, such decoupling does not occur.

Furthermore, let us return to the example of the Lower San Fernando dam. According to the strict sense of factor of safety on force equilibrium, the dam has failed because a substantial part of the upstream side of the dam collapsed and slid into the reservoir. This would be the result if a Newmark type sliding block analysis had been performed implying that the action due to the earthquake loading exceeded the shear resisting capacity of the soil. However, because the motion was arrested, a complete failure did not occur.

The designer can demand that the factor of safety on force equilibrium should never be allowed to fall below unity. However, this could lead to extremely conservative and costly design. On the other hand, the soil-structure system may be so stiff that a brittle failure may occur at a higher level of excitation. By contrast a more flexible ductile system with an adequate energy absorption mechanism which is allowed to move and "fail" may survive a similar level of excitation with minor damages to the structure and its human inhabitants.

In order to model the saturated soil, three main components are needed for the numerical simulation and they are as follows:

a. The establishment of an adequate mathematical framework to describe the phenomenon
b. The establishment of a numerical (discrete) approximation procedure
c. The establishment of an adequate constitutive relationship for the material behaviour.

Each of them is a major topic on its own and involves a certain degree of approximation. There are also different levels of agreement amongst researchers on approaches to be taken in each of them. The mathematical framework, the process of numerical approximation and the definition of an adequate constitutive relationship are first presented. Then the basic equations used in the numerical modelling are introduced together with the basic information for the numerical analyses. Comparison with the results of a three-dimensional centrifuge experiment is then given before the final discussions and conclusions.

## 2.2 Mathematical framework for the dynamic behaviour of saturated soil

In this area, there is almost total agreement on the approach to be taken. Take for instance the VELACS project [1], which stands for the Verification of Liquefaction Analysis by Centrifuge Studies. Most of the numerical predictions for the centrifuge tests, except the test modelling a level ground with a single layer of sand, used formulations [2,3] that can be traced back to the original Biot dynamic formulation [4]. The Biot formulation provides a mathematical description of physical behaviour which can adequately describe the transient behaviour of saturated soil for most geomechanics applications except notably fast pile driving and explosive events.

The basic equations for the Biot formulation are:

d. The equilibrium equation of the soil-pore fluid mixture
e. The equilibrium equation for the pore fluid which is a generalisation of the Darcy's equation to include the acceleration of the soil skeleton
f. The conservation of mass for the pore fluid
g. The concept of effective stress
h. The constitutive equation

The formulation emphasised the two phase nature of saturated soil. The effective stress equation emphasised the disparity between the nature of the two media so the simple distribution of stress between the two media according to the ratio of volume would not be correct. Lastly, the use of two equilibrium equations emphasised that the two phases are acting separately and a fully drained or an undrained analysis would not be appropriate in general [5,6].

In dynamic analysis, a fully drained calculation assumes no change in pore water pressure occurred, usually on the basis that permeability is high. However, a high permeability may lead to a high fluid velocity which, in turn, would be preceded by a high fluid acceleration. The high level of fluid velocity and acceleration would introduce substantial changes in pore water pressure and invalidate the original assumption of the calculation.

On the other hand, a fully undrained calculation during earthquake shaking could be justified for soils with relatively low permeability such as silt and clay but not in the case of sand where drainage can be substantial during the short duration of seismic activity. Furthermore this assumption of no drainage, breaks down during the consolidation stage. In order to obtain a correct spatial distribution of permanent settlement, the build-up, the redistribution and drainage of the pore water pressure has to be as accurate as possible.

These simplified fully drained and fully undrained analyses were employed, in most cases, due to the limited availability of inexpensive computing power. However, the situation is now changing and, for two-dimensional examples, the results could be obtained on an Intel personal computer within reasonable computational time. Nevertheless, the result in this paper is obtained using the BlueBEAR computer cluster at the University of Birmingham [7].

## 2.3 Numerical (discrete) approximation procedure

A slightly more diverse approach has been taken by researchers in this area. Again quoting the VELACS project, of the numerical predictors performing Class A predictions for model 3, the following three strategies, all based on the finite element method, have been used for the numerical discretisation of the Biot equation [3]:

The u-p formulation - where u is the displacement of soil skeleton and p is the pore fluid pressure are the basic variables [8-10].

The u-U or u-w formulation - when U is the displacement of the pore fluid and w is the averaged relative fluid displacement defined in accordance with Darcy's law [11-15].

The u-U-p formulation - which employs both the displacement of the soil skeleton and the fluid as well as the pore fluid pressure as the basic unknowns [16-18].

Although the u-p formulation neglects the averaged relative fluid acceleration, it has been found to be satisfactory for static, consolidating and slow dynamic loading including most seismic calculations and it has been widely adopted. The most popular implementation is the fully Implicit procedure [8-10,19]. Other variants included the Explicit u/Implicit p formulation [20,21] and Implicit-Implicit staggered [22-24].

For faster dynamic conditions, the u-U or the u-w formulation would be more appropriate with the retention of the averaged relative fluid acceleration. Again a number of variants exist for this strategy namely Implicit u-U [12,14,15,25] and the fully Explicit u-w [11]. The fully Explicit u-w formulation has the advantage of simplicity and only two by two matrices are required to be solved even under three-dimensional conditions. Although it suffers from more degrees of freedom per node than the u-p formulation and is only conditionally stable, no iteration is required within a time step. Also the scheme lends itself more easily to vectorisation and parallelisation and requires less storage. This method together with the Explicit u/Implicit p and Implicit-Implicit staggered formulation can be attached to a solid finite element program, to augment its capability to include two phase analysis.

The u-U-p formulation, although being formulated by Zienkiewicz and Shiomi [15] is a relatively new development. With three sets of independent variables, the procedure imposed a higher computational requirement. But as it contains no extra terms in addition to the u-U or the u-w formulation, they should share similar converged results. However as the formulation has a mixed character, it could have a better convergence characteristic but this has yet to be shown.

## 2.4 A adequate constitutive relationship for the material behaviour

There is generally little if any agreement on the way to approach the issue of constitutive modelling. However, there is some agreement on the following major issues:

a. A peak and residual friction angle can be defined for cohesionless soil like sand
b. The dilative characteristic of dense sand leads to cyclic mobility and contractive behaviour of loose sand, and on the other hand, could lead to static or cyclic initial liquefaction
c. The classical elasto-plastic model with single yield surface, associative flow rule and elastic interior is, in general, inadequate for the modelling of the cyclic behaviour of soil.

The authors would not attempt a detailed discussion on this matter in this paper. Many material models have been developed and these material models have varying degrees of success under dynamic loading conditions. There may be more constitutive models than the number of researchers in the field. Mainly, two diametrically opposite approaches have been taken.

The first is to include as many aspects of soil behaviour in the model as possible leading to a sort of Grand Unified Model. This implies a large number of soil model parameters, and thus a number of soil property tests are required to find them. The models should be able to account for all foreseeable aspects of soil behaviour, but usually this would lead to heavy computational overheads. From a scientific point of view, this is a direction which has to be pursued, not least to form a framework for the calibration of simpler models in their own area of applicability.

On the other hand, for engineering applications, simpler models must suffice, for reasons such as computational efficiency, limited test data available, site variability, and uncertainty in input parameters including earthquake input. Simpler models can also enhance understanding and allow extensive parametric studies to be performed. These simpler models have to be used with care to avoid some of their assumptions being violated during the analysis. For example, a simple classical non-associative Mohr Coulomb model may be adequate for static analysis. However, when used in cyclic analysis, appropriate damping must be introduced to account for the hysteretic damping available in the physical material [26]. The material model used in this paper is the Pastor Zienkiewicz mark III model [27].

## 2.5 Validation of the numerical procedure

Having established the numerical procedure, it is imperative to establish the extent of validity of the assumptions made during the three stages of numerical formulation. This can be performed individually for the three stages:

The constitutive model can be compared with laboratory experiments such as the conventional triaxial apparatus, true triaxial apparatus, hollow cylinder and shear box tests. The constitutive model can be tested against other constitutive models. A parametric sensitivity analysis can also be performed. These tests can be conveniently performed on a soil model testing computer program such as SM2D [28]. Single subroutine call is involved for each separate constitutive model implementation and this allows the same subroutine to be used in the soil model testing mode or in various finite element programs such as DIANA-SWANDYNE III [29] and GLADYS-3D [30].

The numerical approximation can be tested against analytical solutions obtained using linear elastic material such as the Terzaghi consolidation and solutions for dynamic (cyclic) steady state [5,6]. But in order for such analytical solutions to be used correctly, the basic assumption of the formulation such as negligible relative average fluid acceleration for the u-p formulation must be respected.

However, in order to test the validity of the assumptions made in the initial mathematical framework and the behaviour of the constitutive model in the boundary value problems, one has to resort to the comparison with physical results from prototype or model experiments. It is generally accepted that centrifuge experiments are able to provide a good representation of the prototype behaviour with proper scaling of stress level within the model [31]. In this paper, the calculated results of displacement, pore pressure and acceleration using the implicit u-p formulation using the computer program DIANA-SWANDYNE III are compared with centrifuge experiment model 12 performed at the Rensselaer Polytechnic Institute for the VELACS project.

# 3 Numerical Modelling and Computer Code

Because of space limitation, only a brief outline is given in this section. Further details can be found in Zienkiewicz *et al.* [8,10,32] and Chan [9]. The computer program DIANA-SWANDYNE III [29] has been developed for the static, consolidating and dynamic analysis of soils. It is based on the fully implicit u-p approximation of the Biot [4] formulation for a saturated porous medium possibly interacting with a solid component *e.g.* a structure.

SWANDYNE III has access to an array of soil models via the soil model library interface and adopts the fully implicit scheme. Using the continuity, momentum and generalised Darcy equations, the complete Biot equation governing deformable porous media can be expressed in three-dimensional Cartesian co-ordinates using indices notation as:

$$\sigma_{ij},_j + \rho b_i - \rho \ddot{u}_i - \rho_f \ddot{w}_i = 0 \tag{1}$$

where $\sigma_{ij}$ is the total stress tensor (tensile positive), $u_i$ and $w_i$ are the displacement of the solid skeleton and average (Darcy) fluid velocity respectively, $b_i$ is the body force per unit mass, $\rho_s$, $\rho_f$ and $\rho$ are the densities of the solid grain, fluid and mixture respectively, with $\rho = (1-n)\rho_s + n\rho_f$ and $n$ being the porosity of the porous media. By including porosity n in the definition of the averaged relative fluid velocity:

$$\dot{w}_i = n(\dot{U}_i - \dot{u}_i) \tag{2}$$

their value conforms with the one used in the original Darcy equation (see *e.g.* [33]) with $U_i$ being the displacement of the fluid. The generalised Darcy equation can be expressed as:

$$\dot{w}_i = \frac{v_k}{v_f \rho_k g_k} k_{ij} \left( -p,_j + \rho_f g_j - \rho_f \ddot{u}_j - \frac{\rho_f \ddot{w}_j}{n} \right) \tag{3}$$

where $k_{ij}$ is the permeability tensor. For material with isotropic permeability k (unit = length/time), $k_{ij} = k\delta_{ij}$, where $\delta_{ij}$ is the Kronecker delta. Furthermore $\nu_k$, $\rho_k$ and $g_k$ are the viscosity of the fluid, density of fluid and gravitational acceleration at which the permeability is measured and $\nu_f$ is the viscosity of the fluid actually being used. It is common to use a substitute fluid with higher viscosity in centrifuge tests in order to maintain the compatibility of the dynamic behaviour and the consolidation behaviour when the actual prototype soil is used in centrifuge experiments. Lastly the continuity equation for the fluid phase can be expressed as:

$$\frac{\dot{p}}{Q} + \dot{u}_i,_i + \dot{w}_i,_i = 0 \tag{4}$$

where Q is the averaged bulk modulus defined with $K_s$ being the bulk modulus of the soil grains and $K_f$ the bulk modulus of the pore fluid as:

$$\frac{1}{Q}=\frac{1-n}{K_s}+\frac{n}{K_f} \tag{5}$$

By eliminating pore pressure p between Equations (3) and (4), the u-w formulation (so-called because u and w are the primary variables) results. On the other hand, if the averaged relative fluid acceleration is neglected from Equations (1) and (3), and $\dot{w}_i$ are eliminated between Equations (3) and (4), the u-p formulation is recovered. By neglecting further the solid acceleration in Equations (1) and (3), the standard consolidation equation is obtained. Lastly, if both the solid and fluid velocities are neglected, the resulting equations are now decoupled and they are the static solid equation and the steady state seepage equation. Using the finite element method [34,35] for spatial discretisation, the discretised u-p formulation is as follow: The dynamic form:

$$\begin{aligned}&\mathbf{M\ddot{u}}+\mathbf{P(u)}-\mathbf{Qp}=\mathbf{f_u}\\&\mathbf{G\ddot{u}}+\mathbf{Q^T\dot{u}}+\mathbf{S\dot{p}}+\mathbf{Hp}=\mathbf{f_p}\end{aligned} \tag{6a}$$

The consolidation form:

$$\begin{aligned}&\mathbf{P(u)}-\mathbf{Qp}=\mathbf{f_u}\\&\mathbf{Q^T\dot{u}}+\mathbf{S\dot{p}}+\mathbf{Hp}=\mathbf{f_p}\end{aligned} \tag{6b}$$

The static and steady state form:

$$\begin{aligned}&\mathbf{P(u)}-\mathbf{Qp}=\mathbf{f_u}\\&\mathbf{Hp}=\mathbf{f_p}\end{aligned} \tag{6c}$$

where M is the mass matrix, G is the dynamic coupling matrix, S is the compressibility matrix, H is the permeability matrix, (sometimes a viscous damping term C is added to the first equation in 6a), u and p are the vectors for the nodal value of u and p respectively, Q is the coupling matrix, P(u) is non-linear internal force vector which is given by:

$$\mathbf{P(u)}=\int_\Omega \mathbf{B}^T \underline{\sigma}' d\Omega \tag{7}$$

where B is the usual displacement-strain transformation matrix for the finite element method, $\underline{\sigma}'$ is vector of all effective stress components at the integration (Gauss) points and Ω is the domain concerned. Equations (5-7) together with the undrained version of Equations (6a) and (6c) formed the basis of the implicit finite element program DIANA-SWANDYNE III. The time discretisation is performed using the generalised Newmark method presented by Katona and Zienkiewicz [36].

# 4 Experimental Set-up for VELACS model 12 and Results

As part of the VELACS experiments, Test No.12 is a dynamic soil-structure interaction test where a model of a structure was placed on a saturated sand layer overlaid by a thin silt layer and subjected to earthquake-like base motion. As shown in Figure 1, the structure is 4m high (in prototype scale, 4cm in model scale), while the soil layer is a 6m deep sand layer overlaid by a 1m deep silt layer. The structure foundation extends 0.5m below the surface of the sand layer and the water surface is 1m above the silt surface. The average pressure of the structure acting on the sand layer is 150kPa. The earthquake input motion is 10 cycles of 2Hz sine wave, with an amplitude of 0.25g (see Figure 8). The tests were performed at 100g centrifugal acceleration at Princeton University, Rensselaer Polytechnic Institute and University of California, Davis.

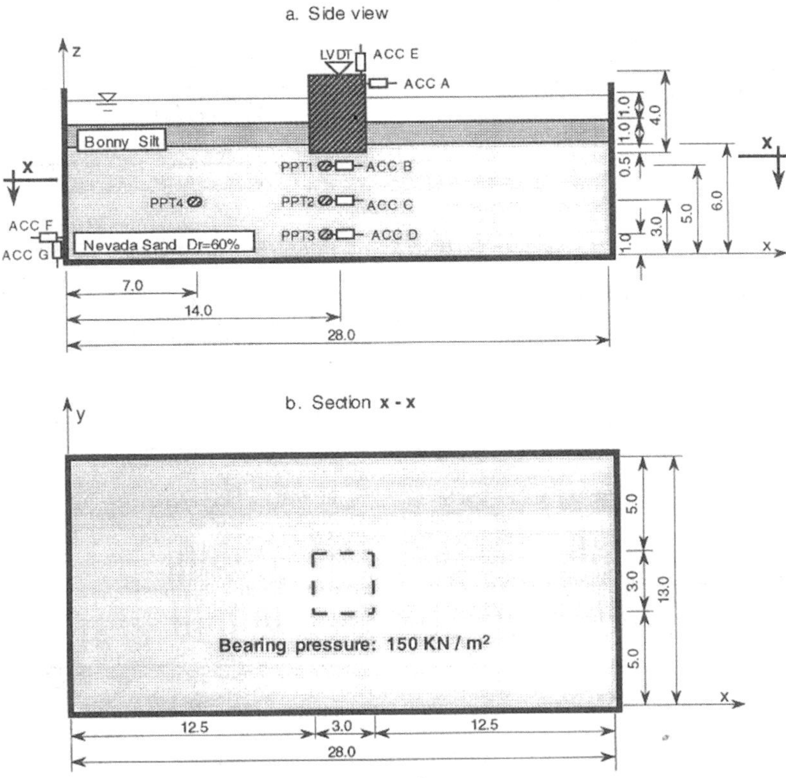


Figure 1: Experimental set up of model No.12 of VELACS [37]

In this test, since the foundation does not extend the entire width of the container, the model cannot be modelled satisfactorily under 2D plane strain conditions. For this reason, 3D numerical modelling was performed in prototype scale for this test.

A finite element mesh with 1107 27-noded Lagrangian brick elements was used in the numerical analysis, and the FEM mesh was shown in Figure 2. A 10 cycles of 2Hz sinusodial acceleration in the x-direction is applied at the bottom of the soil layer (z=0m) and all of the sides (x=0m or x=28m or y=0m or y=13m). The time history of the input acceleration is shown on Figure 3. Constraint on displacement in the y-direction was applied on the boundary at y=0m and y=13m. Pore pressure and total stress of 9.8kPa were applied at the surface of the soil (z=7m) to simulate the pressure created by 1m of water. An associated Mohr Coulomb constitutive model was used to simulate the 1m depth of Bonnie silt and a PZ3 model was used to simulate the behaviour of the saturated sands. As an initial stresses state is required before the dynamic analysis, a static drained analysis was performed. In the static analysis stage, the Associated Mohr Coulomb model was used for both the silt and sands layers, while a linear elastic model was used for the structure block. In order to avoid tensile stress and high stress ratio, the friction angle ϕ' was reduced to 25 degree and cohesion c'=1kPa was used for the Mohr Coulomb model. The Young's modulus for sands was assumed to be E=195MPa and E=4MPa for the silt in order to obtain an initial stress state. The density of the foundation block was assumed to be 3800kg/m$^2$ to obtain the average pressure of 150kPa on the sands layer. The material parameters are listed in Table 1.

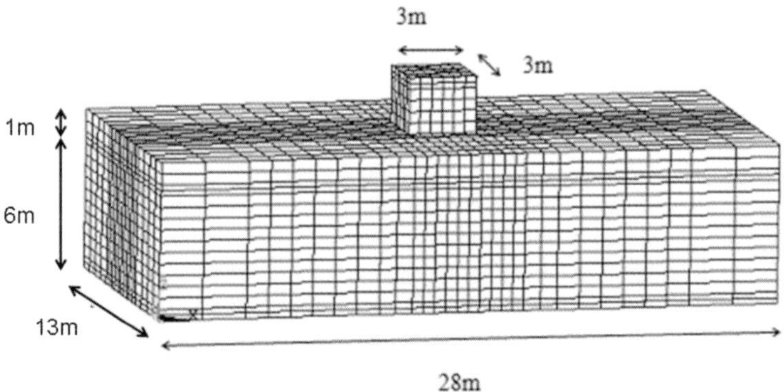


Figure 2: FEM mesh for model No.12 of VELACS with 1107 27-noded Langrangian brick elements

| | $E(MPa)$ | $\nu$ | $c(kPa)$ | $\phi'$ | $e$ | $\rho_s$ (kg/m$^3$) | $\rho_f$ (kg/m$^3$) | $K_f$ (MPa) | $K_s$ (MPa) |
|---|---|---|---|---|---|---|---|---|---|
| Sand | 195 | 0.31 | 1 | 35 | 0.66 | 2670 | 1000 | 1000 | $10^{14}$ |
| Silt | 4 | 0.31 | 1 | 25 | 0.63 | 2670 | 1000 | 1000 | $10^{14}$ |
| Block | 200000 | 0.49 | / | / | 0 | 3800 | / | / | / |

Table1: Soil parameters for static analysis

In Table 1, $\nu$ represents the Poisson's Ratio, $e$ the void ratio, $\rho_s$ the density of solid particles, $\rho_f$ the density of fluid, $K_f$ the Bulk Modulus of the fluid, while $K_s$ is the Bulk Modulus of the solid particles.

In the earthquake analysis, the PZ3 model was adopted for the sands. The constitutive model and parameters were kept unchanged for the silt layer and the foundation block. The permeability $k=5.6x10^{-5}$m/s was adopted for sand and $k=1x10^{-5}$m/s was adopted for silt based on the experimental data. The parameters of PZ3 model were calibrated with the Monotonic Test CIUC Test 6012 [38] by Chan *et al.* [39]. The parameters were shown on the Table 2.

| $K_{evo}$ | 2000 kPa | $K_{eso}$ | 2600 kPa | $p'_o$ | 4kPa |
|---|---|---|---|---|---|
| $M_g$ | 1.32 | $M_f$ | 1.30 | $\alpha_f = \alpha_g$ | 0.45 |
| $\beta_0$ | 4.2 | $\beta_1$ | 0.05 | $H_0$ | 750 |
| $H_{u0}$ | 40000kPa | $\gamma_u$ | 2.0 | $\gamma_{DM}$ | 0 |

Table 2: Soil parameters in earthquake analysis (for sands)

The time step in the earthquake analysis is 0.02s, and a total of 500 steps was performed. The numerical results were compared with the experimental results obtained by Rensselaer Polytechnic Institute [40]. This is because only the experiment performed by Rensselaer Polytechnic Institute gave a relatively complete set of results (including the acceleration at accelerometers ACC A, ACC B, ACC C, ACC D, ACC E, ACC F in Figures 3 to 8 and the pore pressure at PPT1, PPT2, PPT3, PPT4 in Figures 9 to 12 and the vertical settlement at the top of the structure in Figure 13), as mentioned in [37]. One should note that the input motion achieved in the experiment is not exactly the same as the target motion specified for the numerical predictions as shown in Figure 8. Nevertheless the departure is not too serious.

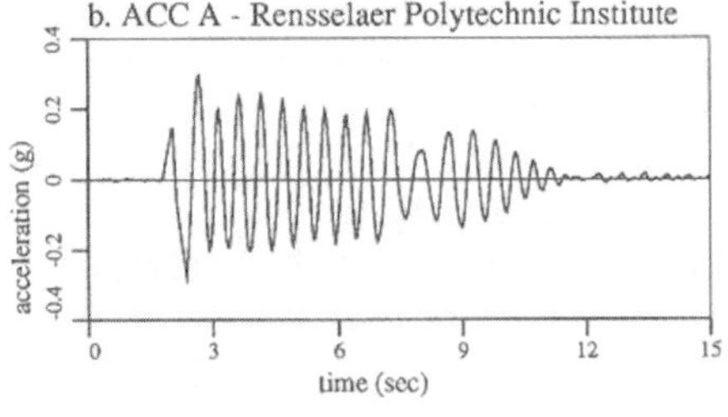


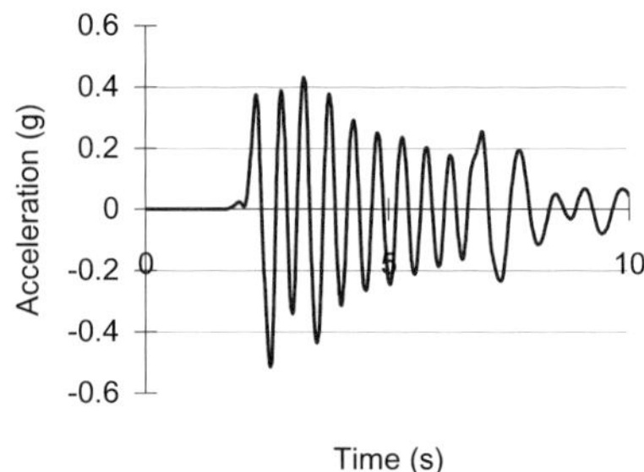


Figure 3: Transducer ACC A - Horizontal acceleration on the structure

At the top of the structure, as shown in Figure 3, the numerical analysis predicts a slightly higher amplitude at first and this is because the initial input shown in Figure 8 was smaller in amplitude than the target. The comparison is excellent after the initial phase. After the passage of the earthquake, the structure goes into free vibration mode although the decay is faster for the numerical prediction than the experimental measurement.

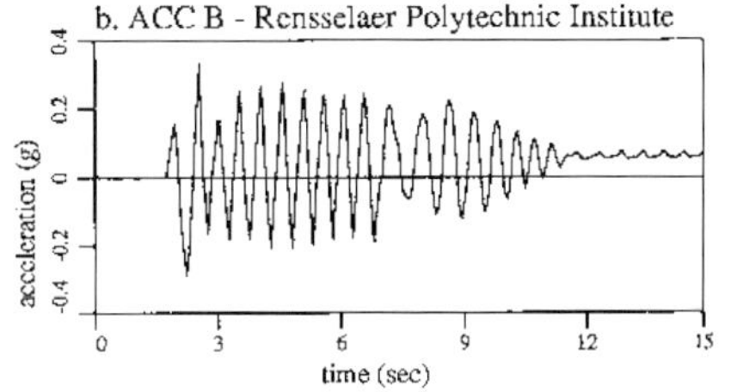


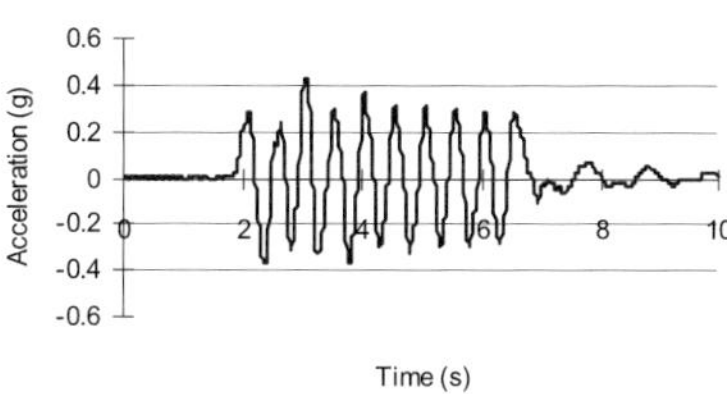


Figure 4 : Transducer ACC B - Horizontal acceleration below the structure

The comparison for ACC B in Figure 4 is very good except the free vibration phase after the passage of the earthquake. In Figure 5, the numerical prediction shows a slightly smaller acceleration amplitude than the experimental result, this could be caused by the greater stiffness degradation in the constitutive model used than the physical behaviour.

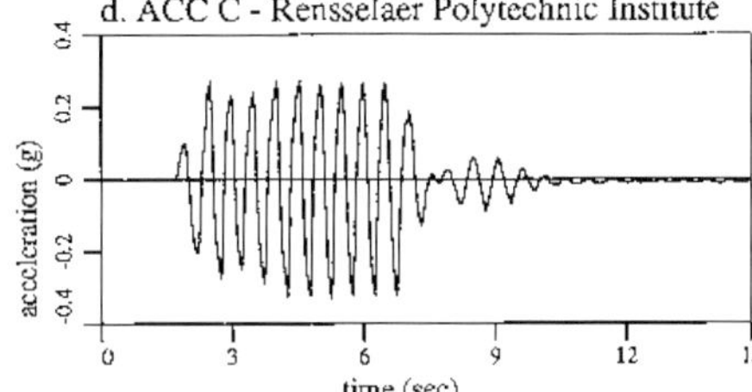


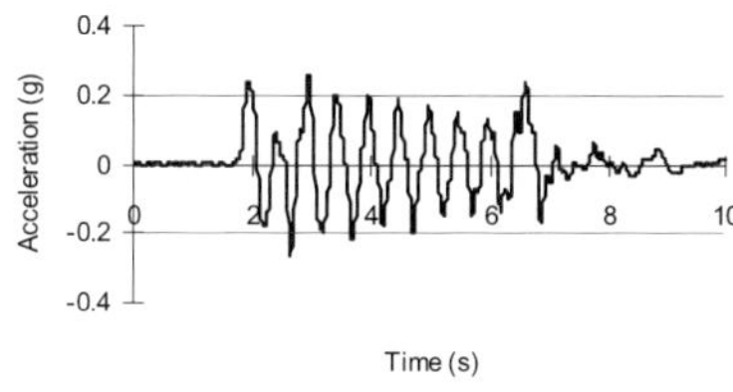


Figure 5: Transducer ACC C – Horizontal acceleration at the middle of the sand layer

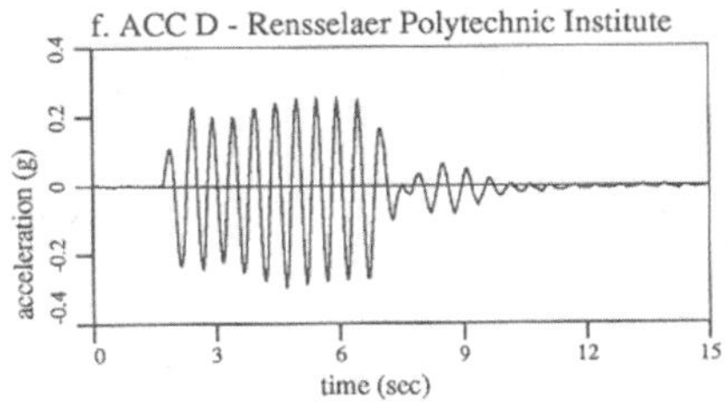


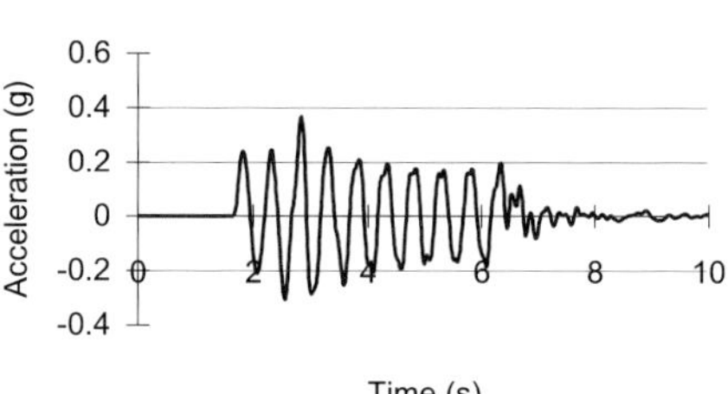


Figure 6: Transducer ACC D – Horizontal acceleration near the bottom of the sand layer

The comparison is very good for ACC D in Figure 6 except for the single higher peak for the numerical prediction. The measured vertical acceleration in Figure 7 shows non-symmetric behaviour between the phase when the structure is moving upwards and the phase when it is moving downward. This is understandable because, as the structure moves up, a slight gap opens at the interface with the soil at the bottom of the structure and this may reduce the amount of upward acceleration being transmitted. However as no slip element is used in the numerical analysis, full contact is always assumed therefore the vertical acceleration prediction is more symmetric.

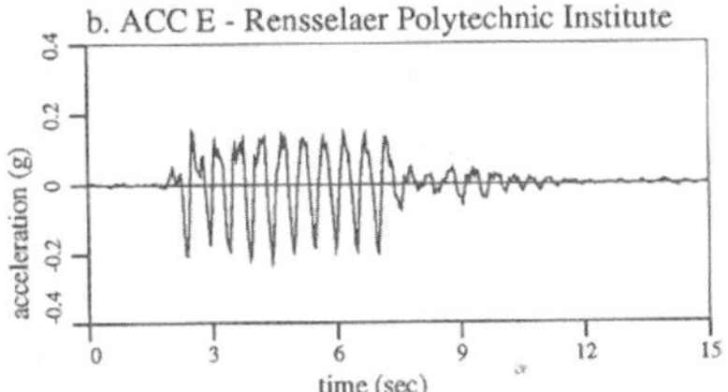


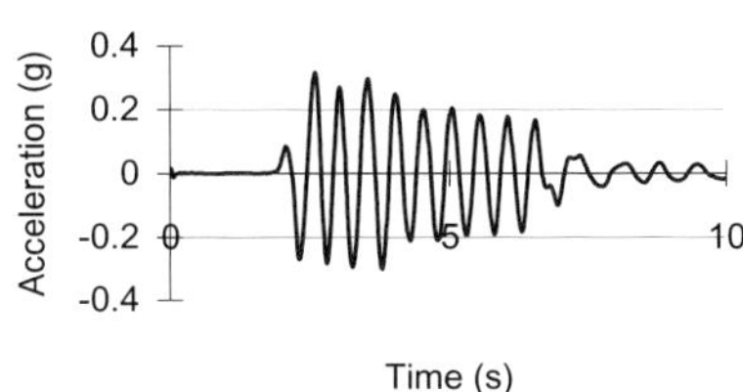


Figure 7: Transducer ACC E – Vertical acceleration of the structure

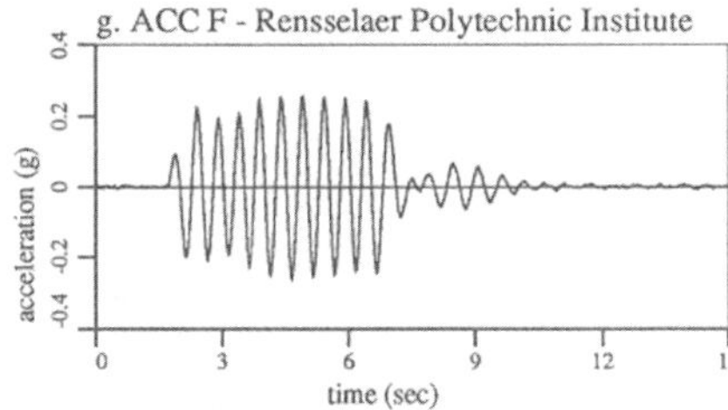


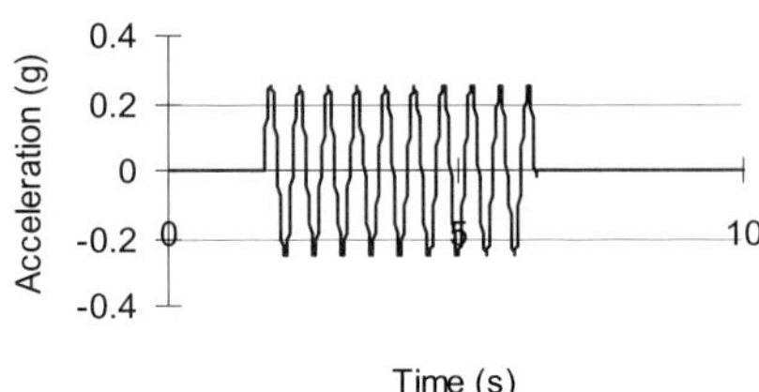


Figure 8: Transducer ACC F – Horizontal acceleration of the box

In order to reduce the computational effort (the analysis required around 9 cpu-hours), only the first ten seconds of the analysis were used in the numerical prediction therefore the time history is only half as long as the experimental results and much of the consolidation phase was not modelled. However, very good comparison is still obtained. In Figure 9, the behaviour is almost identical for the first 10 seconds even though the rise in excess pore water pressure was slightly slower in the numerical prediction and the consolidation in the experiment started before 10 seconds while the peak of excess pore water pressure continued in the numerical prediction up to 10 seconds. The difference could be due to the fact that a smaller elastic modulus was predicted by the numerical analysis therefore the rate of consolidation is much slower. Very good comparison is also obtained for PPT2 in Figure 10 and the behaviour is almost identical for the first 10 seconds even though the rise in excess pore water pressure was slightly faster in the numerical prediction.

Although a rise of 30kPa was seen in the excess pore pressure, no liquefaction occurred as the initial vertical effective stress at this point was at least 35kPa.

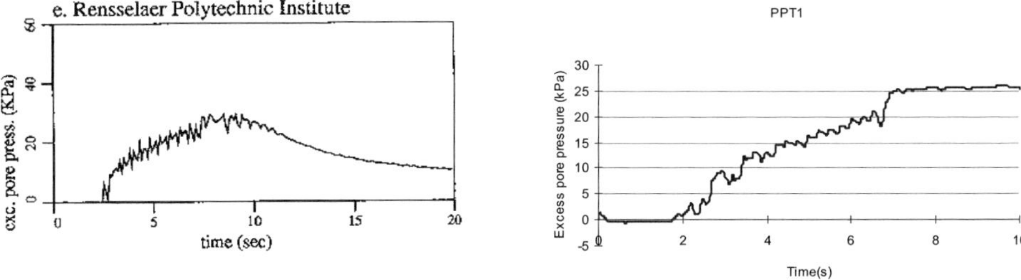


Figure 9: Excess Pore Pressure response at location PPT1

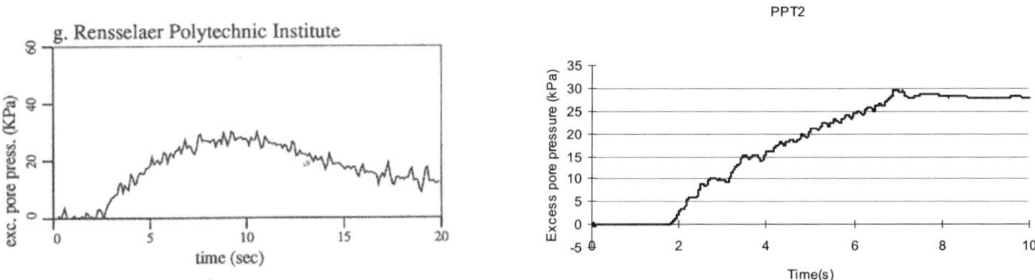


Figure 10: Excess Pore Pressure response at location PPT2

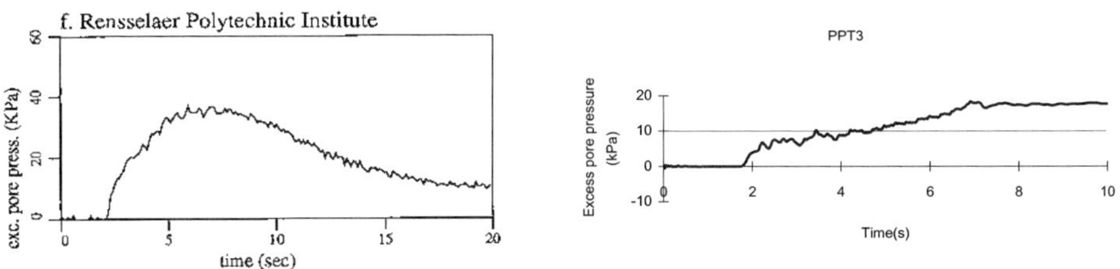


Figure 11: Excess Pore Pressure response at location PPT3

Despite a higher and faster rise in excess pore pressure for both PPT 3 near the bottom of the sand layer and PPT4 in the far field, the general trend of the behaviour is still very comparable especially for the amplitude for the oscillation in the pore pressure trace. This implies that the reversible volume change characteristic is modelled well but the irreversible volume change is less for the numerical prediction.

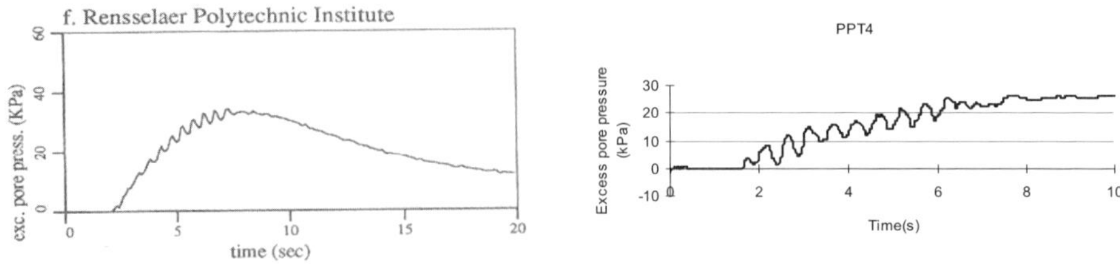


Figure 12: Excess Pore Pressure response at location PPT4

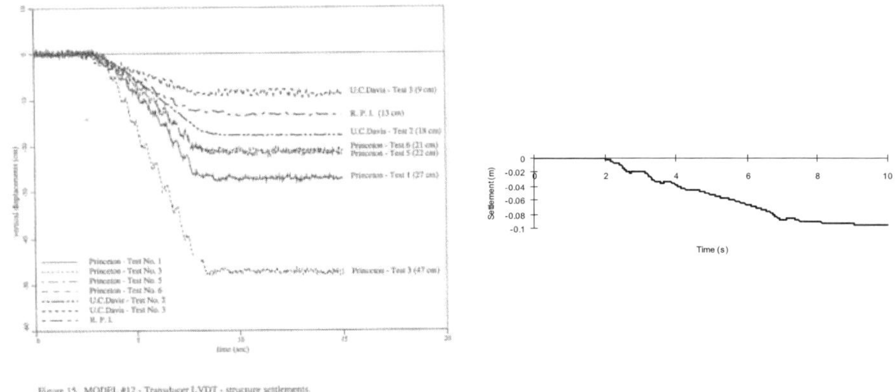


Figure 13: Vertical settlement at the top of the block

Around 10cm settlement was predicted by the numerical analysis which is between the value obtained at UC Davis (9cm) and RPI (13cm). But a very wide spread of the measured vertical settlement was observed therefore the measured final settlement value may not be too trustworthy. However the numerical prediction correctly modelled the trend at which the settlement develops.

# 5 Discussions and Conclusions

In order to calculate the overall behaviour of a non-trivial three-dimensional saturated soil/structure system, careful consideration has to be made for the the behaviour of structure, pore fluid and the soil as well as the soil-structure interface. In this paper, the numerical modelling of saturated soil is discussed in detail together with a comparison with soil-structure physical experiments performed at the RPI centrifuge facility.

In order to model the dynamic behaviour of saturated soil adequately, it is necessary to employ an appropriate mathematical framework, numerical (discrete) approximation and an adequate constitutive relationship. The Biot formulation, the u-p formulation and the Pastor-Zienkiewicz mark III model have been shown to be adequate for the applications shown. Good comparisons have been obtained in general for the physical centrifuges tests.

Furthermore, the numerical procedure did not involve expensive computational resources. Most of the two-dimensional results have been obtained on an Intel personal computer with reasonable computational time and the three-dimensional analyses can be done relatively quickly on a more powerful processor. This computational efficiency would allow parametric studies which are essential in the design process. Further analyses on various three-dimensional examples and further investigations on the use of iterative solvers instead of the direct solver used in this study are under development in order to deliver this powerful design and analysis tool into the hands of practising engineers.

# References

[1] K. Arulanandan, "Why VELACS? (VErification of Liquefaction Analysis using Centrifuge Studies)", Proc. VELACS symp., UC Davis, 17-20 Oct., Vol. 2, 1239-1266, 1994.

[2] A.H.C. Chan, R. Siddharthan, K. Ito, "Overview of the Numerical Predictions for VELACS Model No.3 in Verification of numerical procedures for the analysis of soil liquefaction problems", UC Davis, 17-20 Oct, A.A. Balkema, Rotterdam, 1994.

[3] I.M. Smith, "An overview of numerical procedures used in the VELACS project", Proc. VELACS symp., UC Davis, 17-20 Oct., Vol. 2, 1321-1338, 1994.

[4] M.A. Biot, "Theory of propagation of elastic waves in a fluid-saturated porous solid, part I - low-frequency range", J. Acoust. Soc. Am., 28 (2), 168-178, 1956.

[5] B.R. Simon, O.C. Zienkiewicz, D.K. Paul, "An analytical solution for the transient response of saturated porous elastic solids", Int. J. Num. Anal. Geomech., 8, 381-398, 1984.

[6] O.C. Zienkiewicz, C.T. Chang, P. Bettess, "Drained, undrained, consolidating and dynamic behaviour assumptions in soils", Géotechnique, 30 (4), 385-395, 1980.

[7] A. Reed, "Notes for the participants of the Blue Bear Open Service", University of Birmingham, 2007.

[8] O.C. Zienkiewicz, A.H.C. Chan, M. Pastor, B.A. Schrefler, T. Shiomi, "Computational Goemechanics with special reference to Earthquake Engineering", John Wiley and Sons Ltd, Chichester, February, 1999.

[9] A.H.C. Chan, "A unified Finite Element Solution to Static and Dynamic Geomechanics problems", Ph.D. Dissertation, University College of Swansea, Wales, 1988.

[10] O.C. Zienkiewicz, A.H.C. Chan, M. Pastor, D.K. Paul, T. Shiomi,"Static and Dynamic Behaviour of Geomaterials - A rational approach to quantitative

solutions, Part I - Fully Saturated Problems", Proc. Roy. Soc. Lond., Vol. A429, 285-309, 1990.

[11] A.H.C. Chan, O.O. Famiyesin, D. Muir Wood, "A Fully Explicit u-w Schemes for Dynamic Soil and Pore Fluid Interaction", APCOCM Hong Kong, 11-13 Dec., Vol. 1, 881-887, 1991.

[12] J.H. Prevost, "DYNAFLOW: A Nonlinear transient finite element analysis program", Report 81-SM-1, Department of Civil Engineering, Princeton University, Princeton NJ, 1987.

[13] J.H. Prevost, "Nonlinear dynamic response analysis of soil and soil-structure interacting systems, in Soil Dynamics and Geotechnical Earthquake Engineering (ed.) Pedro Sêco e Pinto, 49-126, A.A.Balkema, Rotterdam, 1993.

[14] T. Shiomi,"Nonlinear behaviour of soils in earthquake" - C/Ph/73/83, Ph.D. Dissertation, Univ. Coll. of Swansea, Wales, 1983.

[15] O.C. Zienkiewicz, T. Shiomi ,"Dynamic Behaviour of saturated porous media: The generalized Biot formulation and its numerical solution", Int. J. Num. Anal. Geomech., 8, 71-96, 1984.

[16] A. Anandarajah, "HOPDYNE - A Finite Element Computer Program for the Analysis of Static, Dynamic and Earthquake Soil and Soil-Structure Systems", Internal Report, Johns Hopkins University, Baltimore, 1990.

[17] A. Gajo, A. Saetta, R. Vitaliani, "Evaluation of three- and two-field finite element methods for the dynamic response of saturated soil", Int. J. Num. Meth. Engrg., 37, 1231-1247, 1994.

[18] R.S. Sandhu, H.L. Shaw, S.J. Hong, "A three-field finite element procedure for analysis of elastic wave propagation through fluid-saturated soils", Soil Dyn. Earth. Engrg., Vol. 9, 58-65, 1990.

[19] E.J. Parra-Colmenares, "Numerical modeling of liquefaction and lateral ground deformation including cyclic mobility and dilation response in soil systems", Ph.D. Dissertation, Rensselaer Polytechnic Institute, Troy, New York, 1996.

[20] O.C. Zienkiewicz, K.H. Leung, E. Hinton, C.T. Chang, "Liquefaction and permanent deformation under dynamic conditions - Numerical solution and Constitutive relations", Chapter 5 in "Soil Mechanics - Transient and cyclic loads", John Wiley, Chichester, 1982.

[21] K.H. Leung, "Earthquake response of saturated soils and liquefaction", Ph.D. Dissertation, University College of Swansea, Wales, 1984.

[22] O.C. Zienkiewicz, D.K. Paul, A.H.C. Chan, "Numerical solution for the total response of saturated porous media leading to liquefaction and subsequent consolidation", NUMETA 87 conference, Swansea, July 6-10, 1987.

[23] O.C. Zienkiewicz, D.K. Paul, A.H.C. Chan, "Unconditionally stable staggered solution procedure for soil-pore fluid interaction problems", Int. J. Num. Meth. Engrg., 26, 1039-1055, 1988.

[24] D.K. Paul, "Efficient dynamic solutions for Single and Coupled multiple field problems, Ph.D. Dissertation, University College of Swansea, Wales, 1982.

[25] S.J. Lacy, "Numerical Procedures for Nonlinear transient analysis of two-phase soil systems", Ph.D. Dissertation, Department of Civil Engineering, Princeton University, Princeton NJ, 1986.

[26] A.H.C. Chan, O.O. Famiyesin, D. Muir Wood, "Numerical modelling of dynamic experiments on dry sand beds using simple models", ECONMIG 94, 249-255, 1994.

[27] M. Pastor, O.C. Zienkiewicz, "A generalised plasticity hierarchical model for sand under monotonic and cyclic loading", NUMOG II, Ghent, April, 131-150, 1986.

[28] A.H.C. Chan, "User manual for SM2D - Soil Model Tester for 2-dimensional application", School of Civil Engineering, University of Birmingham, 1995.

[29] A.H.C. Chan, J. Ou, "User Manual for DIANA SWANDYNE-III", Department of Civil Engineering, University of Birmingham, 2008.

[30] A.H.C. Chan, "User Manual for GLADYS-3D", Department of Civil Engineering, University of Birmingham, 2007.

[31] A.N. Schofield, "Cambridge Geotechnical Centrifuge Operations - 20th Rankine Lecture", Géotechnique, 30(3), 227-268, 1980.

[32] O.C. Zienkiewicz, M. Pastor, A.H.C. Chan, Y.M. Xie, "Computational approaches to the dynamics and statics of saturated and unsaturated soils", Chapter 1 in "Advanced Geotechnical Analysis", Elsevier Applied Science Publishers Ltd, London and New York , 1991.

[33] R.F. Craig, "Soil Mechanics" (5th edn), Chapman & Hall, London, 1992.

[34] O.C. Zienkiewicz, R.L. Taylor, "The Finite Element Method - Volume 1: Basic Formulation and Linear Problems (4th edn)", McGraw-Hill Book Company, London, 1989.

[35] O.C. Zienkiewicz, R.L. Taylor, "The Finite Element Method - Volume 2: Solid and Fluid Mechanics, Dynamics and Non-linearity (4th edn)", McGraw-Hill Book Company, London, 1991.

[36] M.G. Katona, O.C. Zienkiewicz, "A unified set of single step algorithms Part 3: The Beta-m method, a generalisation of the Newmark scheme", Int. J. Num. Meth. Engrg., 21, 1345-1359, 1985.

[37] J.H. Prevost, I. Krstelj, R. Popescu, "Overview of experimental results for centrifuge model No.12, Verification of Numerical Procedures for the Analysis of Soil Liquefaction Problems, Volume 2 , 1619-1634, 1993.

[38] Earth Technology Corporation,"VELACS Laboratory testing program-Soil data report", 1992.

[39] A.H.C. Chan, O.O. Famiyesin, D. Muir Wood, "Numerical Prediction for Model No. 11", Proc. VELACS symp., UC Davis, 17-20 Oct., Vol. 1, 909-931, 1993.

[40] R.A. Carnevale, A.W. Elgamal, "Experimental results of RPI centrifuge Model No 12, Verification of Numerical Procedures for the Analysis of Soil Liquefaction Problems", Volume 1, 1019-1026, 1993.

# Author Index


Aifantis, E.C., 189
Aigner, E., 209
Ammar, A., 247
Askes, H., 189

Barth, F.G., 321
Bazilevs, Y., 1
Bennett, C.R., 151
Bennett, T., 189
Bittnar, Z., 61
Böhm, H.J., 321

Calo, V.M., 1
Cascón, G., 229
Cascón, J.M., 229
Chan, A.H.C., 335
Chinesta, F., 247
Cottrell, J.A., 1
Croft, T.N., 151
Cross, M., 151

De, S., 173
Dechant, H.-E., 321
Doltsinis, I., 295

Eberhardsteiner, J., 209
Eriksson, A., 127
Escobar, J.M., 229
Evans, J., 1
Eyheramendy, D., 17

Gebhardt, J.E., 151
Gitman, I.M., 189

Harlen, O., 81
Hößl, B., 321
Hughes, T.J.R., 1

Innocente, M.S., 103

Jimack, P.K., 81

Lackner, R., 209
Lamari, H., 247
Lipton, S., 1

Mackie, R.I., 41
Macri, M., 173
Mang, H.A., 209
McBride, D., 151
Montenegro, R., 229

Nikishkov, G.P., 271
Nishidate, Y., 271

Ou, J., 335
Oudin-Dardun, F., 17

Patzák, B., 61

Rammerstorfer, F.G., 321
Ranc, N., 247
Rodríguez, E., 229

Scott, M.A., 1
Sederberg, T.W., 1
Sienz, J., 103

Tenchev, R., 81

Walkley, M.A., 81
Wistuba, M., 209

# Keyword Index

adaptive refinement-derefinement, 229
arachnids, 321
arbitrary Lagrangian-Eulerian meshes, 81
artificial intelligence, 103
asphalt, 209
atomic-scale, 271

biomechanics, 321
Biot formulation, 335
bitumen, 209
Brownian dynamics, 247

component oriented, 41
composites, 173
computational fluid dynamics, 151
computer aided design, 1
cracking, 173, 209
curse of dimensionality, 247

deformation processes, 295
distributed computing, 41

earthquake, 335
enrichment, 173

finite elements, 17, 41, 81, 209, 229, 271, 321
finite sums decompositions, 247
fresh concrete casting, 61
fresh concrete rheology, 61

gradient elasticity, 189

heap leaching, 151
higher-order continuum, 189
homogenization, 173
human movements, 127

identification, 209
inelastic solids, 295
interface-capturing, 61
isogeometric analysis, 1

kinetic theory, 247

mechanisms, 127
mesh smoothing, 229
mesh untangling, 229
model reduction, 247
model validation, 151
molecular dynamics, 247
multi-physics, 17
multi-scale, 173, 189, 209, 247
muscular forces, 127

nanostructure, 271
nested meshes, 229
non-linear mechanics, 17
non-Newtonian flow, 61
NURBS, 1

object-oriented, 17, 41
optimization, 103, 127

parallel processing, 17, 41
particle swarms, 103
partition of unity, 173
pavement, 209
polymer melts, 81
process modelling, 151

quantum mechanics, 247

rate sensitivity, 295
reactive porous media, 151
representative volume element, 189

scheduling, 103
self-positioning, 271
sensors, 321
separated representation, 247
soil dynamics, 335
soil-structure interaction, 335
spiders, 321
stability of evolution, 295
statistical mechanics, 247

T-splines, 1
tetrahedral mesh generation, 229
thermal loading, 209
thermomechanical coupling, 295

upscaling, 209

validation, 209
variably saturated flow, 151
viscoelasticity, 81, 209

wave dispersion, 189
wave propagation, 189

SAXE-COBURG
PUBLICATIONS
mmviii